Cytokines

The Practical Approach Series

SERIES EDITORS

D. RICKWOOD
Department of Biology, University of Essex
Wivenhoe Park, Colchester, Essex CO4 3SQ, UK

B. D. HAMES
Department of Biochemistry and Molecular Biology
University of Leeds, Leeds LS2 9JT, UK

★ **indicates new and forthcoming titles**

Affinity Chromatography
Anaerobic Microbiology
Animal Cell Culture
(2nd Edition)
Animal Virus Pathogenesis
Antibodies I and II
★ Basic Cell Culture
Behavioural Neuroscience
Biochemical Toxicology
★ Bioenergetics
Biological Data Analysis
Biological Membranes
Biomechanics — Materials
Biomechanics — Structures
and Systems
Biosensors
★ Carbohydrate Analysis
(2nd Edition)
Cell–Cell Interactions
★ The Cell Cycle
★ Cell Growth and Apoptosis
Cellular Calcium

Cellular Interactions in
Development
Cellular Neurobiology
Clinical Immunology
Crystallization of Nucleic Acids
and Proteins
★ Cytokines (2nd Edition)
The Cytoskeleton
Diagnostic Molecular Pathology
I and II
Directed Mutagenesis
★ DNA Cloning 1: Core Techniques
(2nd Edition)
★ DNA Cloning 2: Expression
Systems (2nd Edition)
★ DNA Cloning 3: Complex
Genomes (2nd Edition)
★ DNA Cloning 4: Mammalian
Systems (2nd Edition)
Electron Microscopy in Biology
Electron Microscopy in
Molecular Biology
Electrophysiology

Enzyme Assays

Essential Developmental Biology

Essential Molecular Biology I and II

Experimental Neuroanatomy

★ Extracellular Matrix

★ Flow Cytometry (2nd Edition)

Gas Chromatography

Gel Electrophoresis of Nucleic Acids (2nd Edition)

Gel Electrophoresis of Proteins (2nd Edition)

★ Gene Probes 1 and 2

Gene Targeting

Gene Transcription

Glycobiology

Growth Factors

Haemopoiesis

Histocompatibility Testing

★ HIV Volumes 1 and 2

Human Cytogenetics I and II (2nd Edition)

Human Genetic Disease Analysis

Immunocytochemistry

In Situ Hybridization

Iodinated Density Gradient Media

★ Ion Channels

Lipid Analysis

Lipid Modification of Proteins

Lipoprotein Analysis

Liposomes

Mammalian Cell Biotechnology

Medical Bacteriology

Medical Mycology

★ Medical Parasitology

★ Medical Virology

Microcomputers in Biology

Molecular Genetic Analysis of Populations

★ Molecular Genetics of Yeast

Molecular Imaging in Neuroscience

Molecular Neurobiology

Molecular Plant Pathology I and II

Molecular Virology

Monitoring Neuronal Activity

Mutagenicity Testing

★ Neural Cell Culture

Neural Transplantation

Neurochemistry

Neuronal Cell Lines

NMR of Biological Macromolecules

★ Non-isotopic Methods in Molecular Biology

Nucleic Acid Hybridization

Nucleic Acid and Protein Sequence Analysis

Oligonucleotides and Analogues

Oligonucleotide Synthesis

PCR 1

★ PCR 2

★ Peptide Antigens

Photosynthesis: Energy Transduction

★ Plant Cell Biology

★ Plant Cell Culture (2nd Edition)

Plant Molecular Biology

★ Plasmids (2nd Edition)
Pollination Ecology
Postimplantation Mammalian
  Embryos
Preparative Centrifugation
Prostaglandins and Related
  Substances
★ Protein Blotting
Protein Engineering
Protein Function
Protein Phosphorylation
Protein Purification
  Applications
Protein Purification Methods
Protein Sequencing
Protein Structure
Protein Targeting

Proteolytic Enzymes
★ Pulsed Field Gel
  Electrophoresis
Radioisotopes in Biology
Receptor Biochemistry
Receptor–Ligand
  Interactions
RNA Processing I and II
Signal Transduction
Solid Phase Peptide
  Synthesis
Transcription Factors
Transcription and Translation
Tumour Immunobiology
Virology
Yeast

Cytokines

A Practical Approach
SECOND EDITION

Edited by

FRANCES R. BALKWILL

Imperial Cancer Research Fund
Lincoln's Inn Fields
London

IRL PRESS
——at——
OXFORD UNIVERSITY PRESS
Oxford New York Tokyo

Oxford University Press, Walton Street, Oxford OX2 6DP
Oxford   New York
Athens Auckland Bangkok Bombay
Calcutta Cape Town Dar es Salaam Delhi
Florence Hong Kong Istanbul Karachi
Kuala Lumpur Madras Madrid Melbourne
Mexico City Nairobi Paris Singapore
Taipei Tokyo Toronto
and associated companies in
Berlin Ibadan

Oxford is a trade mark of Oxford University Press

Published in the United States
by Oxford University Press Inc., New York

© Oxford University Press, 1995
First edition published 1991
Second edition 1995

Users of books in the Practical Approach Series are advised that prudent
laboratory safety procedures should be followed at all times. Oxford
University Press makes no representation, express or implied, in respect of
the accuracy of the material set forth in books in this series and cannot
accept any legal responsibility or liability for any errors or omissions
that may be made.

A catalogue record for this book is available from the British Library

Library of Congress Cataloging in Publication Data
Cytokines : a practical approach / edited by Frances R. Balkwill.–
2nd ed.
(The Practical approach series; 155)
Includes bibliographical references and index.
1. Cytokines.   2. Cytokines–Research–Methodology.   I. Balkwill,
Frances R.   II. Series.
QR185.8.C95C982   1995   612'.01575–dc20   95–4092
ISBN 0 19 963567 6 (Hbk)
ISBN 0 19 963566 8 (Pbk)

Typeset by Footnote Graphics, Warminster, Wilts
Printed in Great Britain by Information Press Ltd, Eynsham, Oxon.

Preface

Cytokines are of central importance in the regulation of immunity, inflammation, tissue remodelling, and embryonic development. They comprise a group of low molecular weight proteins that are produced by many different cell types and generally act in a paracrine or autocrine fashion, their production being transient and subject to complex and strict control. Cytokines are rarely released singly. An individual cytokine is able to stimulate the production of many others generating a network that interacts with other cell regulators such as hormones and neuropeptides.

An understanding of the cytokine network and its manipulation in disease states has considerable clinical potential. As soluble mediators of immunity, inflammation, and cell growth, cytokines are being used with increasing success in malignant disease, chronic virus and parasitic infections, and some rare congenital immune deficiencies. Paradoxically, uncontrolled or excessive cytokine production may contribute to the pathophysiology of some acute and chronic infections, autoimmune disease, and neoplasia. Thus cytokine antagonists, be they antibodies or peptide mimetics, have therapeutic potential in many diseases.

In the past 15 years cytokine research has become increasingly relevant to many different areas of the biological sciences and medicine. Because of the multidisciplinary nature of cytokine research, and the overlapping biologic properties of cytokines *in vitro*, it would be impossible to devote one chapter of this 'recipe book' to each cytokine currently identified. Instead, the book has been designed with three main aims: first, to select a number of biologic models and outline the ways of assaying cytokine activity in these systems; second, to provide a spectrum of techniques for assaying cytokines at the level of messenger RNA and protein, whether the sample be single cells, tissue sections, tissue cultures, culture supernatants, or clinical specimens; and third, to discuss strategies for purifying, and cloning new cytokine genes and defining the actions of their protein products.

On receipt of all these detailed and fascinating protocols for the successful first edition of this book, I realised that this volume of the Practical Approach series unintentionally fulfilled another purpose—it provided a guide to some of the most important cell biology, immunology, biochemical, and molecular techniques in use today, a guide that is not exclusively for those interested in cytokines, but of interest to scientists from many different disciplines in medical research and the biological sciences.

In this second edition, the majority of techniques have been revised and updated by their original authors, and important new techniques have been added. In particular, the book now contains a detailed guide to genetic manipulation of murine cytokine genes, leading to the production of cytokine transgenic and 'knock-out' mice, assays for soluble cytokine receptors, and assays for the most recently discovered cytokines (up to and including IL-14). A quick scan through the list of contents reveals that almost two hundred different techniques are now described!

For those readers who are using these 'Practical Approaches' for studies with cytokines, I would like to make the following comments. First, concerning the assay of cytokines. The regulation of cytokine production is complex with transcriptional and translational controls, soluble receptor proteins that inhibit biologic activity, and naturally occurring receptor antagonists. It is thus advisable, as several authors in this volume have pointed out, to use a number of complementary assay systems. Moreover *ex vivo* manipulation of cells and tissues may be enough to stimulate cytokine production: heparin can be a powerful inducer, as can adhesion to solid substrates. And all such responses are amplified in the presence of endotoxin which is, in itself, a powerful cytokine inducer. Indeed results obtained using the majority of techniques described in this book can be influenced strongly by contamination with endotoxin. Frequent checks of all reagents with commercially available endotoxin assays are recommended.

Another important point to remember is that most of the cytokine activities described in this book are based on *in vitro* assays. The relevance of these *in vitro* activities to the endogenous function of cytokines is probably best defined at this moment in time by construction of mice transgenic for individual cytokines, or 'knock-out' mice unable to make a particular cytokine.

There is still much to learn about the cytokine network both *in vitro* and *in vivo*, there are new cytokines to be discovered and characterised, new clinical applications for cytokines and their antagonists, and many pathways of cytokine gene activation to be illuminated. I hope that this description of techniques will help. Finally, I would like to thank all the authors of this second edition for communicating their techniques and experience so clearly and concisely.

London   F.R.B.
April 1995

Contents

List of contributors   xxiii

Abbreviations   xxix

1. Cloning and expression of cytokine genes   1
  R. Contreras, J. Demolder, and W. Fiers

  1. Introduction   1

  2. Synthesis of cDNA libraries   2
     Preparation of mRNA   2
     Conversion of mRNA to cDNA   3

  3. Screening of the cDNA libraries for cytokine genes   4
     Screening on the basis of biological activity   4
     Screening on the basis of protein sequence data   9
     Screening on the basis of nucleic acid sequence data   9
     Differential screening   9

  4. Heterologous expression of cytokine genes   10
     Expression in mammalian cells   10
     Expression in lower eukaryotes   11
     Expression in bacterial systems   13

  5. Conclusion   15

  Acknowledgements   15

  References   16

2. Southern blotting of cytokine genes   19
  M. O. Diaz and T. W. McKeithan

  1. Introduction   19

  2. Basic technique   19
     Preparation of DNA   19
     Restriction enzyme digestion of DNA   21
     Gel electrophoresis   22
     Transfer of DNA to membranes   24
     Preparation of radioactive probes   27
     Membrane hybridization   29

Contents

3. Common artefacts and interpretation   30
Contamination with plasmid DNA   30
Partial digests   31
Degraded DNA: incomplete transfer   31
Background sources: repetitive sequences   31
Quantitation   33

References   33

3. Northern analysis, ribonuclease protection, and *in situ* analysis of cytokine messenger RNA 35

M. Stuart Naylor, Michele Relf, and Frances R. Balkwill

1. Introduction   35

2. General precautions for handling RNA   36

3. Northern analysis   36
Electrophoresis of RNA for Northern blot   38
Probe labelling   40

4. Ribonuclease protection   42
Controls   42
Preparation of template DNA   43
Analysis on sequencing gel   45

5. *In situ* hybridization   46
Sample handling   46
Pre-hybridization   47
Probe choice, labelling, and hybridization   47
Controls   48
Sample preparation   49
Interpretation   54

References   55

4. Polymerase chain reaction to quantitate cytokine mRNA 57

Cornelia Platzer and Thomas Blankenstein

1. Introduction   57

2. General remarks   58

3. Cytokine RT-PCR   59

4. Control fragment design and construction   60

Contents

5. Quantitative PCR 62

Acknowledgements 67

References 68

5. Comparative polymerase chain reaction 69
Sylvie de Kossodo

1. Introduction 69

2. Methods 70

References 76

6. Receptor-binding studies 77
Thomas M. Mariano, Lara Izotova, and Sidney Pestka

1. Introduction 77

2. Methods 78
Cell culture 78
Interferon radiolabelling 78
Binding of [^{32}P]Mu-IFN-γ to tissue culture cells 79
Binding of [^{32}P]Mu-IFN-γ to *X. Laevis* oocytes 81
Covalent cross-linking of [^{32}P]Mu-IFN-γ to cells and oocytes 82

3. Cloning of the Mu-IFN-γ receptor (Mu-IFN-γR) 85
Introduction 85
Procedures for isolating the Mu-IFN-γR cDNA 85

4. Cloning functional type I and II interferon receptors 86

5. Concluding remarks 88

Acknowledgements 89

References 89

7. Measurement of cytokine induction of PC-specific phospholipase C and sphingomyelinases 93
Stefan Schütze and Martin Krönke

1. Phosphatidylcholine-specific phospholipase C 95
Detection of 1,2-DAG production from PC in metabolically labelled cells 95
Analysis of phosphatidylcholine (PC) 99
Detection of phosphorylcholine (Pchol) 100
Estimation of combined phospholipase D (PLD)/phosphatidic acid-phosphatase (PA-Ptase) activities in DAG-formation 101
Inhibitors of PC-PLC 102

2. Sphingomyelinases   102
   Detection of SM hydrolysis in metabolically labelled cells   103
   Detection of ceramide   104
   Estimation of SMase activity *in vitro*   105

Acknowledgements   108

References   109

8. Two-dimensional gel electrophoresis for purification, sequencing, and identification of cytokine-induced proteins in normal and malignant cells   111
Diana M. Smith, Huu M. Tran, and Lois B. Epstein

1. Introduction   111

2. Preparation of cell lysates   112
   Samples for analytical gels   112
   Samples for preparative gels   114

3. Isolation of proteins from cell lysates by 2D PAGE   115
   Resolving proteins on 2D analytical gels   115
   Resolving proteins on preparative 2D gels   118

4. Identification of proteins by immunoblot analysis   119

5. Microsequencing of proteins isolated by preparative
   2D PAGE   121

6. Analysis of the gels   123
   Visual approach to analysis   123
   Computer based analysis   123

7. Analytical 2D PAGE applications   123
   Analytical gels for detecting alterations in gene/protein expression   123
   Work from our laboratory employing analytical 2D gels   124

8. Preparative 2D PAGE applications   124
   Preparative gels for detecting alterations in protein expression 124
   Work from our laboratory employing preparative 2D gels and
      microsequencing techniques   125

9. Concluding remarks   126

Acknowledgements   127

References   127

9. Antiviral activity of cytokines   129
John A. Lewis

1. Introduction   129
   Choice of an assay system   129

2. Assaying for antiviral effects   131
    Multi-well assay procedure   132
    Micro-titre plate assay   135
    Virus yield reduction assay   137

3. Preparation of virus stocks   140

References   141

10. Assays for human B cell responses to cytokines   143

Robin E. Callard, Karolena T. Kotowicz, and
Susan H. Smith

1. Introduction   143

2. Reagents used for *in vitro* experiments with human B cells   144
    Culture medium   144
    Holding medium   145
    Lymphocyte separation medium   145
    Sheep red blood cells (SRBC)   146
    Phorbol ester   146
    Anti-IgM   146
    Staphylococcus aureus Cowan 1 strain (SAC)   147
    Influenza virus   147
    ^{3}H-thymidine   147
    FITC-goat anti-mIg   147
    Recombinant cytokines   147
    Monoclonal antibodies   148

3. Preparation of human B cells   148
    Mononuclear cell preparation   148
    Depletion of T cells by E-rosetting   150
    Depletion of monocytes by adherence   151
    Preparation of heavy and light B cells   153
    Separation of B cell subpopulations with monoclonal antibodies   153
    Immunofluorescence analysis of B cell preparations 156
    Human B cell lines for assaying B cell growth and differentiation
      factors   158

4. B cell activation   159
    Measurement of RNA synthesis   160
    Measurement of B cell activation by expression of surface
      activation antigens such as CD23 and surface IgM   160

5. Assays for B cell proliferation in response to cytokines   162
    Cytokine stimulation of B cell proliferation   162
    Cytokine stimulation of proliferation by B cell lines   165

6. Assays for cytokines which stimulate human B cell
differentiation (Ig secretion)   166
    Co-stimulation assays for cytokines which stimulate
      immunoglobulin secretion (B cell differentiation factors)   166
    Assays for B cell differentiation factors using B cell lines   166
    Assays for cytokine regulation of Ig isotype responses   168
    T cell replacing factors   169

7. Enzyme immunoassays for immunoglobulin secreted
by human B cells   171
    Measurement of Ig (IgM, IgG, IgA, and IgE) secretion   171
    Measurement of IgG subclasses by ELISA   173
    ELISA for specific antibody production   174

References   176

11. T cells and cytokines 179

S. B. A. Cohen, J. Clayton, M. Londei, and M. Feldmann

1. Introduction   179

2. Isolation of T cells for analysis of cytokine production   179
    T cell separation using a one-step Ficoll gradient   180
    Negative B cell depletion using immunomagnetic beads   182

3. Analysis of T cell cytokine production   183

4. Cytokine production by T cells of different phenotypes   186
    Cytokine production by CD4$^+$ cells   186
    Cytokine production by CD8$^+$ T cells   187
    Cytokine production by CD4$^-$ CD8$^-$ $\alpha\beta$ T cells   187
    Cytokine production by $\gamma\delta$ T cells   187
    Optimization of cytokine production in response to various
      stimuli   187

5. Cytokine production of antigen specific T cells   188

6. Effect of cytokines on T cells   188
    Regulation of T cell growth   188
    Variations on cloning technique   192
    Isolation of autoreactive T cells   192
    Regulation of activation   193
    Measurement of T cell growth in response to cytokines and
      antigen stimulation   193
    T cell proliferation—direct [^{3}H]thymidine incorporation   193
    IL-2 release bioassay   194

7. Conclusions   195

References   195

12. The generation and quantitation of cell-mediated cytotoxicity 197

Elizabeth Grimm and William Loudon

1. Introduction 197

2. Techniques for the isolation of lymphocyte populations 199
Isolation of PBMC, PBL, or LGL 199
Isolation of specific lymphoid populations based upon
phenotypic markers or light-scatter characteristics 201
Isolation of tumour infiltrating lymphocytes 203

3. Generation of lymphokine-activated killers 204
The generation of IL-2 activated killers 204
Anti-CD3/IL-2 activated 'T cell' killers 205
Alternative cytokine activation strategies 206

4. Cryopreservation of lymphocytes and tumours 207

5. Cytotoxicity assays 208
^{51}Cr-release cytotoxicity assay 208
Single-cell CMC assay 211
Alternative approaches for measuring CMC, including non-
radioactive assays 212

References 212

13. Assays for chemotaxis 215

Jo Van Damme and René Conings

1. Introduction 215

2. Chemotaxis under agarose 215
Preparation of test samples 218
Assay procedure 218

3. Chemotaxis in micropore filters 220

4. *In vivo* chemotactic responses to cytokines 222

Acknowledgements 224

References 224

14. Cytokine regulation of endothelial cells 225

*P. Allavena, E. Dejana, F. Bussolino, A. Vecchi, and
A. Mantovani*

1. Introduction 225

2. Endothelial cells   226
    HUVEC   226
    Mouse PmT-transformed lines   227

3. Chemotaxis   229
    Boyden chamber assay   230
    Micro-chemotaxis method   231
    Comments   233

4. Procoagulant activity (PCA) or thromboplastin activity 233
    Comments   234

5. Platelet-activating factor (PAF)   235
    Cell cultures used for PAF assay   235
    Extraction of PAF   236
    Identification of PAF   237
    Quantitation of PAF by bioassay   237
    Quantification of PAF by HPLC-tandem mass spectrometry 239
    Comments   240

6. Leukocyte adhesion and transmigration   241
    Comments   242

Acknowledgements   244

References   244

15. Biological assays for haemopoietic growth factors   247

N. G. Testa, C. M. Heyworth, B. I. Lord, and E. A. de Wynter

1. Introduction   247

2. Preparation of cell suspensions   248
    Murine bone-marrow   248
    Human bone-marrow   249
    Separation of excess red cells from human bone-marrow samples 250

3. Standard clonal assays   250
    Murine bone-marrow   250
    Human bone-marrow or peripheral blood 252
    General comments   252

4. Serum-free cultures   253
    General comments   253
    Preparation of the reagents   254
    Preparation of serum-free medium   254
    Comments   254

5. Growth factor-dependent cell-lines   254

6. Enrichment of progenitor cell populations from murine bone-marrow 256

Contents

General comments   256
Murine cells   256
Human cells   261

References   267

16. Assays for macrophage activation by cytokines   269

Anthony Doyle, Michael Stein, Satish Keshav, and Siamon Gordon

1. Introduction   269

2. Assay for respiratory burst activity   270

3. Assay for secretion of nitric oxide   271

4. MHC class II expression   272

5. Macrophage mannosyl receptor (MMR) assays   273
Single cell assays   276

6. Macrophage cytokine assays   276

7. Conclusions   277

References   277

17. Measurement of proliferative, cytolytic, and cytostatic activity of cytokines   279

Frances Burke, Enrique Rozengurt, and Frances R. Balkwill

1. Introduction   279

2. Effects of cytokines on cell growth *in vitro*   279
Direct cell counting   279

3. Determination of effects on growth of cytokines using indirect methods   282
The alkaline phosphatase assay   282
The MTT-type assay   283

4. Measurement of cell lysis   283
Measurement of cell lysis using crystal violet   283
Measurement of cell lysis using the lactate dehydrogenase assay   284

5. Cell cycle studies   286
Measurement of [^{3}H]thymidine incorporation into acid-insoluble material   286
Autoradiography of labelled nuclei   288
Cytofluorimetry   290

Measurement of bromodeoxyuridine incorporation to assess cell
cycle changes *in vivo*   291
Mitotic index   293

6. DNA fragmentation   293

References   295

18. Production of cytokine transgenic and
knockout mice   297
Manolis Pasparakis and George Kollias

1. Introduction   297

2. Production of transgenic mice by pronuclear injection
of DNA   297
Equipment   297
Animals   298
Preparation of DNA for microinjection   298
Preparation of the pipettes used for pronuclear injections   299
Recovery of oocytes   300
Microinjection procedure   301
Integration and expression of injected genes   304
Identification of transgenic progeny   304

3. Production of knockout mice   308
Embryonic stem cells   308
Culturing ES cells   308
Gene targeting in ES cells   311
Transfection of ES cells   313
Picking and expansion of ES cell colonies   314
Storage and recovery of ES cell clones   315
Identification of targeted ES cell clones   317
Generation of chimaeric mice   318

4. Transgenic and knockout mice in cytokine research   322

Acknowledgements   324

References   324

19. Development of antibodies to cytokines   327
S. Poole

1. Introduction   327

2. Choice of immunogen   327
Recombinant cytokines   327
Purified 'natural' cytokines   327
Conjugates of cytokines   328

Small peptide homologues 328
Enzymatic digests of cytokines 332

3. Choice of animal 332

4. The adjuvant 332

5. Immunization schedules 333
Mice 333
Rabbits 334
Guinea-pigs 335

6. Screening 336

7. Purification of antibodies 337

Acknowledgements 337

References 338

20. **Visualizing the production of cytokines and their receptors by single human cells** 339
C. E. Lewis and A. Campbell

1. Introduction 339

2. Immunolocalization of cytokines 341
Alkaline phosphatase anti-alkaline phosphatase (APAAP) method 341
Streptavidin–biotin (AB) method 343

3. Detection of cytokine release by individual cells 347
Reverse haemolytic plaque assay (RHPA) 347
Elispot assay 351

4. Discussion 353

Acknowledgements 355

References 355

21. **Quantitative biological assays for individual cytokines** 357
Meenu Wadhwa, Christopher Bird, Lisa Page, Anthony Mire-Sluis, and Robin Thorpe

1. Introduction 357
Bioassays compared with immunoassays and receptor binding assays 357

2. Biological assays for interleukins 358
Bioassay of interleukin-2 (IL-2) 358

Contents

Bioassay of interleukin-1 (IL-1)   359
Bioassay of interleukin-4 (IL-4)   360
Bioassay for interleukin-5 (IL-5)   364
Bioassay for interleukin-6 (IL-6)   365
Bioassay for interleukin-7 (IL-7)   367
Bioassay for interleukin-9 (IL-9)   368
Bioassay for interleukin-10 (IL-10)   368
Bioassay for interleukin-11 (IL-11)   369
Bioassay for interleukin-12 (IL-12)   370
Bioassay for interleukin-13 (IL-13)   371
Bioassay for interleukin-14 (IL-14)   371

3. Bioassays for colony stimulating factors—interleukin-3,
granulocyte colony stimulating factor (G-CSF),
macrophage colony stimulating factor (M-CSF),
granulocyte macrophage colony stimulating factor (GM-
CSF), stem cell factor (SCF), and leukaemia inhibitory
factor (LIF)   372
   Maintenance of MO-7e/TF-1 cell-lines   374
   Maintenance of GNFS-60 and MNFS-60 cell-lines 376
   Bioassay for leukaemia inhibitory factor (LIF) 377
   Bioassay for stem cell factor (SCF)   378

4. Bioassay for interleukin-8 (IL-8) and other chemokines   378

5. Bioassays for other cytokines   381
   Bioassay for tumour necrosis factors   381
   Bioassay for interferons (IFNs)   382
   Bioassay for TGF-β   384
   Bioassay for oncostatin-M   386

6. Analysis of results   388

Acknowledgements   389

References   390

22. **RIA, IRMA, and ELISA assays for
cytokines and their soluble receptors**   393
A. Meager

1. Introduction   393

2. Radioimmunoassays (RIA)   394

3. Immunoradiometric assays (IRMA)   397
   IRMA assays for cytokines   397
   IRMA assays for soluble cytokine receptors   399

4. Enzyme-linked immunosorbent assays (ELISA)   400

Contents

5. Problems   402
    Calibration   402
    Mixtures of cytokine and soluble receptors   403
    Assay diluent or matrix   403
    Samples from different sources   403

References   404

A1. *Suppliers of specialist items*   405

Index   411

Contributors

P. ALLAVENA
Istituto di Ricerche Farmacologiche 'Mario Negri', Milano, Italy.

FRANCES R. BALKWILL
Biological Therapy Laboratory, Imperial Cancer Research Fund, PO Box 123, Lincoln's Inn Fields, London WC2A 3PX, UK.

CHRISTOPHER BIRD
Division of Immunobiology, National Institute for Biological Standards and Control, Blanche Lane, South Mimms, Potters Bar, Herts EN6 3QG, UK.

THOMAS BLANKENSTEIN
Max-Delbrück-Centrum für Molekulare Medizin, Robert-Rössle-Str. 10, 13122 Berlin, Germany.

FRANCES BURKE
Biological Therapy Laboratory, Imperial Cancer Research Fund, PO Box 123, Lincoln's Inn Fields, London WC2A 3PX, UK.

F. BUSSOLINO
Dipartimento di Genetica, Biologia, Chimica Medica, Università di Torino, Torino, Italy.

ROBIN E. CALLARD
Cellular Immunology Unit, Institute of Child Health, 30 Guilford Street, London WC1N 1EH, UK.

A. CAMPBELL
Nuffield Department of Pathology and Bacteriology, John Radcliffe Hospital, Headington, Oxon OX9 3DU, UK.

J. CLAYTON
Nature, Macmillan Journals, 4 Little Essex Street, London WC2R 3LF, UK.

S. B. A. COHEN
Kennedy Institute of Rheumatology, Sunley Division, 1 Lurgan Avenue, Hammersmith, London W6 8LW, UK.

RENÉ CONINGS
Rega Institute for Medical Research, University of Leuven, Minderbroeders-straat 10, 3000 Leuven, Belgium.

R. CONTRERAS
Laboratory of Molecular Biology, State University, Gent, Belgium.

Contributors

SYLVIE DE KOSSODO
Biological Therapy Laboratory, Imperial Cancer Research Fund, PO Box 123, Lincoln's Inn Fields, London WC2A 3PX, UK.

E. A. DE WYNTER
Paterson Institute for Cancer Research, Christie Hospital NHS Trust, Wilmslow Road, Manchester M20 9BX, UK.

E. DEJANA
CEA, HEM/DBMS, INSERM U217, CEN-G, Grenoble, France.

J. DEMOLDER
Laboratory of Molecular Biology, State University, Gent, Belgium.

M. O. DIAZ
Loyola University, Oncology Institute, 2160 South 1st Avenue, Maywood, Illinois 60153, USA.

ANTHONY DOYLE
Queensland Institute of Medical Research, Post Office, Royal Brisbane Hospital, Queensland 4029, Australia.

LOIS B. EPSTEIN
Cancer Research Institute and Department of Pediatrics, University of California, San Francisco, CA 94143, USA.

M. FELDMANN
Kennedy Institute of Rheumatology, Sunley Division, 1 Lurgan Avenue, Hammersmith, London W6 8LW, UK.

W. FIERS
Laboratory of Molecular Biology, State University, Gent, Belgium.

SIAMON GORDON
University of Oxford, Sir William Dunn School of Pathology, South Parks Road, Oxford OX1 3RE, UK.

ELIZABETH GRIMM
Department of Tumor Biology, University of Texas, M. D. Anderson Cancer Center, Houston, TX 77030, USA.

C. M. HEYWORTH
Paterson Institute for Cancer Research, Christie Hospital NHS Trust, Wilmslow Road, Manchester M20 9BX, UK.

LARA IZOTOVA
Department of Molecular Genetics and Microbiology, UMDNJ-Robert Wood Johnson Medical School, 675 Hoes Lane, Piscataway, New Jersey 08854–5635, USA.

Contributors

SATISH KESHAV
University of Oxford, Sir William Dunn School of Pathology, South Parks Road, Oxford OX1 3RE, UK.

GEORGE KOLLIAS
Department of Molecular Genetics, Hellenic Pasteur Institute, 127 Vas Sofias Av., Athens 11521, Greece.

KAROLENA T. KOTOWICZ
Cellular Immunology Unit, Institute of Child Health, 30 Guilford Street, London WC1N 1EH, UK.

MARTIN KRÖNKE
Institut für Immunologie, Christian-Albrechts-Universitat, Keil, Brunswiker Str.4, 24105, Kiel, Germany.

C. E. LEWIS
Nuffield Department of Pathology and Bacteriology, University of Oxford, John Radcliffe Hospital, Headington, Oxon OX9 3DU, UK.

JOHN A. LEWIS
Department of Anatomy and Cell Biology, SUNY Health Science Centre, 450 Clarkson Avenue, Brooklyn, New York 11203, USA.

M. LONDEI
Kennedy Institute of Rheumatology, Sunley Division, 1 Lurgan Avenue, Hammersmith, London W6 8LW, UK.

B. I. LORD
Paterson Institute for Cancer Research, Christie Hospital NHS Trust, Wilmslow Road, Manchester M20 9BX, UK.

WILLIAM LOUDON
Department of Tumor Biology, University of Texas, M. D. Anderson Cancer Center, Houston, TX 77030, USA.

ALBERTO MANTOVANI
Istituto di Ricerche Farmacologiche 'Mario Negri', Milano, Italy.

THOMAS M. MARIANO
Department of Molecular Genetics and Microbiology, UMDNJ-Robert Wood Johnson Medical School, 675 Hoes Lane, Piscataway, New Jersey 08854-5635, USA.

T. W. McKEITHAN
Department of Pathology, University of Chicago, Box 420, Chicago, IL 60637, USA.

ANTHONY MEAGER
Division of Immunobiology, National Institute for Biological Standards and Control, Blanche Lane, South Mimms, Potters Bar, Herts EN6 3QG, UK.

Contributors

ANTHONY MIRE-SLUIS
Division of Immunobiology, National Institute for Biological Standards and Control, Blanche Lane, South Mimms, Potters Bar, Herts EN6 3QG, UK.

M. STUART NAYLOR
Institute of Cancer Research, Haddow Laboratories, 15 Cotswold Road, Sutton, Surrey SM2 5NG, UK.

LISA PAGE
Division of Immunobiology, National Institute for Biological Standards and Control, Blanche Lane, South Mimms, Potters Bar, Herts EN6 3QG, UK.

MANOLIS PASPARAKIS
Department of Molecular Genetics, Hellenic Pasteur Institute, 127 Vas Sofias Av., Athens 11521, Greece.

SIDNEY PESTKA
Department of Molecular Genetics and Microbiology, UMDNJ-Robert Wood Johnson Medical School, 675 Hoes Lane, Piscataway, New Jersey 08854–5635, USA.

CORNELIA PLATZER
Klinikum der Friedrich-Schiller Universität, Klinik für Innere Medizin II, Erlanger Allee 101, 07740 Jena, Germany.

S. POOLE
Division of Endocrinology, National Institute for Biological Standards and Control, Blanche Lane, South Mimms, Potters Bar, Herts EN6 3QG, UK.

MICHELE RELF
Biological Therapy Laboratory, Imperial Cancer Research Fund, PO Box 123, Lincoln's Inn Fields, London WC2A 3PX, UK.

ENRIQUE ROZENGURT
Growth Regulation Laboratory, Imperial Cancer Research Fund, PO Box 123, Lincoln's Inn Fields, London WC2A 3PX, UK.

STEFAN SCHÜTZE
Institut für Immunologie, Christian-Albrechts-Universität Kiel, Brunswiker Str. 4, 24105, Kiel, Germany.

DIANA M. SMITH
Cancer Research Institute, University of California, San Francisco, CA 94143, USA.

SUSAN H. SMITH
Cellular Immunology Unit, Institute of Child Health, 30 Guilford Street, London WC1N 1EH, UK.

Contributors

MICHAEL STEIN
University of Oxford, Sir William Dunn School of Pathology, South Parks Road, Oxford OX1 3RE, UK.

N. G. TESTA
Paterson Institute for Cancer Research, Christie Hospital NHS Trust, Wilmslow Road, Manchester M20 9BX, UK.

ROBIN THORPE
Division of Immunobiology, National Institute for Biological Standards and Control, Blanche Lane, South Mimms, Potters Bar, Herts EN6 3QG, UK.

HUU M. TRAN
Cancer Research Institute, University of California, San Francisco, CA 94143, U.S.A.

JO VAN DAMME
Rega Institute for Medical Research, University of Leuven, Minderbroeders-straat 10, 3000 Leuven, Belgium.

A. VECCHI
Istituto di Ricerche Farmacologiche 'Mario Negri', Milano, Italy.

MEENU WADHWA
Division of Immunobiology, National Institute for Biological Standards and Control, Blanche Lane, South Mimms, Potters Bar, Herts EN6 3QG, UK.

Abbreviations

AET	aminoethylisothiouronium bromide hydrobromide
AML	acute myeloid leukaemia
AP	alkaline phosphatase
APAAP	anti alkaline phosphatase
A-SMase	acidic, endosomal sphingomyelinase
ATCC	American tissue culture collection
BCDF	B cell differentiation factor
BCG	bacillus Calmette-Guerin
BCGF	B cell growth factor
BDT	bis-diazotolidine
BFU-E	erythroid colonies
BHK	baby hamster kidney
Bo	bovine
BSA	bovine serum albumin
BTG	bovine thyroglobulin
CD40L	CD40 ligand
CFC	colony forming cells
CFU-E	erythroid progenitors
CHEF	contour-clamped horizontal electrical field
CHO	Chinese hamster ovary cells
CID	collision-induced dissociation
CMC	cell-mediated cytotoxicity
CPER	cytopathic effect reduction
$CrCl_3$	chromium chloride hexahydrate
CSF	colony stimulating factor
CTL	cytotoxic T lymphocytes
DAG	diacylglycerol
DME	Dulbecco's modified Eagle's medium
DMEM	Dulbecco's modified essential medium
DMSO	dimethylsulfoxide
dpm	disintegrations per minute
DSS	disuccinimidyl suberate
E^+	E-rosette forming T cell
E^-	non-rosette forming lymphocyte
EBV	Epstein–Barr virus
EC	endothelial cells
ECGS	endothelial cell growth supplement
EDA	eosinophil differentiation assay
EDTA	ethylene diamine tetra acetic acid

EIA	enzyme immunoassays
ELAM-1	endothelial leukocyte adhesion molecule-1
ELISA	enzyme-linked immunoabsorbent assay
EMCV	encephalomyocarditis virus
EMEM	Eagle's minimum essential medium
Eo-CFC	eosinophil colony forming cell
EPO	eosinophil peroxidase
Epo	erythropoietin
EqS	horse serum
ES	embryonic stem cells
FACS	fluorescence activated cell sorter
FALS	forward angle light scatter
FBS	fetal bovine serum
FCS	fetal calf serum
FDA	fluorescent dye fluoroscein diacetate
FGF	fibroblast growth factor
FITC	fluorescein isothiocyanate
Fmin	fluorescence at very low calcium
Fmax	fluorescence at very high calcium
fMLP	formylmethionyl-leucylphenylalanine
G-CFC	granulocyte colony forming cells
G-CSF	granulocyte colony stimulating factor
GAM	goat anti-mouse
GCP	granulocyte chemotactic protein
GM-CFC	granulocyte and macrophage colony forming cells
GM-CSF	granulocyte macrophage colony stimulating factor
Gp	G protein
GuSCN	guanidinium thiocyanate
hCG	human chorionic gonadotropin
HBSS	Hank's balanced salt solution
HPTLC	high performance thin layer chromatography
HRP	horseradish peroxidase
HSA	human serum albumin
^{3}HTDR	tritiated thymidine
Hu	human
HUIFN	human IFN
HUTNF	human TNF
HUVEC	human umbilical venous endothelial cells
ICAM-1	intercellular adhesion molecule-1
IEF	isoelectric focusing
IFN	interferon
IL	interleukin
IMDM	Iscove's modified Dulbecco's medium
IP$_3$	inositol 1,4,5-trisphosphate

IRMA	immunoradiometric assay
KLH	keyhole limpet haemocyanin
LAB	labelled strepavidin–biotin
LAK	lymphokine-activated killer cell
LCR	locus control region
LDH	lactate dehydrogenase
LFA-1	lymphocyte function associated antigen 1
LGL	large granular lymphocyte
LIF	leukaemia inhibitory factor
LPBA	liquid phase binding assays
LR	ligand/receptor
LSIMS	liquid secondary ion mass spectrometry
LT	lymphotoxin
LU	lytic units
Mø	macrophage
M-CFC	macrophage colony forming cell
M-CSF	macrophage colony stimulating factor
MAb	monoclonal antibodies
MAC	monoclonal anti-cytokine
MACS	magnetic cell sorter
MAR	mouse anti-rabbit
MBS	maleimidobenzoic acid N-hydroxy succinimide
MCP	monocyte chemotactic protein
Meg-CFC	megakaryocyte colony forming cell
MEM	minimal essential medium
MFI	median fluorescence intensity
MF	mating factor
MFR	mannosyl fucosyl receptor
MHC	major histocompatibility complex
MLP	major late promoter
M-CSF	macrophage colony stimulating factor
MMLV	Molony murine leukaemia virus
MMR	macrophage mannosyl receptor
MOPS	3-N-morpholinopropanesulfonic acid
M_r	molecular weight
MTT	(3-(4,5-dimethylthiazol-2-ys)-2,5-diphenyl tetrazolium salt
Mu	murine
4-MUH	4-methylumbelliferyl heptanoate
MuTNF	mouse TNF
NADG	N-acetyl-D-glucosamine
NBB	naphthol blue black
NBS	N-bromosuccinimide
NBT	nitrobluetetrazolium
NCI	National Cancer Institute

NGS	normal goat serum
NK	natural killer
NHS	normal human AB serum
NIBSC	National Institute for Biological Standards and Controls
NPP	*p*-nitrophenyl phosphate
N-SMase	neutral plasma membrane bound sphingomyelinase
OD	optical density
OLB	oligo-labelling buffer
OPD	orthophenylenediamine
PA	phosphatidic acid
PAF	platelet activating factor
PA-Ptase	phosphatidic acid-phosphatase (PA-Ptase)
PBA	phosphate buffered saline + 0.1% bovine serum albumin
PBL	peripheral blood lymphocytes
PBMC	peripheral blood mononuclear cells
PBS	phosphate buffered saline
PCA	procoagulant activity
Pchol	phosphorylcholine
PC-PLC	phosphatidylcholine-specific phospholipase C
PC-PLD	phosphatidylcholine-specific phosphalipase C
PCR	polymerase chain reaction
PdB	phorbol dibutyrate
PDQUEST	protein databases quantitative electrophoresis standardized
PEG	polyethylene glycol
PFU	plaque forming units
$PG1_2$	prostacyclin
PHA	phytohaemagglutinin
p*I*	isoelectric point
PI	phosphoinositol
PIP_2	phosphatidyl inositol bisphosphate
PKC	protein kinase C
PL	phospholipase
PLC	phospholipase C
PLD	phospholipase D
PMA	phorbol myristate acetate
PMN	polymorphonuclear cells
PMS	pregnant mare serum
pNPP	*p*-nitro-phenylphosphate
PPD	purified protein derivative
PrA	protein A
PrA-sRBC	PrA-conjugated sRBC
PRP	platelet-rich plasma
PVP	polyvinyl pyrrolidone
R	recombinant

Abbreviations

Ra	rat
RAC	rabbit anti-cytokine
RALS	right angle light scatter
RAM	rabbit anti-mouse IgG
RFLP	restriction fragment length polymorphism
RHPA	reverse haemolytic plaque assay
RIA	radioimmunoassay
RT-PCR	reverse transcriptase-polymerase chain reaction
SAC	*Staphylococcus aureus* Cowan
SCF	stem cell factor
SDS	sodium dodecyl sulfate
SF	synovial fluid
SFV	semlike forest virus
SM	sphingomyelin
SMase	sphingomyelinase
SPBA	solid phase binding assays
SRBC	sheep red blood cells
TCA	trichloroacetic acid
TG	triglyceride
TGF-α	transforming growth factor α
TGF-β	transforming growth factor β
ThP	T helper precursor cells
TIL	tumour-infiltrating lymphocytes
TLC	thin-layer chromatography
TMC	tonsillar mononuclear cells
TNF	tumour necrosis factor
TPO	thyroid peroxidase
TRF	T cell replacing factor
VCAM	vascular cell adhesion molecule
VEGF	vascular endothelial growth factor
VSV	vesicular stomatitis virus
WGA-FITC	fluoresceinated wheat germ agglutinin

1

Cloning and expression of cytokine genes

R. CONTRERAS, J. DEMOLDER, and W. FIERS

1. Introduction

Cytokines are proteins secreted by a cell into the extracellular fluid where they exert their effects on the same cells (autocrine activity) or on neighbouring cells (paracrine activity) by interacting with specific receptors. Most cytokine proteins have, on a molar basis, very high biological activities (for instance, colony stimulating factors are active at 10^{-11} to 10^{-13} M concentrations) which usually can be assayed in rapid, sensitive and fairly specific *in vitro* test systems. These fast and simple detection systems have played a crucial role in the cloning of the corresponding cytokine genes by providing tools for either purification of the protein, the mRNA, or the direct detection of biologically active clones in cDNA expression libraries. Other cytokines, such as the strongly inducible chemokine family (the mRNA for these molecules can represent 1% of the total poly(A)$^+$ RNA after stimulation), act in much higher concentrations (10^{-3} to 10^{-9} M) and have often been cloned by different techniques such as induction-specific differential hybridization or subtracted libraries. An important factor in a successful cloning procedure is the presence of an abundant source of protein and mRNA. The expression level of the desired gene in this source determines the frequency with which the genes will be found in the cDNA libraries.

After cloning and sequencing the gene, efficient expression of the cDNAs has been obtained for all cytokines in prokaryotic, yeast or fungal, or mammalian systems. The most important steps concerning cloning and expression of cytokine proteins, except for the techniques used for purification of the natural or recombinant (r) cytokine proteins, will be described in this chapter. A general guide for purification of proteins has appeared recently (1). A more detailed description of the biological assays used for detection of individual cytokines is given in Chapter 21.

2. Synthesis of cDNA libraries

2.1 Preparation of mRNA

A prerequisite for the synthesis of a large, full-length cDNA library is an excellent preparation of mRNA. Several procedures for the preparation of eukaryotic mRNA have been described in detail in previous volumes of this series or in several laboratory manuals (2–5). In summary, two major procedures have been used. In the first, extraction of total RNA is obtained under strong denaturing conditions with guanidinium thiocyanate followed by centrifugation in CsCl buffer (6) and a sucrose gradient (7). This method is especially advised for the extraction of mRNA from primary cells such as spleen or peripheral blood lymphocytes. A second, shorter procedure for the extraction of cytoplasmic RNA by lysis of the cells in NP40 buffer, has been used routinely in our laboratory (2). A detailed description of the latter procedure is given in *Protocol 1*.

Protocol 1. Preparation of cytoplasmic RNA from cultured cells[a,b]

Equipment and reagents

- PBS: 0.14 M NaCl, 2.7 mM KCl, 6.5 mM Na_2HPO_4, 1.5 mM KH_2PO_4, pH 7.2; sterile RNase-free
- NP40 lysis buffer: 0.5% NP-40, 10 mM Tris–HCl, pH 8.0, 140 mM NaCl, 1.5 mM $MgCl_2$, 1000 U/ml RNasin (Promega)
- HB4 Sorvall rotor
- 2 × RNA extraction buffer: 0.2 M Tris–HCl, pH 7.5, 25 mM EDTA, 0.3 M NaCl, 2% SDS
- Proteinase K

- TE buffer: 10 mM Tris–HCl, pH 8.0, 1 mM EDTA
- Phenol/TE
- Phenol/chloroform/isoamyl alcohol (50:49:1, v/v)
- Chloroform/isoamyl alcohol (96:4, v/v)
- Ethanol
- Falcon blue-cap tube (50 ml)
- 1% agarose gel

Method

1. Wash the cells twice with ice-cold, sterile RNase-free PBS.

2. Resuspend the cells in 10 ml ice-cold NP40 lysis buffer. Shake carefully and place on ice for 3 min.

3. Remove the nuclei by centrifugation (2500 r.p.m., HB4 Sorvall rotor, 10 min at 4°C). Carefully recover 9 ml supernatant.

4. Add 1 vol. of 2 × RNA extraction buffer and proteinase K to 200 μg/ml. Incubate at 37°C for 30 min.

5. Extract once with phenol/TE, once with 50% phenol/49% chloroform/ 1% isoamyl alcohol, and once with 96% chloroform/4% isoamyl alcohol. Precipitate the RNA with 2 vol. of ethanol for at least 2 h at −20°C in a 50 ml Falcon blue-cap tube.

6. Collect the RNA by centrifugation at 3500 r.p.m. for 30 min at 4°C. Wash

the pellet with 70% ethanol and dry under vacuum. Dissolve in 2 ml TE pH 8.0.

7. Measure the yield and quality of the RNA by determining its absorbance at 260 and 280 nm. Also test its quality by electrophoresis on a 1% agarose gel.

---

[a] Adapted from ref. 7.
[b] Protocol for 10^8 lymphoid suspension culture cells.

Classically, the poly(A)$^+$ fraction of the total RNA preparation is purified by chromatography on oligo(dT)-cellulose or poly(U)-Sepharose (5). Recently, rapid and efficient procedures, using oligo(dT$_{30}$) bound on latex (8), or magnetic (9) beads have been introduced and used successfully for the preparation of several cDNA libraries (our unpublished results).

Certain studies of cDNA libraries of small numbers of cells require scaled-down versions of protocols for the isolation of RNA: a modification of the guanidinium thiocyanate technique (6) was published by Hahnel *et al.* (10), and the acidic guanidinium thiocyanate–phenol–chloroform method was modified by Weng *et al.* (11) and Belyavsky *et al.* (12).

More recently, several companies have commercialized very useful kits for the preparation of mRNA from mammalian cells or tissues.

2.2 Conversion of mRNA to cDNA

The conversion of mRNA to cDNA has become a standard technique for which several commercial kits are available. The procedure can be modified according to the way the cDNA will be inserted in the vector. The choice of the vector, in turn, is dependent on the way the library will be screened (see Section 3). A good overview of the different routes for the preparation of cDNA, and of the vectors which can be used for insertion, is given by Kimmel and Berger (13). In summary:

(a) If only screening by nucleic acid probes or antibodies is envisaged, all protocols giving a sufficiently large library will suffice.

(b) Screening for very rare genes, which require libraries of $> 10^6$ clones, with antibodies in an expression library will be easier in a λgt11 vector than in a plasmid.

(c) It is easier to screen large phage libraries (the best choice is λgt10) than large plasmid libraries with oligonucleotide probes. A detailed description for the synthesis of λgt10 and λgt11 libraries is given by Huynh *et al.* (14).

(d) If screening by functional expression is to be used, the choice is again between a phage or a plasmid system. Both systems have been used successfully. The λZAP-II vector system is particularly attractive since the expression cassette is easily recovered for subsequent transcription

3

experiments in a plasmid background by co-infection with a helper phage (15, 19). In several respects it is, however, simpler to work with plasmid than with phage libraries. The subcloning of the phage inserts into a plasmid expression vector, which is usually desirable for obtaining maximal expression levels, is also avoided.

(e) If cDNA libraries have to be made from limiting amounts of material, adapted techniques for mRNA preparation and cDNA synthesis are to be used (16).

(f) Since the transformation frequencies of *Escherichia coli* cells obtained by improved protocols (17) and by rather simple electroporation procedures routinely vary between 10^8 and 10^9 transformants/μg DNA (ref. 18 and our own experience), it is possible to construct libraries in plasmids encompassing several million clones starting from 5 μg poly(A)$^+$ mRNA.

3. Screening of the cDNA libraries for cytokine genes

Historically, the first cytokine genes were isolated on the basis of biological signals obtained by injection into *Xenopus laevis* oocytes of mRNA released from filters containing pools of cDNA plasmids (hybrid release selection) (7, 20–27). This procedure requires an abundant source of highly active mRNA. Later on, by improvement of sensitive protein sequencing procedures, several cytokine genes were isolated by hybridization selection with oligonucleotide probes deduced from the amino acid sequence (28–32). More recently, elegant selection strategies were used to isolate the genes for cytokines by direct expression in *in vitro* (33–35) or *in vivo* systems (36–38). Very recently, cytokine gene clones have been isolated by selection procedures which identified the biological activity after the clone was selected (39, 40).

Screening procedures on the basis of biological activities are described in Section 3.1, screening procedures based on data obtained from the purified protein are described in Section 3.2, screening on the basis of nucleic acid sequence data is described in Section 3.3 and screening by comparison of different cDNA libraries is described in Section 3.4.

3.1 Screening on the basis of biological activity

3.1.1 Hybrid selection

The first cytokines were cloned on the basis of a selection procedure which is called 'hybrid selection'. Briefly, the procedure is as follows:

(a) Groups of a limited number of plasmid DNAs (derived from a fraction of the total library) are fixed on solid supports (nitrocellulose or nylon membrane, DBM paper) and hybridized to authentic (mostly sucrose gradient-enriched) mRNA preparations isolated from the cell lines or organs producing the desired cytokine.

(b) The hybridized mRNA fraction is then eluted from the filters and tested for the presence of mRNA coding for the desired cytokine by injection into *X. laevis* oocytes.

This procedure is only applicable if mRNA preparations are available which produce high biological signals after translation in *X. laevis* oocytes. Indeed, a certain loss of biological activity of the mRNA preparation usually occurs during the hybridization procedure. Although technically demanding, the procedure has permitted the isolation of the genes coding for interferons (20–23) and the first cloned cytokines (7, 24–26). The technique of mRNA translation by injection in *X. laevis* oocytes is described in detail by Colman (41), although in our laboratory much simpler equipment is preferred to that available commercially (42).

Lomedico *et al.* (43) used an elegant selection technique to isolate the cDNA clone for murine (mu)IL-1 based on the availability of a goat anti-(mu)IL-1 antiserum. Groups of cDNA clones were screened by hybrid selection using mRNA derived from IL-1-producing cells. The released mRNA fractions were then translated in reticulocyte lysates, and the proteins were immunoprecipitated by means of the specific antiserum and analysed on denaturing polyacrylamide gels. In this way, they could identify a 33 kDa polypeptide, which corresponds with the precursor to the heterogeneous collection of lower molecular weight IL-1 polypeptides found in the culture supernatant of a stimulated macrophage-like tumour cell line (P388D1). It was estimated that only 0.005% of the poly(A)$^+$ RNA from superinduced cells coded for this IL-1 precursor protein, showing the great potential of the screening technique.

3.1.2 Direct expression by *in vitro* transcription

The simplicity and high efficiency of an *in vitro* expression system for the expression of a cytokine was first illustrated by injecting into *X. laevis* oocytes transcripts obtained from the human (h) IFN-β gene transcribed from the λ phage P$_L$ promoter by means of *E. coli* RNA polymerase (35). The complicated, but necessary preparation of 'capped' transcripts for this system, previously carried out by post-transcriptional enzymatic conversion of the 5'-terminal triphosphate to a cap structure (44), can be avoided by a simple addition of pre-synthesized, commercially available cap structures during the *in vitro* transcription reaction under appropriate conditions (35). Similar observations using a promoter derived from a *Salmonella* phage have also been reported (45).

The cDNA clone for murine (Mu) B cell growth factor (mIL-4) was the first to be isolated using an *in vitro* selection technique based on translation in *X. laevis* oocytes of SP6 polymerase-produced transcripts (33) (*Figure 1*). The source of biologically active mRNA was insufficient to use the hybrid selection procedure. Transcripts of the total cDNA library (45 000 clones),

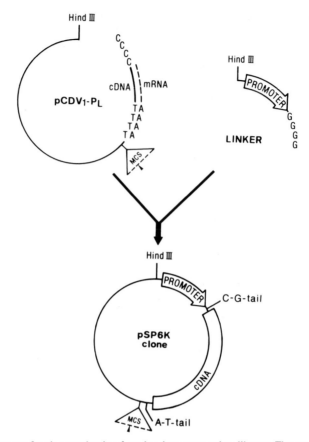

Figure 1. Strategy for the synthesis of an *in vitro* expression library. The construction of the vector primer and the linker fragment are described by Noma *et al.* (33). The cDNA is synthesized according to the Okayama and Berg method (51), whereby the T-tailed vector is used to prime the first-strand synthesis, then the first strand is tailed with CTP and terminal transferase (top left), and the construction is closed by annealing with the linker containing the promoter (top right) and second-strand synthesis. Transcription of the clones is carried out after linearization of the plasmid with a restriction enzyme cutting in the multi-cloning site MCS (arrowhead in the bottom diagram).

as well as from an enriched library (4000 clones) produced upon translation a biological signal, and allowed the isolation of an IL-4 clone. The same procedure was used to isolate the cDNA clone for IL-5 (34). *In vitro* expression systems have also been very useful for cloning the genes of transporter (46) and receptor molecules (47).

3.1.3 Direct expression in bacterial cells
Although direct expression of eukaryotic cDNA in bacterial cells as such is practically impossible, eukaryotic coding sequences can be expressed in

bacteria after fusion to bacterial proteins. Protein synthesis in the heterologous bacterial host is initiated at the bacterial initiation signal present in front of the homologous protein fusion partner to which the eukaryotic polypeptide is further added during the elongation process. This fusion protein can then be detected if an antibody is available. Immunological screening for proteins is carried out most frequently with recombinant expression libraries constructed in either plasmid (48) or bacteriophage λgt11 vectors (49). In this way a human high-molecular-weight B-cell growth factor was cloned from PHA (phyto-haemagglutinin) stimulated Namalva cells using the λZAP expression vector (15, 50, Stratagene). As the process of preparing antibodies for a cytokine is lengthy, this cloning procedure has not been followed frequently.

3.1.4 Direct expression in mammalian cells

The possibility of identifying mammalian genes by direct expression of their cDNAs in mammalian cell lines was first documented by Okayama and Berg in 1982 (51). They designed the pCD mammalian expression vector which contains the SV40 early region promoter, the late splicing junction and the origin of replication (*ori*; see *Figure 2*). The first cytokine identified in a mammalian expression library was murine mast-cell growth factor, which is identical to mIL-3 (36). After selection of a number of possible candidates by the hybrid release technique (see Section 3.1.1), a full-size clone was isolated by transient expression of the plasmids in COS-7 monkey cells. COS-7 cells were used as the recipient for this plasmid, because the expressed SV40 large T antigen drives the replication of the pCD DNA from the SV40 *ori*.

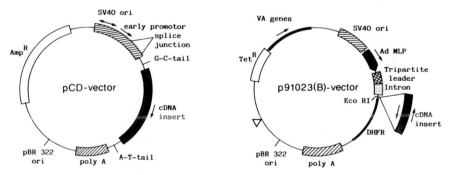

Figure 2. Structure of two mammalian expression vectors pCD and p91023 (B). The pCD vector system was developed by Okayama and Berg (51). The cDNA is synthesized in an oriented way in the vector. Transcription of the cDNA is from the SV40 early promoter clockwise towards the SV40 polyadenylation signal. The p91023 (B) vector is an optimized mammalian expression vector (52). It contains: (i) an SV40 origin and enhancer segment for replication in COS cells; (ii) the adenovirus major late promoter (Ad-MLP) coupled to the adenovirus tripartite leader; (iii) a hybrid intron; (iv) the SV40 polyadenylation signal; (v) the adenovirus VA I and VA II gene region; and (vi) a dhfr cDNA insert. The cDNA is inserted, after ligation to *Eco*RI adaptors, in the *Eco*RI site. The orientation of the insert may be sense or anti-sense.

The pCD-transfected cells were able to express the interleukin to levels which were 300-fold above the background. Plasmid pools containing 1% of full-length clones transfected in COS-7 cells produced readily detectable IL-3 levels. The same group used the above described expression plasmid for the selection of human full-size GM-CSF cDNAs in a library of 10 000 individual colonies. Screening of the culture medium of COS-7 cells transfected with 40 groups of 48 individually grown clones for colony-stimulating activity revealed four positive pools from which the individual clone could easily be identified (37). A similar technique, using an optimized expression vector p91023(B) (see *Figure 2*), was applied by a group at the Genetics Institute to isolate the genes for hGM-CSF and hIL-3 (52, 53). The cloning of the latter proved to be a complicated task (53). Indeed, the usual technique of cross-hybridization between murine and human genes was inapplicable because of low nucleotide sequence homology. The problem was further complicated by a low expression level of the cytokine by activated human peripheral blood lymphocytes. Using an assay developed for the detection of GM-CSF, gibbon T cell lines were identified which secreted a growth factor non-neutralized by anti-GM-CSF serum. Hence the multi-CSF (IL-3) clone was isolated from the gibbon cDNA library by screening transfected COS-1 cells for a biological activity in the hGM-CSF assay in the presence of antiserum to GM-CSF. The human clone was then identified by cross-hybridization with the gibbon clone. The human and gibbon clones were found to differ in amino acid sequence at only 11 positions.

Direct expression strategies in mammalian systems have proved to be extremely powerful for the isolation of receptor protein genes (54, 55). However, this approach requires high-quality (i.e. full-length) cDNA libraries. At present, numerous receptor genes have been cloned and sequence comparison has allowed the identification of receptor families such as the TNF/NGF receptor superfamily. For certain receptors, found on cells of the immune system, ligands are unknown and, therefore, they constitute ideal candidates for use in a search for novel immunoregulating cytokines. Recently, the ligand for an inducible T cell antigen was cloned by expression cloning in COS-7 cells (38). It was found to be a membrane glycoprotein with an extracellular domain showing clear sequence homology to corresponding regions of TNF, LT-α (lymphotoxin), LT-β and ligands for the CD40 and CD27 receptors. The interaction of the 4-1BB antigen with its ligand, induces proliferation of activated thymocytes and splenic T cells. In this case, growth of one cell, apparently, is activated by contact with an antigen present on another cell, unlike the case with classical cytokines, which are secreted, migrating molecules.

Besides transient expression systems, as described earlier, the potential use of more permanent expression systems for cloning new cytokines was illustrated by Rayner and Gonda (56). They synthesized a directional cDNA library in a retroviral vector. By functional screening in the appropriate cell lines they could isolate the cDNAs for IL-3 and GM-CSF. The more general use of retroviral vectors for persistent expression *in vivo* was described by Naviaux and Verma (76).

The use of retroviral vectors is also described in systems where gene therapy by cytokines is envisaged (57).

3.2 Screening on the basis of protein sequence data

When a portion of the amino acid sequence of a protein is known, an oligonucleotide can be designed based on this information. Due to the degeneracy of the genetic code, the selection of the protein region for reverse translation into nucleic acid sequence is somewhat complicated. Several aspects, such as length, G+C content, self-complementarity and complexity have to be considered. A detailed analysis of these considerations is described in the literature (58–60). The technique of oligonucleotide screening was used successfully for a number of lymphokines [e.g. hM-CSF (29), hIL-6 (30), hG-CSF (31), mGM-CSF (32), hTNF (26), mTNF (27), etc.]. In some cases, a special technique of overlapping oligonucleotide probes may be advantageous to obtain high-radioactivity probes by a simple filling-in polymerization using Klenow enzyme (27).

More recently, hybridization probes were generated by using oligonucleotides derived from the protein sequence, in an *in vitro* DNA amplification reaction (PCR). In this way, larger probes are generated with complete homology to the desired gene. This technique was used to clone a ninth member of the fibroblast growth factor (FGF-9) family (61).

3.3 Screening on the basis of nucleic acid sequence data

Once the cDNA sequence for one organism (often human or mouse) has been cloned, it is often desirable to isolate the cDNA for the same gene function in other organisms (degree of relatedness at the molecular level, functional cross-reactivity between species, etc.). For this heterologous hybridization techniques are often applied. Before screening heterologous libraries, the usefulness of the probe is tested on genomic DNA digests or mRNA preparations of cells which are known to express the gene of interest.

Hybridization probes can also be synthesized by PCR technology starting from oligonucleotides derived from more conserved regions in the proteins. The number of such successful gene isolations is so large that it is impossible to even make a reference list.

3.4 Differential screening

In differential screenings, cDNA sub-libraries from activated cytokine producing cells (e.g. a T lymphocyte) are analysed by a primary screening criterion. This criterion is inducible upon a certain stimulation of the producing cell. In a second screening the activity of the cloned protein is determined. In this way, interleukin-13 was discovered after stimulation of the T cell CD28 antigen by the B7/BB1 B lymphocyte surface antigen (39). The sub-library can be formed by substraction (39) or screening with degenerate oligonucleotides derived from a homologous coding region found for certain groups of cytokines. Additional criteria, such as the presence of destabilizing TATTT sequences in the 3′ untranslated regions can also be taken into consideration to make the decision to further study a selected clone (40).

The identification of tissue-specific or cell-type specific genes by the powerful differential display technology, first described by Liang and Pardee (64), will also, most likely, contribute to selection of interesting sub-groups of cDNA libraries from immune-active cell types. These techniques represent a new approach to previously described procedures in which the known biological activity was used for purification of a protein, or for isolation of cDNA by expression cloning. Here, the biological activity is determined after the gene has been cloned.

4. Heterologous expression of cytokine genes

After a putative cytokine cDNA clone is isolated by a technique which is not based on direct expression, it is most important to demonstrate that the clone obtained can produce biologically active cytokine. To this end, several expression systems are available. However, a mammalian system will often be chosen instead because these systems guarantee a more faithful translation and processing from a minimally manipulated clone, although the expression levels are often rather low. Indeed, secondary modifications will be carried out and secretion will almost certainly be observed if the complete information for the signal sequence is present in the cDNA clone. Efficient vectors for transient expression in African green monkey cells, based on SV40 replication and expression signals, were developed by Gheysen and Fiers (65), and later optimized by Huylebroeck *et al.* (66). Often, similar vectors, from which the coding information for the large T antigen has been deleted, are used to express genes in COS cells (67).

Once a biologically active clone is identified, a strategy has to be selected for the preparative synthesis of the desired cytokine in larger quantities. Three main options are open: expression in mammalian cells; expression in lower eukaryotes; or bacteria.

4.1 Expression in mammalian cells

Mammalian expression has a number of advantages which arise from the fact that the heterologous gene is expressed in a host cell which most closely resembles its natural milieu. The proteins fold in a correct way and they form the appropriate disulphide bridges. Furthermore, post-translational processing and modifications (especially glycosylation) will be more faithful. Two examples of mammalian expression systems will be briefly discussed: a transient expression system and a permanent expression system.

4.1.1 Transient expression

Transient expression of a gene is obtained by the transfection of a transgenome (usually a manipulated viral genome; mostly SV40) in a host cell which allows replication of the transfected DNA to high copy numbers, ensuring a high-level expression of the heterologous gene (65). A second system is the SV40 vector/COS cell system. COS cells are African green monkey kidney cells, the permissive hosts of the SV40 virus, which express constitutively SV40 T antigen required for SV40-DNA replication. The high episome DNA copy number, combined with the high transfection efficiency of these cells, results in a very efficient level of heterologous gene expression (e.g. around 1 µg IL-2/10^6 cells) at 48–72 h post-transfection (67).

Although being non-mammalian, insect cells behave very much like mammalian cells. They have been used for heterologous production of several cytokines (68) after cloning the genes in a baculovirus vector. They also carry out post-translational modifications like *N*- and *O*-glycosylation, but the carbohydrate structures obtained do not prevent a rapid clearing of such proteins after injection into the bloodstream (69). Transient expression systems may be advantageous for the synthesis of gene products which show toxicity to the host cells (66).

4.1.2 Permanent expression

For larger scale production purposes, massive transfection of mammalian cells is an unpractical procedure. However, it is possible to construct cell lines which constitutively secrete certain cytokines fairly efficiently. For example, stable cell lines were obtained secreting up to 10 µg/ml IFN-γ in the culture medium by co-transfection of CHO (Chinese Hamster Ovary) dhfr$^-$ cells with the plasmid pAdD26SV(A)$^{-3}$ (70) containing a selectable dhfr marker, and the plasmid pSV2-IFN-γ, containing an expression cassette of the gene of interest, in this case hIFN-γ (71). Because the genes (selective marker and expression cassette) often become linked during the co-transfection process, there is the possibility of co-amplification of the expression cassette by selection for amplification of the dhfr gene during growth in increasing concentrations of methotrexate (67). In certain cases, however, additional amplification efforts did not increase the initially obtained (rather low) production levels (72). The major drawbacks of the animal cell culture approach are the high cost of media, the requirement of highly skilled labour and the possibility of contamination by viruses, other cytokines, etc.

4.2 Expression in lower eukaryotes

Because cytokines are all secreted proteins, their native structure is formed during a delicate and complicated translocation process through the secretion pathway. This folding process is often difficult to mimic in test-tube conditions.

(A) pUDT₂ secretion vector

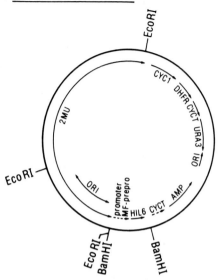

(B) **Sequence of the secretion vector gene fusions**

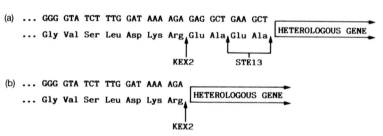

Figure 3. Structure of the yeast pUDT2 secretion vector. (**A**) The pUDT2 vector contains: (i) the total 2-mu plasmid for replication in cir° strains; (ii) the dhfr amplification cassette (63); (iii) the ura3 selective marker; (iv) the bacterial selection and replication sequences (Amp and *ori*), and (v) the secretion cassette (promoter, MF-prepro, CYCC terminator). (**B**) The precise structure of the fusion between the mating factor prepro-sequence and the heterologous gene is shown. Depending on the type of fusion, the KEX2 and STE13 proteases (a) or only KEX2 enzyme (b) are involved in the processing of the fusion precursor protein.

Therefore, the heterologous production of cytokines preferentially is best achieved using a host with an active secretion system. An elegant secretion system has been worked out for the yeast *Saccharomyces cerevisiae* (73). The heterologous protein is fused to the mating factor prepro-leader region (see *Figure 3B*) which contains processing signals for the yeast proteases KEX2 and STE13. The KEX2 protease cleaves the precursor at the carboxyl-

Table 1. Expression levels obtained for cytokines using different promoters in the yeast secretion system

Heterologous gene	Promoter	Yield (μg/ml) [a]
hIL-2	α-MF [b]	6.0
mIL-2	α-MF	0.1
	GAL1	3.0
mIL-5	α-MF	0.3 [c]
hIL-6	α-MF	5.0
	GAL1	30.0

[a] Biological activity measured in the non-purified medium of a yeast culture grown to an $OD_{600} \approx 6$.
[b] α-mating factor.
[c] Data taken from ref. 72.

terminal side of Lys-Arg, and the STE13 gene product, the dipeptidyl amino-peptidase A, removes the Glu–Ala dipeptides at the amino terminus of the excised protein in case they are present. The presence of the Glu–Ala dipeptides is optional. They will be inserted only in fusion situations where defective KEX2 cleavage is encountered.

We have used the mating factor secretion system for the expression of hIL-2 and mIL-2, mIL-5 and hIL-6 (see *Figures 3* and *4*). The amounts of cytokines secreted by the yeast cells is shown in *Table 1*. The use of the strong and well-regulated GAL1 promoter is often beneficial or even essential when expression of the heterologous gene confers counter-selective pressure. Worth noting is the observation that the formation of homodimeric IL-5, which is crucial for biological activity by oxidation of two conserved cysteine residues, occurs faithfully in the yeast secretion system (72). It should also be mentioned that the glycosyl groups synthesized by the yeast cells, are different from their mammalian counterparts. This may be a limitation when *in vivo* use of the recombinant protein is considered.

Filamentous fungi have also been used as hosts for expression of cytokines. They were selected as promising hosts for their extremely high capacity to secrete certain proteins such as cellulases and amylases. The secretion of heterologous mammalian cytokines was, however, much lower than the secretion of homologous proteins (74). Recent studies (75) have shown that it was possible to improve the secretion of the heterologous protein drastically by fusing it to such a highly expressed fungal protein (e.g. *A. niger* glucoamylase).

4.3 Expression in bacterial systems

From the very early days of cytokine cloning, bacterial expression systems have been most helpful in providing tools for the cheap production of abundant

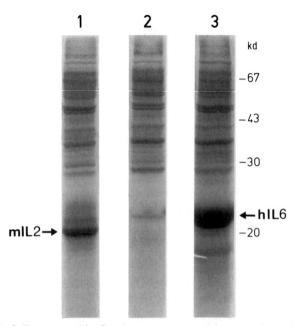

Figure 4. PAGE of rILs secreted in *Saccharomyces cerevisiae* growth medium. *S. cerevisiae* cells were grown in SDC (0.67% yeast nitrogen base, 2% dextrose, 0.2% casamino acids) to an $OD_{600} \approx 4$, and induced by transfer to a galactose-containing medium SGC (0.67% yeast nitrogen base, 2% galactose, 2% casamino acids) for 15 h. The proteins in 1 ml medium, from which the cells were removed by centrifugation, were precipitated by adding 100 μl deoxycholate (1 mg/ml), 100 μl 100% TCA and incubation for 30 min on ice. The protein pellet was dissolved in 40 μl sample buffer (78), neutralized with 4 μl 1 M Tris base, boiled for 2 min, and loaded on a denaturing 15% polyacrylamide gel (78). **Lane 1** shows the proteins from a strain transformed with a plasmid containing an expression cassette for mIL-2 (the promoter is GAI1; arrow: mIL-2 product). **Lane 2** is a non-transformed strain. **Lane 3** is a strain with an expression cassette for hIL-6 (arrow: hIL-6 product).

amounts of mammalian protein (7, 77). For this purpose, the mature cytokine gene (i.e. without the signal sequence information) is placed downstream after a powerful promoter ribosome-binding region. More than 40% of the total protein in the culture may correspond to the protein of interest. A complication which often occurs, however, is that this heterologous protein is present in the bacterial cell as insoluble inclusion bodies. Procedures must then be followed to dissolve these in chaotropic media, followed by renaturation and extensive purification. One of the exceptions is the bacterial expression of hTNF from the heat-inducible P_L promoter, where a production of 44 μg/ml of soluble, fully active, homotrimeric protein was obtained (ref. 28; see *Figure 5*).

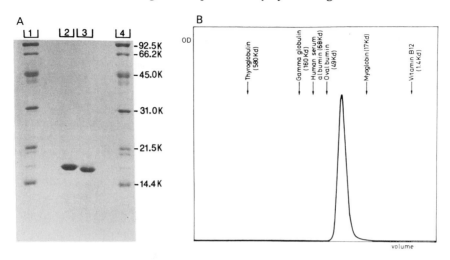

Figure 5. Characterization of *E. coli* rhTNF and mTNF. **(A)** SDS-PAGE of purified rhTNF (**lane 2**) and rmTNF (**lane 3**). **Lanes 1** and **4** represent marker proteins. **(B)** Gel filtration profile of purified rhTNF using matrix TSK G3000. Chromatography was carried out in PBS at a flow rate of 1 ml/min. The elution positions of the mol. wt markers are indicated by arrows. The true mol. wt of trimeric rhTNF is 52 kDa, but the 'apparent' mol. wt is only ≈35 kDa due to retardation by interaction with the matrix. (Figure reproduced from ref. 62 with permission.)

5. Conclusion

A wide range of elegant cloning techniques have been used for the isolation of cytokine genes. Due to improvements in mRNA preparation methods, cDNA synthesis technology and transfection procedures, direct expression methods become more and more attractive for the cloning of new biological activities. In fact, they even allow one to screen for hypothetical functions (e.g. natural antagonists or synergists).

The abundant synthesis of protein in heterologous systems is possible for all cytokines. Depending on the purpose for cloning the gene or for *in vivo* or *in vitro* experiments, and the requirement for secondary modification, one can choose from the more faithful, but more expensive, mammalian expression systems, yeast or fungal secretion systems or bacterial expression.

Acknowledgements

Research in the authors' laboratory was supported by the Belgian Fund for Medical Research (FGWO) and the Belgian and Flemish Departments of Science Policy (OOA and Emerging Technology Centre). R. C. is a Research Director with the NFWO.

References

1. Deutscher, M. P. (ed.) (1990). *Methods in Enzymology*, **Vol. 182.** Academic Press, New York.
2. Clemens, M. J. (1984). In *Transcription and translation: a practical approach* (ed. B. D. Hames, and S. J. Higgins). IRL Press, Oxford.
3. Berger, S. L. and Kimmel, A. R. (ed.) (1987). *Methods in enzymology*, **Vol. 152.** Academic Press, New York.
4. Janssen, K. (1994). *Current protocols in molecular biology.* John Wiley and Sons, Boston.
5. Sambrook, J., Fritsch, E. F., and Maniatis, T. (1989). *Molecular cloning, a laboratory manual.* Cold Spring Harbor Laboratory Press, New York.
6. Chirgwin, J. M., Przybyla, A. E., MacDonald, R. J., and Rutter, W. J. (1979). *Biochemistry*, **18**, 5294.
7. Devos, R. *et al.* (1983). *Nucleic Acids Res.*, **11**, 4307.
8. Kurybayashi, K., Hikata, M., Hiraoka, O., Miyamoto, C., and Furuichi, Y. (1988). *Nucleic Acids Res*, Symposium Series **19**, 61.
9. Hornes, E. and Korsner, L. (1990). In *Basic and Applied Research*, February 10–14, pp. 85, Sweden.
10. Hahnel, A. C. *et al.* (1990). *Development*, **110**, 555.
11. Weng, C. *et al.* (1989). *Mol. Reprod. Dev.*, **1**, 233.
12. Belyavsky, A. *et al.* (1989). *Nucleic Acids Res.*, **17**, 2919.
13. Kimmel, A. R. and Berger, S. (1987). In *Methods in enzymology* (ed. A. R. Kimmel and S. Berger), **Vol. 152**, p. 307, Academic Press, New York.
14. Huynhn, T. V., Young, R. A., and Davis, R. W. (1985). In *DNA cloning Volume I. a practical approach* (ed. D. M. Glover), p. 49. IRL Press, Oxford.
15. Alting-Mees, M. *et al.* (1992). *Strategies*, **5**, 58.
16. McCarrey, J. and Williams, S. (1994). *Curr. opinions in biotechnology*, **5**, 34.
17. Hanahan, D. (1985) In *DNA cloning: Volume I. a practical approach* (ed. D. M. Glover), p. 109. IRL Press, Oxford.
18. Dower, W. J., Miller, J. F., and Ragsdale, C. W. (1988). *Nucleic Acids Res.*, **16**, 6127.
19. Short, J. M., Fernandez, J. M., Sorge, J. A., and Huse, W. D. (1988). *Nucleic Acids Res.*, **16**, 7583.
20. Derynck *et al.* (1980). *Nature*, **285**, 542.
21. Nagata, S., Mantei, N., and Weissmann, C. (1980). *Nature*, **287**, 401.
22. Gray, P. *et al.* (1982). *Nature*, **295**, 503.
23. Devos, R. *et al.* (1982). *Nucleic Acids Res.*, **10**, 2487.
24. Taniguchi, T. *et al.* (1983). *Nature*, **302**, 305.
25. Fung, M. C. *et al.* (1984). *Nature*, **307**, 233.
26. Pennica, D. *et al.* (1984). *Nature*, **312**, 724.
27. Fransen, L. *et al.* (1985). *Nucleic Acids Res.*, **13**, 4417.
28. Marmenout, A. *et al.* (1985). *Eur. J. Biochem.*, **152**, 515.
29. Kawasaki, E. *et al.* (1985). *Science*, **230**, 291.
30. Hirano, T. *et al.* (1986). *Nature*, **324**, 73.
31. Nagata, S. *et al.* (1986). *Nature*, **319**, 415.
32. Gough, N. M. *et al.* (1984). *Nature*, **309**, 763.

33. Noma, Y. *et al.* (1986). *Nature*, **319**, 640.
34. Kinashi, T. *et al.* (1986). *Nature*, **324**, 70.
35. Contreras, R. *et al.* (1982). *Nucleic Acids Res.*, **10**, 6353.
36. Yokota, T. *et al.* (1984). *Proc. Natl. Acad. Sci. USA*, **81**, 1070.
37. Lee, F., *et al.* (1985). *Proc. Natl. Acad. Sci. USA*, **82**, 4360.
38. Goodwin, R. G. *et al.* (1993). *Eur. J. Immunol.*, **23**, 2631.
39. Minty, A. *et al.* (1993). *Nature*, **362**, 248
40. Samal, B. *et al.* (1994). *Mol. Cell. Biol.*, **14**, 1431.
41. Colman, A. (1984). In *Transcription and translation: a practical approach* (ed. B. D. Hames and S. J. Higgins), p. 271. IRL Press, Oxford.
42. Contreras, R., Cheroutre, H., and Fiers, W. (1981). *Anal. Biochem.*, **113**, 185.
43. Lomedico *et al.* (1984). *Nature*, **312**, 458.
44. Paterson, B. and Rosenberg, M. (1979). *Nature*, **279**, 692.
45. Krieg, P. and Melton, D. (1987). In *Methods in enzymology* (ed. R. Wu), **Vol. 155**, 397. Academic Press, New York.
46. Hediger, M. *et al.* (1987) Nature, **330**, 379.
47. Masu, Y. *et al.* (1987). *Nature*, **329**, 836.
48. Helfman, D. *et al.* (1983). *Proc. Natl. Acad. Sci. USA*, **80**, 31.
49. Young, R. A. and Davis, R. D. (1983). *Proc. Natl. Acad. Sci. USA*, **80**, 1194.
50. Ambrus, J. L. *et al.* (1993). *Proc. Natl. Acad. Sci. USA*, **90**, 6330.
51. Okayama, H. and Berg, P. (1982). *Mol. Cell. Biol.*, **2**, 161.
52. Wong, G. G. *et al.* (1985). *Science*, **228**, 810.
53. Yang, Y.-C. *et al.* (1986). *Cell*, **47**, 3.
54. Aruffo, A. and Seed, B. (1987). *Proc. Natl. Acad. Sci. USA*, **84**, 8573.
55. Izotova, L., Mariano, T. M., and Pestka, S. Chapter 6 of this book.
56. Rayner, J. R. and Gonda, T. J. (1994). *Mol. Cell. Biol.*, **14**, 880.
57. Hwu, P. *et al.* (1993). *J. Immunol.*, **150**, 4104.
58. Wallace, R. B. and Miyada, C. G. (1987). In *Methods in Enzymology* (ed. S. Berger, and A. Kimmel), **Vol. 152**, 432.
59. Lathe, R. (1985). *J. Mol. Biol.*, **183**, 1.
60. Martin, F. H. and Castro, M. M. (1985). *Nucleic Acids Res.*, **13**, 8927.
61. Miyamoto, M. *et al.* (1993). *Mol. Cell. Biol.*, **13**, 4251.
62. Tavernier, J. *et al.* (1987). In *Lymphokines 13* (ed. D. R. Webb, and D. V. Goeddel), p. 181.
63. Zhu, J.-D. *et al.* (1985). *Bio/Technology*, **3**, 451.
64. Liang, P. and Pardee, A. (1992). *Science*, **257**, 967.
65. Gheysen, D. and Fiers, W. (1982). *J. Mol. Appl. Genet.*, **1**, 385.
66. Huylebroeck, D. *et al.* (1988). *Gene*, **66**, 163.
67. Cullen, B. R. (1987). In *Methods in Enzymology* (ed. S. L. Berger and A. R. Kimmel), **Vol. 152**, 684.
68. Morelle, C. (1993). *Biofutur*, N° **125**, 1.
69. Sareneva *et al.* (1993). *J. Interferon Res.*, **13**, 267.
70. Kaufman, R. J. and Sharp, P. A. (1982). *Mol. Cell. Biol.*, **2**, 1304.
71. Scahill, S. J., Devos, R., Van der Heyden, J., and Fiers, W. (1983). *Proc. Natl. Acad. Sci. USA*, **80**, 4654.
72. Tavernier, J. *et al.* (1989). *DNA*, **8**, 491.
73. Brake, A. *et al.* (1984). *Proc. Natl. Acad. Sci. USA*, **81**, 4642.
74. Carrez, D. *et al.* (1990). *Gene*, **94**, 147.

75. Contreras, R. *et al.* (1991). *Bio/Technology*, **9**, 378.
76. Naviaux, R. K. and Verma, I. M. (1992). *Current Opinion in Biotechnology*, **3**, 540.
77. Derynck, R. *et al.* (1980). *Nature*, **287**, 193.
78. Laemmli, U. K. (1970). *Nature*, **227**, 680.

2

Southern blotting of cytokine genes

M. O. DIAZ and T. W. McKEITHAN

1. Introduction

Southern blot analysis of genomic sequences (1) is one of the basic tools of molecular genetics. This technique is principally used to determine the physical organization of particular DNA sequences for which a probe is available. Its power of resolution has recently been extended up into the megabase (10^6 base-pair) level by the technique of pulsed field gel electrophoresis (2). In many cases, the organization of large gene families, and the distance between particular cytokine genes have been determined using this technique.

The chromosomal assignment of particular genes can be made by hybridization to Southern blots of DNA from panels of somatic cell hybrids that contain diverse subgenomic complements of chromosomes from the species of interest. Genomic rearrangements close to a gene, due to chromosomal rearrangement or viral insertion, as well as amplification or deletion of a gene, can be detected by Southern blot analysis.

A particularly useful application is the detection of restriction fragment length polymorphisms (RFLPs), which can be used for tracing the allele of a certain disease gene through several generations of a family, or to determine the recombination frequency between linked loci to construct genetic linkage maps. RFLPs can also be used to detect loss of heterozygosity at certain loci, caused by deletion, chromosome loss, or mitotic recombination. Finally, an application that has become very important in the modern strategies to identify genes by reverse genetics is the detection of gene coding or regulatory sequences by the analysis of conserved sequences between different species (zoo blots).

2. Basic technique

2.1 Preparation of DNA

DNA can be prepared from animal cells by a number of techniques that involve lysing the cells, dissociating the nucleoproteins with the use of detergents, and precipitating the proteins while the DNA is maintained in solution. The most efficient methods use enzymes to digest protein and RNA, and a final deproteinization step using phenol and chloroform is usually included (3). The method in *Protocol 1* can be used to prepare DNA from a suspension of animal cells. The DNA obtained will have a molecular weight in excess of 100 kilobases (kb).

Protocol 1. Isolation of DNA from animal cells

Equipment and reagents

- NTE: 100 mM NaCl, 10 mM Tris–HCl, 1 mM EDTA, pH 8.0
- Proteinase K
- 0.5 M EDTA
- 20% SDS
- Phenol
- Chloroform/isoamyl alcohol (24:1, v/v)
- 7.5 M ammonium acetate
- Plastic pipettes

Method

1. Suspend the cells in 1 ml of NTE per $20–50 \times 10^6$ cells.

2. Per ml, add 11 µl of 100 µg/ml proteinase K and 55 µl of 0.5 M EDTA. Swirl to mix. Add 55 µl of 20% SDS to lyse the cells. Mix gently.

3. Digest at 50°C for 2 h to overnight.

4. Add 1 vol. of phenol equilibrated with NTE. Shake gently for 15–60 min.

5. Centrifuge at 3000 r.p.m. at room temp. for 15 min.

6. Transfer top aqueous layer to a new tube, avoiding the precipitate at the interface. Optionally, to increase the yield, additional NTE can be added and mixed, and the centrifugation repeated.

7. Add an equal volume of a 24:1 mixture of chloroform/isoamyl alcohol. Shake gently for 15–60 min.

8. Centrifuge as before and transfer aqueous layer to a new tube. [a]

9. Per milliliter, add 0.5 ml of 7.5 M ammonium acetate and 3 ml of ethanol. Shake gently until the DNA has precipitated.

10. Fish out the precipitated DNA with a plastic pipette and transfer to an excess of 70% ethanol. Shake 15–60 min, depending on the size of the precipitate.

11. Remove precipitate and partially dry. Do not dry completely or it will be very difficult to dissolve. Dissolve in TE at 50°C. This may require several hours.

[a] If it has been impossible to avoid the precipitate at the interface, perform an additional centrifugation underlaying 25% sucrose in NTE.

For many purposes, DNA molecules of a molecular weight in the order of megabases are needed. In order to obtain DNA of this length it is necessary to avoid shearing during the preparation. This is achieved by immobilizing the cells before lysis in a gel matrix, into which the enzymes and other chemicals are allowed to diffuse (2, 4, 5) as described in *Protocol 2*. The DNA in solution or in agarose plugs can be stored for years in TE buffer (10 mM Tris–HCl, 1 mM EDTA) at 4°C without degradation. This technique is also useful when the sample is very limited, since standard methods using extraction in solution may give poor recovery if relatively few cells are available.

Protocol 2. Isolation of DNA in agarose plugs

Equipment and reagents

- Plug solution: 0.1 M NaCl, 0.1 M EDTA, 10 mM Tris–HCl, pH 8.0
- Low gelling temperature agarose
- Plug mould
- N-laurylsarcosine
- Proteinase K
- 100 mM phenylmethylsulfonyl fluoride (**Caution: TOXIC!**)
- Waterproof adhesive tape

Method

1. Count nucleated cells and pellet. Calculate the number of plugs to be made; 1.6×10^6 cells are typically used in each plug.

2. Resuspend in 25 µl of plug solution (0.1 M NaCl, 0.1 M EDTA, 10 mM Tris, pH 8.0 with HCl) per plug. Maintain at 37°C.

3. Add an equal volume of melted 1% low-gelling-temperature agarose in plug solution at 37°C. Mix. Maintain at 37°C.

4. Pipette 50 µl into each well of plug mould; chill at 4°C for 30 min.[a]

5. For each 10 plugs, prepare a tube containing 3.5 ml of plug solution, 35 mg N-laurylsarcosine and 3.5 mg proteinase K. Remove the tape from the plug mould. Squirt the plugs into the tubes with a rubber bulb.

6. Digest at 48–50°C (NO MORE!) for two days.

7. Remove the liquid and transfer the plugs to 10–15 ml of TE/10 plugs. Add 0.01 vol. of 100 mM phenylmethylsulfonyl fluoride (CAUTION!: toxic) to this first wash. Place the tubes on a rocker at 4°C for 4 h.

8. Wash the plugs at least six times in TE over a two-day period. The plugs are very fragile and difficult to see. They are most easily seen with a black background.

[a] Acrylic plug moulds are made with 2×5 mm rectangular holes. Waterproof adhesive tape should cover the bottom of the holes.

2.2 Restriction enzyme digestion of DNA

The digestion of the DNA by restriction enzymes produces fragments of specific length from the sequences containing a certain gene. The length of a restriction fragment is a genetic character that can be changed by mutations or chromosome rearrangements, but, with some exceptions (RFLPs), it is constant for the species. Restriction enzymes cut DNA readily in solution when used in the optimal conditions specified by the manufacturers, but for some purposes, simultaneous digestion with several enzymes may be performed in a universal buffer, or a set of a few basic buffers (6). The digestion of DNA contained in agarose plugs is slower, because the enzymes need time to diffuse from the surface to the centre of the plug. It is necessary to presoak the plug in the enzyme buffer before adding the enzyme. Inhibitors for some restriction enzymes may be present in DNA preparations, especially those derived from nucleated cells in blood that are contaminated with erythrocytes. The addition of 1–4 mM spermidine to the digestion buffer can relieve the inhibition in many cases (7). Note, however, that spermidine may cause the DNA to precipitate from low salt buffers. If inhibitors are present, digestion at lower DNA concentration may also be useful.

After digestion, the enzymes can be inactivated by heating at 65°C after addition of EDTA. Sometimes, concentration of the digestion products is necessary, to reduce the volume before loading the digestion mix in the gel. If precipitation with ethanol is used at this step, the presence of spermidine may slow down the redissolution of the DNA, which may require several hours at 37°C to complete.

2.3 Gel electrophoresis

2.3.1 Standard agarose gel electrophoresis

The separation of the DNA fragments is performed by electrophoresis through an agarose gel (typically 0.8–1.0%) (8). Standard agarose gel electrophoresis, performed at constant voltage, can resolve DNA fragments between 70 and 80 000 base-pairs, but different agarose concentrations must be used to resolve different sizes within this range (8). Also, the higher end of this range can be resolved only in very low concentrations of agarose, making the gels difficult to handle. Separation by standard electrophoresis can be performed in any of a number of cells either home-made or commercially available. Horizontal gels are easier to cast and handle and do not require refrigeration. Vertical gels prepared between glass plates need some support to prevent the gel from sliding down through the bottom of the sandwich. A frosted glass plate can be used to prepare the gel sandwich, or a strip of porous material (plastic foam sponge) at the bottom of the buffer reservoir can be used to hold the gel in place.

The running buffer is an electrolyte solution providing conductivity and pH buffering. The most commonly used electrophoresis buffers are solutions of Tris-borate or Tris-acetate, with EDTA to chelate divalent ions (e.g. 40 mM Tris-acetate, 1 mM EDTA, pH 8.0).

Casting of the gels is done either in the cell itself, or in a separate casting device that provides a mould of the right size. Moulds for horizontal gels can be made very easily by wrapping a waterproof sticky tape around the edge of a rectangular glass or plastic plate of appropriate dimensions. The tape protruding under the edge is folded and stuck to the underside of the plate. The agarose is melted in the appropriate volume of the running buffer, cooled down to 55–60°C (the temperature at which the flask wall can be touched without burning the skin), and let stand for a few minutes until all air-bubbles disappear. Then it is poured into the mould, with the well-forming comb in place and allowed to cool until gelling.

The wells are made by inserting a comb with prismatic teeth during casting of the gel. When the gel has hardened, the comb is removed, and the apparatus is filled with buffer. The digestion mix containing the DNA fragments is made denser than the buffer by adding to it 1/5th vol. of a solution of glycerol or Ficoll plus marker anionic dyes [e.g. 0.3% Bromophenol blue, 0.3% xylene cyanol, and 15% Ficoll (Pharmacia) in water], which serve as visual indicators of the migration rate of the DNA in the gel. Typically 6–8 μg of DNA is loaded for a 6 mm-width lane. The denser solution can be pipetted into the well through the lighter buffer. It is important to start the run as soon as possible after loading the samples to avoid diffusion of the DNA out of the wells.

The electrophoresis can be run at 1–2 V/cm. It is best to use a regulated power supply capable of maintaining constant amperage. Recirculation of the buffer between the cell chambers may be necessary during long runs. The resolution of the fragment separation is generally best when gels are run at low voltage, overnight.

2.3.2 Pulsed field gel electrophoresis

When DNA fragments larger than 20 kb have to be resolved it is better to use one of the forms of pulsed-field gel electrophoresis in which the field orientation is switched at regular intervals (2, 5, 9–11). There are several commercial devices for these procedures, all of them considerably expensive, but with some ingenuity it is possible to build the necessary equipment, buying an electronic laboratory timer and a high-voltage switching relay. There are several different geometric and electronic arrangements of electrodes for pulsed field electrophoresis, which we cannot discuss here, but, in our experience, the most efficient one is the contour clamped horizontal electrical field (CHEF) system (10). For any of the commercial systems, detailed instructions come included, and the essentials of the assembly and loading of the gels are similar.

The gels are cast in a mould, as for a regular electrophoresis, with insertion of a comb forming wells slightly wider than the size of the DNA agarose plugs. The plugs are inserted in the wells before covering the gel with buffer, and are sealed in place with molten agarose. The gel is then set on its platform inside the buffer, and the field is applied. In this kind of separation, the resolution depends not on the agarose concentration, but on the length of the alternating pulses. If a large range of DNA fragment sizes has to be resolved, the length of the pulses can be varied either in steps or continuously over the length of the run.

2.3.3 Monitoring the separation

The quality of the separation can be monitored under UV illumination after staining the gel in a solution of ethidium bromide (see *Protocol 3*). For documentation, place the gel on transparent plastic film (Saran wrap) on a transilluminator, a box containing a UV source and topped by a UV transparent filter which blocks the visible emission from the lamp, and photograph the pattern of bands through a red or yellow filter, together with a graduated ruler as a length reference.

2.3.4 Controls and size markers

Depending on the purpose of the Southern blot, it may be necessary to include certain controls and size markers. When looking for possible genomic re-arrangements, a germline genomic DNA control is usually included. This may be DNA from any somatic tissue that is known not to rearrange the sequences of interest during development. For human DNA Southerns, human placental DNA is frequently used as a control (*Figure 2*). When studying the hybridization of a probe to a panel of somatic cell hybrids, a control of DNA from the parental cell line(s) or species of the hybrids is convenient.

Appropriate size markers for regular electrophoresis can be obtained by digesting phage λ or M13 DNA with different restriction enzymes. These DNAs will not hybridize to plasmid vector DNAs. One of the most-used standards is a *Hind*III restriction digest for phage λ DNA (*Figure 2*). Multimer standards for different fragment size ranges are available from different companies.

For PFGE electrophoretic separations, where the size of the fragments resolved is larger, uncut yeast chromosomes (9) and multimers of phage λ (12) are used.

The markers can be radioactively end labelled with Klenow fragment of *Escherichia coli* polymerase I and a [^{32}P]-deoxyribonucleotide triphosphate, in which case they will appear as bands in the autoradiogram, or can be used cold, in which case they are visualized after ethidium bromide staining of the gel. In the last case, a photograph of the stained gel must be taken, with a length reference that can be correlated to some feature of the autoradiogram; for example, a graduated ruler with the zero mark matched to the level of the wells, or the upper edge of the gel.

2.4 Transfer of DNA to membranes

The DNA in the gel must be denatured and transferred to a porous membrane on to which it attaches irreversibly after dehydration. In order for the DNA fragments to move out of the gel in a reasonable time, they must have a maximum size of about 2 kb. Larger fragments must be nicked to reduce their length. Nicking is obtained either by exposing the ethidium-bromide-stained DNA to UV light (13), or by partial acid depurination followed by exposure to alkali. Denaturation is achieved by soaking the gel in alkali and then neutralizing it with a buffer. The DNA fragments can be moved out of the gel and on to the membrane by mechanical means, allowing a saline solution to flow through the gel pulled either by capillary forces (1), or by pressure gradients, or by electrophoresis. A procedure for DNA transfer, using capillary driven liquid flow, is described in *Protocol 3*.

Protocol 3. Southern transfer of DNA

Equipment and reagents

NB: Caution: ethidium bromide is mutagenic, wear gloves to handle the gel, add hypochlorite to ethidium bromide solutions 30 min before discarding down the sink.

- 1 μg/ml ethidium bromide
- UV transilluminator with rulers (preferably fluorescent)
- UV protective clothing (goggles, long sleeves, face shield, gloves)
- 0.4 M NaOH + 0.6 M NaCl
- 0.5 M Tris–HCl, pH 7.4, 1.5 M NaCl
- Charged nylon membrane
- 10 × SSC buffer: 1.5 M NaCl, 0.15 M Na citrate, pH 8.0

- Whatman grade 540 SFC (or equivalent) filter paper
- Sponge
- Paper towels
- Parafilm or equivalent
- Thin plastic or glass plate
- 250–500 g weight
- 0.4 M NaOH
- 0.2 M Tris + 2 × SSC

Method

1. With gentle shaking at room temp., stain the gel in 1 μg/ml ethidium bromide for 45 min

2. Photograph the gel on a UV transilluminator with rulers (preferably fluorescent) placed along the sides.

3. Expose the gel to the UV light for several minutes (typically 5–10 min) (CAUTION! UV light may damage your skin and eyes: wear UV protective clothing). For a given transilluminator, different lanes in a test gel should be exposed to UV for various times to determine the optimum UV exposure. [a]

4. Shake in 0.4 M NaOH, 0.6 M NaCl, for 45 min at room temp. to denature the DNA.

5. Shake in 0.5 M Tris–HCl, pH 7.4, 1.5 M NaCl for 30–45 min.

6. Cut a charged nylon membrane to a size a few millimeters greater than the dimensions of the gel. Use gloves and flat tipped forceps to handle the membrane. Any traces of oil will interfere with transfer.

Protocol 3. *Continued*

Label the membrane in pencil, and wet in water for a few seconds. Equilibrate in 10 × SSC buffer. Also cut three sheets of filter paper (Whatman grade 540SFC or equivalent) to the dimensions of the gel, and equilibrate in buffer.

7. Soak a clean sponge in 10 × SSC buffer and place in a tray which is filled with buffer almost to the level of the top of the sponge. (A tray covered with several sheets of filter paper with wicks extending into the buffer can substitute for the sponge.) Then, in order, place over it two wetted sheets of filter paper, the agarose gel, the nylon membrane, the three sheets of filter paper cut to the dimensions of the gel, and stack of dry paper towels several inches thick, as in *Figure 1*. A non-permeable membrane (e.g. Parafilm) should be placed around the gel to prevent a 'short circuit' of the capillary flow. While assembling the stack, it is critical to exclude air-bubbles.

8. On top of the stack place a thin plastic or glass plate, and on top of this a small weight (typically 250–500 g). Allow the transfer to take place overnight, and occasionally change the towels if possible.

9. The next day, disassemble the stack, wash the membrane for 1 min in 0.4 M NaOH, neutralize in 0.2 M Tris, 2 × SSC, and dry completely.

10. The extent of transfer can be assessed by restaining the gel in ethidium bromide and examining it under UV light.

Variation: An alternative to the classical capillary transfer here described, is the downward capillary transfer or descending Southern blot (14), in which a stack similar to the one described, is assembled in the inverted order, with the following exceptions: (a) the sponge is omitted, (b) the tray filled with transfer buffer is located at the side of the stack, elevated to the level of the gel, and connected to the top of the stack by a wick made of filter paper, (c) a glass plate is put at the top of the stack. This method avoids crushing the gel, which reduces the efficiency of the transfer in the ascending Southern blot, and is considerably faster, taking from 2.5 to 4 h to complete the transfer.

[a] Alternatively, to nick the DNA before the transfer, shake the gel in 1 M HCl at room temp. for 30 min, neutralize shaking in 1 M Tris–HCl, pH 8.0 for 30 min, and then proceed with step 4.

The porous membranes used to transfer the DNA are either nitrocellulose, or neutral or charged nylon membranes. Nylon membranes are flexible and less breakable than the nitrocellulose ones, allowing for a longer life when subject to multiple hybridizations and washes; however, reinforced nitrocellulose membranes are now available. Nylon membranes (especially charged ones) generally give a stronger signal, but a higher background than

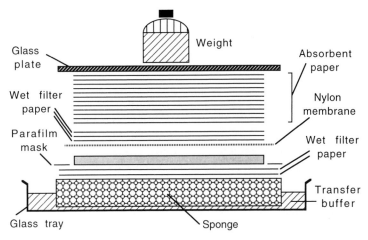

Figure 1. Vertical section of the assembly used for Southern transfer by capillary flow of buffer.

nitrocellulose membranes. After transfer, the membranes are dried in the air at room temperature or under an infrared light bulb. To improve the irreversible binding of the DNA to the membrane, the nitrocellulose or un-charged nylon membranes are further dehydrated in a vacuum oven at 80°C. Be sure to let the oven cool down before releasing the vacuum, because the nitrocellulose may otherwise ignite. The DNA can be cross-linked to nylon membranes by exposure to UV light (15). The same transilluminator used for observation of the gels or a special box containing a timed UV source can be used for this purpose. The DNA cross-linked to the membrane will withstand a larger number of hybridization-stripping cycles than DNA attached only by adsorption to the membranes; nevertheless, too extensive cross-linking may interfere with hybridization.

2.5 Preparation of radioactive probes

Probes for hybridization can be prepared from any kind of polynucleotide. Short synthetic oligonucleotides can be end-labelled by using T4 poly-nucleotide kinase and ^{32}P-ATP (16), but most DNA sequences obtained by cloning or PCR, are labelled by nick-translation (17) or random oligo-nucleotide primed replication with the Klenow fragment of *E. coli* DNA polymerase I (18). The last procedure is preferred since it gives higher specific activities and can be applied to DNA contained in low-melting-temperature agarose (19). There are several commercial kits available for labelling DNA probes by random oligonucleotide primed replication, all based on the method of Feinberg and Volgestein (19) described with slight modifications in *Protocol 4*.

27

Protocol 4. Preparation of radioactive DNA probes by oligo-labelling

Equipment and reagents

- Restriction enzyme(s) appropriate to the plasmid
- Low-melting-temperature agarose gel
- Tris acetate buffer: 50 mM Tris-acetate, 1 mM EDTA (pH 8.0)
- Ethidium bromide (see *Protocol 3*)
- 5 × oligo-labelling buffer (OLB): 250 mM Tris–HCl, pH 8.0, 25 mM MgCl$_2$, 0.05 mM β-mercaptoethanol, 0.1 mM dATP, dTTP, and dGTP, 1 M Hepes (adjusted to pH 6.6 with NaOH), 27 OD units per ml of random hexa-deoxyribonucleotides

- Bovine serum albumin (10 mg/ml)
- DNA probe
- [^{32}P]dCTP (3000 Ci/mmol, 10 μCi/μl)
- Klenow fragment of *E. coli* DNA polymerase I
- Stop solution: 20 mM NaCl, 20 mM Tris–HCl (pH 7.5), 2 mM EDTA, 0.25% SDS, 1 μM dCTP
- Sephadex G-50 column (see Chapter 3)

Method

1. Cut the plasmid with the appropriate restriction enzyme to separate the insert from the vector sequences. Separate the fragments by electrophoresis through a low-melting-temperature agarose gel using Tris-acetate buffer. Stain the gel in 0.2 μg/ml ethidium bromide in water for 30 min, and visualize the fragments under UV illumination.

2. Cut out the region of the gel containing the insert band, put it in a microcentrifuge tube, and weigh it. Add 3 vol. of water, put the tube at 100°C for 7 min, and store it at −20°C.

3. When ready for labelling, reboil the gel for 3 min, and maintain it at 37°C for at least 10 min, then add in the following order to an empty tube:
   (a) 10 μl of 5 × oligo-labelling buffer (OLB)[a];
   (b) 2 μl of bovine serum albumin (10 mg/ml);
   (c) Up to 32 μl of agarose containing 20–50 ng of DNA[b];
   (d) 5 μl of [^{32}P]dCTP (3000 Ci/mmol, 10 μCi/μl);
   (e) 2 units of Klenow fragment of *E. coli* DNA polymerase I;
   (f) Water to 50 μl.

4. Incubate at room temp. for at least 5 h.[c]

5. Terminate the reaction by adding 200 μl of stop solution.

6. Use the labelled probe without further purification, or separate it from the unlabelled nucleotides using a Sephadex G-50 column (for details of how to make this column, see Chapter 3).

[a] Store at −20°C. Reaction buffers from the commercial kits can be substituted.
[b] The length of the probe will vary with the template/primers ratio. If this ratio is very small, the probe fragments will be too short, and will not form stable hybrids with their target after normal stringency hybridization conditions. If the ratio is too large, the probe fragments will be too large, and will give high background-labelling after hybridization.
[c] Some commercial kits provide faster protocols.

For special purposes, one can prepare labelled single-stranded RNA probes using *in vitro* transcription with bacteriophage SP6, T4, or T7 DNA polymerases, from DNA templates cloned in plasmid vectors that contain the appropriate promoters (20).

Chemiluminescence-based detection methods are now as efficient as radioactivity-based methods (21, 22). Commercial kits are available for this purpose, avoiding the inconveniences of using radioactivity, and providing probes with a long shelf life. Detection is made with the same autoradiographic film used to detect radioactivity.

2.6 Membrane hybridization

Pre-treatment and hybridization can be performed in a sealed plastic bag immersed in a shaking water-bath, or in a cylindrical rotating container (23). Both devices allow for minimum volumes of hybridization buffer, and maximum concentrations of the probe. The effective concentration of the probe may be increased by including 5–10% dextran sulphate in the hybridization buffer. Pre-treatment of the membranes is necessary to block sites that could bind the probe unspecifically. SDS, sonicated DNA, and sometimes several other polymers are included in the pre-hybridization mix for this purpose. When a more complete blocking is necessary, the use of dried non-fat milk may be useful (24).

Hybridization is performed at relatively high ionic strength in a buffered saline solution containing a chelating agent for divalent ions. Hybridization is performed overnight, and then the membrane is washed in a saline solution at higher temperature. The degree of similarity between the sequences that remain hybridized (stringency) depends on the ionic strength and temperature of the washings. The lower the ionic strength and the higher the temperature, the higher is the stringency of the hybridization.

After washing the membrane, keep it damp, wrapped in a thin non-permeable plastic membrane (such as Saran Wrap), and laid down on X-ray film inside a light-proof cassette at −70°C. Fluorographic intensifying screens can be used. A film adequate for autoradiography of β emissions is Kodak X-Omat AR radiographic film.

Protocol 5. Hybridization of Southern blots to radioactive probes

Equipment and reagents

- Sealable plastic bags (e.g. Seal-a-Meal)
- Pre-hybridization solution (for 20 ml): 10 ml deionized formamide, 4 ml 50% dextran sulphate, 4 ml 5 M NaCl, 1 ml 20% SDS, 2 mg sonicated single-stranded DNA (e.g. salmon sperm), water to 20 ml
- Radioactive probe
- 1 × SSC, 1% SDS
- Plastic film (e.g. Saran wrap)
- Buffer A: 2 mM Tris-base, pH 7.9, 1 mM EDTA, 0.1% SDS

Protocol 5. *Continued*

Method

1. Place the blot in a sealable plastic bag (e.g. Seal-a-Meal) and seal all edges, except for leaving a spout at one corner for adding reagents.

2. Prepare the pre-hybridization solution ($\sim$0.1–0.2 ml/cm^2).

3. Add the solution to the bag, remove air-bubbles, seal, and incubate at 42°C for 1–2 h.

4. Boil the radioactive probe ($\sim$10^6 d.p.m./ml of mix) for 5 min; chill. Open the bag (if hybridization mix different from pre-hybridization mix, substitute) add the probe; mix well; remove air-bubbles; and reseal the spout.

5. Incubate overnight in a water bath at 42°C, preferably with shaking.

6. Remove the hybridization mix. Carefully open the bag, and wash the blot with an excess of 1 × SSC, 1% SDS at room temperature for 30 min. Wash the blot twice (30 min each time) at 65°C with the same solution.

7. Blot excess liquid, but leave moist. Sandwich between layers of plastic film (e.g. Saran wrap), and expose to radiographic film at −70°C.

8. The blot can be stripped of probe in preparation for rehybridization by incubating at 75°C with buffer A.

Variations: Formamide can be omitted if hybridization is performed at 65°C. Some probes, for example, those with an unusually high C + G content, may require a wash at higher stringency (e.g. 0.1 × SSC, 1% SDS at 65°C or a slightly higher temperature). In particularly difficult cases, it may help to increase the temperature of hybridization. Recently, several companies have offered hybridization accelerators which can be used as an alternative to dextran sulphate and which purportedly reduce the time required for hybridization.

3. Common artefacts and interpretation

Several common problems may result in confusion in interpreting Southern blots.

3.1 Contamination with plasmid DNA

Since radioactive probes are frequently prepared from plasmids, hybridization to contaminating plasmid DNA may be an important source of error. Most commonly used plasmid vectors share DNA sequences and will cross-hybridize to one another. Traces of DNA from plasmids commonly handled in the laboratory frequently contaminate the tips or stems of automatic pipettors, and can then easily make their way into solutions. The signal generated from traces of a linearized or circular plasmid contaminant may be similar in size and intensity to the ones generated by DNA sequences of a unique gene in a complex genome. The use of excised inserts as probes does not solve the problem, because it is difficult to clean the insert completely from vector DNA sequences. Sometimes, the only way to tell if a suspect autoradiographic band is generated by a contaminating plasmid is to hybridize the blot to a plasmid-vector-only probe. The best way to prevent this problem is to maintain separate sets of pipettors and solutions for the preparation and handling of plasmid and of genomic DNA to be used in the Southern blot, and avoiding reusable glassware.

3.2 Partial digests

Additional bands in an autoradiogram may also be generated by partial digestion products. In some cases the new band has a length that is the sum of the lengths of two smaller bands. These bands are of a higher molecular weight than the complete digestion bands. Frequently, but not always, the bands due to partial digestion are of a lower intensity than the complete digestion products. Sometimes the only way to determine if a new band in an autoradiogram is the product of partial digestion, is to repeat the Southern blot, improving the digestion conditions.

3.3 Degraded DNA: incomplete transfer

In addition to new bands, there may be artefactual loss of bands in a Southern blot. One source of band loss is a low molecular weight of the initial DNA preparation. A restriction fragment that exceeds the average size of starting collection of DNA fragments will be represented by a smear of fragments of different sizes that will be either too weak to be seen, or will show like a wide fuzzy band at the top of the autoradiogram. Large DNA fragments may also be transferred incompletely or not at all to the membrane, if not nicked properly.

Incomplete transfer of DNA may also be due to air-bubbles interposed between the gel and the membrane during the transfer, or to non-permeable spots on the membrane resulting from manufacturing defects, or from films of grease deposited on the membrane by ungloved hands or other sources.

3.4 Background sources: repetitive sequences

A frequent problem in Southern blots is heavy background or spurious spots in the autoradiograms. One source of background may be incomplete washing of the unspecifically bound probe. Generalized background all over the membrane may be due to a high molecular weight of the probe. This can be dealt with by preparing a lower molecular weight probe (decreasing the proportion of template to primers), more completely blocking of the membrane during pre-hybridization (24), and increasing the SDS concentration during the washings (up to 4%). Background restricted to the DNA-containing lanes may be due to high G-C base composition of the probe. This can be dealt with by increasing the temperature of the hybridization and the stringency of the washings. Lane-restricted background may also be due to DNA degradation, and will show as a smear that extends down from specific bands in the autoradiogram. Lane-restricted background may be due to repetitive sequences represented in the probe. This problem can be solved by pre-annealing the probe and the membrane with sonicated total genomic DNA.

When dealing with gene families, depending on the stringency of the hybridization, one can see a number of different bands in the autoradiograms (*Figure 2*). This may be used to advantage when the organization of the whole gene family is to be studied. At maximum stringency, the gene of origin of the probe will give the strongest signal.

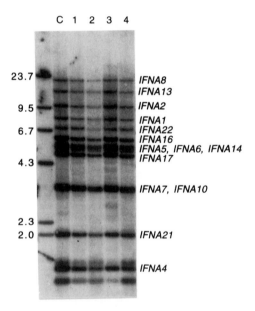

Figure 2. Autoradiograph of a Southern blot of human genomic DNA from the leukaemia cells from four patients with acute lymphoblastic leukaemia cut with the restriction enzyme HindIII. The resulting fragments were separated by standard electrophoresis in a 0.8% agarose gel slab, transferred to a charged nylon membrane, and hybridized to a ^{32}P-labelled probe for the interferon alpha gene family, under low stringency conditions. DNA from human placenta has been run as a control: *lane C*. DNA from phage lambda cut with HindIII and end-labelled with ^{32}P has been run in the first lane as a size marker. The size of the marker fragments is given on the left margin in kb. The interferon genes contained in some of the DNA fragments are indicated on the right margin. Diffuse weak background in the human DNA lanes is due to the unspecific hybridization of GC-rich sequences in the probe.

3.5 Quantitation

Sometimes, the dosage of a gene is of interest, and some kind of quantitation is required. Densitometry of the autoradiograms can provide such quantitation, provided adequate controls are included. Densitometric quantitation of gene dosage is more accurately done on dot or slots blots, where the DNA is loaded directly on to the membrane, and therefore, the variables introduced by the electrophoresis and transfer of the DNA are eliminated. Controls with varying amounts of placental DNA can be included on the same blot. A probe for a gene whose dose is known not to be altered, should be hybridized to the same blot as a standard against which one can measure variations in the dose of the gene under investigation. It is convenient to make several autoradiograms from the same blot with different times of exposure to select one that is within the range of response of the emulsion. Direct or indirect imaging of the hybridized Southern blot membrane, can be obtained with special devices, beta-imagers, or phosphor-imagers, that have linear responses over a larger scale than the autoradiographic film. These devices require a considerable initial investment, but provide more accurate, faster and convenient measurements of the radioactive signal. Phosphor-imagers can also read chemoluminescent signals.

References

1. Southern, E. M. (1975). *Journal of Molecular Biology*, **98**, 503.
2. Schwartz, D. C. and Cantor, C. R. (1984). *Cell*, **37**, 67.
3. Blin, N. and Stafford, D. W. (1976). *Nucleic Acids Res.*, **3**, 2303.
4. Bernards, A., Kooter, J. M., Michels, P. A. M., Roberts, R. P. M., and Borst, P. (1986). *Gene*, **42**, 313.
5. Carle, G. F. and Olson, M. V. (1984). *Nucleic Acids Res.*, **12**, 5647.
6. McClelland, M., Hanish, J., Nelson, M., and Patel, Y. (1988). *Nucleic Acids Res.*, **16**, 364.
7. Bouché, J. P. (1981). *Analytical Biochemistry*, **115**, 42.
8. Sealy, P. G. and Southern, E. M. (1982). In *Gel electrophoresis of nucleic acids: a practical approach* (ed. D. Rickwood and B. D. Hames), p. 77. IRL Press, Oxford.
9. Carle, G. F. and Olson, M. V. (1985). *Proc. Natl. Acad. Sci., USA*, **82**, 3756.
10. Vollrath, D. and Davis, R. W. (1987). *Nucleic Acids Res.*, **15**, 7865.
11. Gardiner, K., Laas, W., and Patterson, D. (1986). *Somatic Cell and Molecular Genetics*, **12**, 185.
12. Waterbury, P. G. and Lane, M. J. (1987). *Nucleic Acids Res.*, **15**, 3930.
13. Lee, H., Birren, B., and Lai, E. (1991). *Analytical Biochemistry*, **199**, 29.
14. Lichtenstein, A. V., Moiseev, V. L., and Zaboikin, M. M. (1990), *Analytical Biochemistry*, **191**, 187.
15. Church, G. M. and Gilbert, W. (1984). *Proc. Natl. Acad. Sci., USA*, **81**, 1991.
16. Richardson, C. C. (1971). In *Procedures in nucleic acid research* (ed. G. L. Cantoni and D. R. Davies), Vol. **2**, p. 815. Harper & Row, New York.

17. Rigby, P. W. J., Dieckmann, M., Rhodes, C., and Berg, P. (1977). *Journal of Molecular Biology*, **113**, 237.
18. Feinberg, A. P. and Volgestein, B. (1983). *Analytical Biochemistry*, **132**, 6.
19. Feinberg, A. P. and Volgestein, B. (1984). *Analytical Biochemistry*, **137**, 266.
20. Melton, D. A., Kreig, P. A., Rebagliati, T., Maniatis, T., Zinn, K., and Green, M. R. (1984). *Nucleic Acids Res.*, **12**, 7035.
21. Martin, R., Hoover, C., Grimme, S., Grogan, C., Holtke, J., and Kessler, C. (1990) *Bio Techniques*, **9**, 762.
22. Engler-Blum, G., Meier, M., Frank, J., and Muller, G. A. (1993). *Analytical Biochemistry*, **210**, 235.
23. Bennhamias, S. (1989). *American Biotechnology Laboratory*, **7**, 10.
24. Johnson, D. A., Gautsch, J. W., Sportsman, J. R., and Elder, J. H. (1984). *Gene Analysis Techniques*, **1**, 3.

3

Northern analysis, ribonuclease protection, and *in situ* analysis of cytokine messenger RNA

M. STUART NAYLOR, MICHELE RELF, and
FRANCES R. BALKWILL

1. Introduction

Northern analysis of RNA allows the quantitation of specific mRNA sequences and the determination of their size. The RNA sample is separated according to size by electrophoresis through a denaturing agarose gel followed by capillary transfer to a membrane to which the RNA is subsequently covalently linked. The messenger RNA (mRNA) of interest is located by probing the membrane with a labelled complementary nucleotide sequence which, in the case of a radioisotopic label, is revealed by autoradiography.

Ribonuclease protection also allows the quantitation of specific RNA sequences. It has the advantage of being more sensitive than Northern analysis and many of the problems of non-specific binding are eliminated. However, the technique does not allow sizing of the transcript. It is possible to distinguish several splice variants by careful selection of the RNA probe. The expression of several different messages may be determined simultaneously providing the probes used are of different sizes and give a characteristic band pattern on a sequencing gel.

In situ hybridization (ISH) to cellular messenger RNA allows the analysis of gene expression in single cells. The availability of good immunohistochemistry techniques for cytokine detection is limited. This is in part due to the nature of these molecules; they are secreted extracellularly, rapidly diffuse from site of secretion and generally have a short biological half-life. Existence of membrane bound forms and ubiquitously expressed receptors serve to confuse the identification of a secreting cell from a target cell from an 'innocent bystander' by antibody detection methods. Therefore ISH is a useful technique for identifying cells expressing cytokine genes, though expression should obviously not been taken as an indicator of translation/ secretion of the biologically active protein.

The reverse transcriptase PCR reaction is another powerful tool for analysing cytokine mRNA. Assay methods are described in Chapters 4 and 5.

2. General precautions for handling RNA

- Wear gloves and change them frequently throughout.
- Wash all glassware, spatulas, forceps, etc., rinse in alcohol and then bake.
- Alternatively fill glassware with 0.1% diethyl-pyrocarbonate (DEPC) in distilled water and allow to stand for 2 h at 37°C.
- Treat all buffers (with the exception of Tris-based buffers where the DEPC reacts with amines in Tris) with DEPC as follows; add DEPC to 0.1% v/v, incubate at 37°C for at least 12 h, then autoclave. Prepare Tris buffers and alcohols in DEPC treated distilled water.
- Wash gel tanks, combs, etc. in soap and water, followed by an ethanol rinse. Treat with 3% hydrogen peroxide for 10 min at room temperature, rinse with DEPC treated distilled water. Beware of damage to gel trays by these treatments.

3. Northern analysis

Protocol 1. A rapid method for isolation of total RNA (approx. 4 h)

This is based on a modification of Chirgwin's protocol developed by Chomczynski and Sacchi (1). Steps 7–10 are a further modification by Puissant and Houdebine (2).

Equipment and reagents

- Denaturing solution (Solution D): 4 M guanidinium thiocyanate (Fluka), 25 mM sodium citrate pH 7, 0.5% sarcosyl, 0.1 M 2-mercaptoethanol. Prepared as follows: 250 g guanidinium thiocyanate dissolved in bottle supplied with; 293 ml water, 17.6 ml 0.75 M sodium citrate, pH 7, and 26.4 ml 10% sarcosyl at 65°C. Store up to 3 months at room temperature. Add 0.36 ml 2-ME/50 ml of stock.
- Phenol (nucleic acid grade, Gibco/BRL)
- 2 M sodium acetate, pH 4
- Chloroform/isoamyl alcohol (49:1)
- 75% ethanol
- Chloroform

- Isopropanol
- 0.5% sodium dodecyl sulphate (SDS)
- 10 mM Tris pH 7.5, 1 mM EDTA, 0.5% SDS
- 4 M LiCl
- Sorvall or Beckman standard centrifuge
- Tissue homogenizer (Polytron, Ultraturrax)
- Microcentrifuge
- Spectrophotometer
- Dry ice
- 1.5 ml capped microcentrifuge tubes (Eppendorf)
- 4 ml or 15 ml polypropylene tubes (Falcon, Sarstedt)

Method

For 100 mg tissue (scale up accordingly. Published method tested in range of 3 mg tissue or 10^6 cells up to 30 g). For cultured cell-lines use 100 μl solution D/10^6 cells.

---

1. Mince tissue on ice.
2. Homogenize in 1 ml of solution D at room temperature, transfer to 4 ml polypropylene tube.
3. Add in order:

   (a) 0.1 ml 2 M sodium acetate, pH 4

   (b) 1.0 ml phenol (water saturated)

   (c) 0.2 ml chloroform/isoamyl alcohol (49:1)

   Mix after each addition. Shake 10 sec, cool on ice 15 min.
4. Centrifuge 10 000 g, 20 min, 4°C
5. Transfer aqueous phase (clear of interface) to fresh tube and mix with 1 ml of isopropanol. Put to −20°C or dry ice for 1 h to precipitate RNA.
6. Centrifuge 10 000 g, 20 min, 4°C.
7. Resuspend in 0.2 ml of 4 M LiCl, microcentrifuge to pellet insoluble RNA.
8. Dissolve pellet in 0.2 ml 10 mM Tris pH 7.5, 1 mM EDTA, 0.5% SDS.
9. Add 0.2 ml chloroform, vortex and microcentrifuge 10 min.
10. Collect upper phase and precipitate with 0.2 ml isopropanol in 0.2 M sodium acetate, pH 5.0.
11. Vacuum or carefully air dry and dissolve in 50 µl 0.5% SDS at 65°C.
12. Determine the RNA quantity and purity as follows: For RNA an optical density of 1.0 is 40 µg/µl.

    (a) Take 4 µl of the sample, add to 995 µl of distilled water

    (b) Transfer the diluted sample to a quartz microcuvette, read optical density (OD) at 260 nm.

    (c) Concentration of the sample will be 10 × OD reading in µg/µl
    e.g. if OD reading = 0.15 at 260 nm
    RNA conc. of sample = 1.5 µg/µl.
13. To assess the purity of the sample read the OD at 280 nm, calculate the ratio of $OD_{260nm}:OD_{280nm}$. If the ratio is 2.0 the sample is extremely pure, if the ratio is below 1.6 it is likely that the sample is contaminated by protein and should therefore be re-extracted with phenol/chloroform and ethanol precipitated.

Purification of mRNA from whole RNA can be simply and rapidly performed using the numerous commercial kits now available. Most of these kits are based on the use of oligo(dT) to pull out polyadenylated RNA. The Promega product 'Poly-ATtract' is a rapid isolation method based on oligo(dT)-conjugated magnetic beads.

Using poly-A mRNA rather than total RNA offers more sensitivity in Northern analysis. 2 µg of poly-A + RNA is typically run per track. One disadvantage of purifying on the basis of the poly-A tail is that it is often the first target for degradation and may be lost in impure preparations or extractions from whole resected tissues.

3.1 Electrophoresis of RNA for Northern blot

It is important to include positive controls of RNA extracted from cell lines that are known to express high levels of mRNA for the cytokine of interest.

For IL-1α, β, and TNF α, we use RNA extracted from HL60 cells stimulated with 50 ng/ml PMA; for IL-6 we use RNA from human foreskin fibroblast cells; for IFN$_\gamma$ and IL-2 we use RNA from Jurkat cells stimulated with 1 μg/ml PHA and 50 ng/ml PMA for 8 h.

Membranes should be probed with beta actin or glyceraldehyde phosphate dehydrogenase (GAPDH) as a control for loading consistency and RNA integrity.

Protocol 2. Electrophoresis of RNA for Northern blot

Equipment and reagents

NB: Caution: ethidium bromide is mutagenic, wear gloves to handle the gel, add hypochlorite to ethidium bromide solutions 30 min before discarding down the sink.

- Agarose
- 10 × MOPS buffer (pH 5.5–7.0): 0.2 M MOPS (3-*N*-morpholinopropanesulphonic acid), 0.05 M sodium acetate, 0.01 M EDTA pH8
- 37% formaldehyde
- 10 mg/ml ethidium bromide
- Deionized formamide
- 80% glycerol
- Bromophenol blue (saturated solution)

- Northern load buffer:
deionized formamide	0.72 ml
10 × MOPS buffer	0.16 ml
37% formaldehyde	0.26 ml
distilled water	0.18 ml
80% glycerol	0.1 ml
Bromophenol blue (sat. solution)	0.08 ml
- Horizontal gel electrophoresis apparatus
- Power supply
- Fume hood
- UV transilluminator and camera

Method

1. Prepare a 1% gel (if target mRNA is less than 2 kb use a higher percentage of agarose e.g. 1.4%) as follows:

   (a) 3 g agarose

   (b) 30 ml 10 × MOPS

   (c) 225 ml DW

2. Microwave to dissolve agarose then cool to 50°C.

3. Add 50 ml of 37% formaldehyde in a fume hood then mix.

4. Add 15 μl of 10 mg/ml ethidium bromide (this can be omitted if it is thought to interfere with transfer), mix and pour on to a levelled gel tray with an appropriate comb fitted.

5. Allow the gel to set in a fume hood.

6. Remove the comb and any tape if used. Place the gel in an electrophoresis tank then cover with 1 × MOPS buffer.

---

7. Take 15 μg samples of total RNA, dry down (if volume > 5 μl), and resuspend in 15 μl of load buffer. Alternatively use 2–5 μg of poly-A + RNA.

8. Run the gel at 90 V for approximately 4–5 h or until dye front reaches the end of the gel. If practical, cool the gel bed with running water and recirculate the buffer.

9. Photograph the gel on a UV transilluminator.

Protocol 3. Northern blotting (see *Figure 1*)

Equipment and reagents

- 20 × SSC: 3M NaCl, 0.3 M Na$_3$ citrate.2H$_2$O, 800 ml distilled water, pH to 7.0, adjust to 1 litre
- Whatman 3 MM filter paper
- Glass plate
- Reservoir tray
- Paper towels
- Nylon membrane e.g. Bio-Rad Zeta-probe
- UV cross-linker
- Saran wrap
- Parafilm or similar

Method

1. Place a glass plate across a large tray filled with 20 × SSC.

2. Overlay the plate with a wick prepared from Whatman 3 MM, soaked in 20 × SSC, cut slightly wider than gel and long enough to drape into reservoir of 20 × SSC in tray.

3. Smooth out any bubbles in the 3 MM wick.

4. Place the gel on top of the wick and surround with Parafilm or suitable impermeable membrane to prevent a short circuit.

5. Place the nylon membrane or equivalent (cut to size) on to gel again avoiding bubbles (check manufacturer's instructions for pre-wetting requirements).

6. Cover the gel with two pieces of pre-wetted 3 MM.

7. Place a stack of paper towels on top (to form a layer approximately 10 cm or more thick) then cover with a glass plate and place a small weight on top (e.g. a filled 400 ml bottle).

8. Leave overnight or for approximately 12 h to transfer.

9. Remove and discard paper towels, 3 MM etc. Orientate the membrane on the gel by making pin marks in the centre of each slot and by cutting off the top right-hand corner of the membrane.

10. Remove the membrane from the gel and wrap in Saran wrap, UV cross-link or bake membrane to fix transferred RNA on to the membrane (check membrane manufacturer's instructions as to cross-linking procedures). Do not let the membrane dry out.

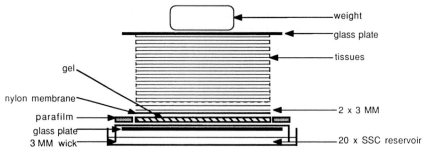

Figure 1. Apparatus for Northern blotting.

3.2 Probe labelling

The protocol used is based on the random priming technique developed by Feinberg and Vogelstein (3, 4) and is outlined in Chapter 2 of this volume. Unincorporated radioactive precursors can be removed on a Sephadex G-50 column. Columns are widely available commercially (e.g. Clontech make a range of columns with different exclusion sizes).

Protocol 4. Spun column (see *Figure 2*)

Equipment and reagents

- 1 ml disposable syringes
- Sterile siliconized glass wool
- Sephadex G-50 (fine)

- TE buffer
- 15 ml disposable tube
- 1.5 ml capped microcentrifuge tubes

Method

1. Plug the bottom of a 1 ml disposable syringe barrel with a small amount of sterile siliconized glass wool.

2. Fill barrel with Sephadex G-50 (fine) equilibrated in TE buffer.

3. Insert prepared column into a 15 ml disposable tube and spin at 1600 *g* for 4 min.

4. Top up the column with Sephadex G-50 and re-spin until the packed volume is approximately 1 ml.

5. Add 0.1 ml of TE buffer and respin as before; 1600 *g* for 4 min.

6. Place the column into a fresh 15 ml tube containing a decapped 1.5 ml microcentrifuge tube. Load the sample obtained from the random priming reaction in a total volume of 100 μl and spin at 1600 *g* for 4 min.

7. Carefully recover the eluted sample in the microcentrifuge tube with forceps and transfer to a new capped tube.

8. Store at −20°C until required.

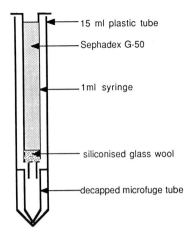

15 ml plastic tube

Sephadex G-50

1ml syringe

siliconised glass wool

decapped microfuge tube

Figure 2. Diagram of a spun column.

Protocol 5. Hybridization

Equipment and reagents

- 1 M sodium phosphate buffer pH 7.2: stock 134 g Na$_2$HPO$_4$.7H2O, 4 ml of 85% H$_3$PO$_4$, water to 1 litre or use a suitable ratio of molar monosodium and disodium phosphates
- 0.1 M EDTA pH 8
- Bovine serum albumin
- Sodium dodecyl sulphate
- Formamide
- Distilled water

- Autoradiography film e.g. XAR 5
- Hybridization oven/bottle system such as the Techne hybridizer
- Bag sealer
- Rocker
- X-ray film developer
- Incubator/waterbath
- TE buffer
- Saran wrap

The following protocol was devised by Church and Gilbert (5) for filter hybridization studies requiring high sensitivity and reprobing.

Method

1. Prepare the hybridization buffer as follows:

   - 1 M sodium phosphate buffer pH 7.2 (see 3.1.7.i)   10 ml   0.2 M
   - 0.1 M EDTA pH 8   0.5 ml   1 mM
   - Bovine serum albumin   0.5 g   1%
   - Sodium dodecyl sulphate   3.5 g   7%
   - Formamide   22.75 ml   45%
   - Distilled water   17 ml   34%

      Total volume   50 ml

Protocol 5. *Continued*

2. Place the UV cross-linked filter in a suitable bag then add 15 ml of the above hybridization mix, expel any bubbles, seal and place at 42°C with gentle rocking to pre-hybridize. Alternatively use the roller bottle system. Pre-hybridization should take place for a minimum of 15 min.

3. Replace with 15 ml of fresh hybridization mix with probe added to a concentration of approximately 10^6 cpm/ml. (Heat probe for 5 min at 100°C to denature, quench on ice then add to hybridization mix prior to addition to the bag.) Place the bag at 42°C, hybridize overnight with gentle rocking.

4. Remove the filter from the bag. Wash filter three times for 30 min each time at 65°C with agitation in the following wash buffer:
   - 1 M sodium phosphate pH 7.2   40 ml   40 mM
   - 0.1 M EDTA pH 8   10 ml   1 mM
   - Sodium dodecyl sulphate   10 g   1%
   - Distilled water   950 ml
   - Drain and wrap in Saran wrap.

5. Store filter moist at 4°C.

6. Expose filter overnight to X-ray film at −70°C in a suitable cassette with intensifying screens. Adjust the exposure time if required.

7. Prior to reprobing strip the filter in boiling TE buffer for 5 min.

4. Ribonuclease protection

A ^{32}P-labelled anti-sense RNA probe is hybridized to target mRNA in the sample, forming double-stranded RNA, which is resistant to digestion by single-strand specific ribonucleases such as RNase A. After purification, the protected fragments are sized on a sequencing gel and can be quantitated by autoradiography and image-analysis or phospho-imaging. A ^{32}P-labelled anti-sense GAPDH probe added to each sample makes a good loading control. It is essential that this is a different length to the test riboprobe.

This ribonuclease protection method is based on that described in ref. (6). The authors also recommend the use of commercially available kits for this method, e.g. the RPAII ribonuclease protection assay kit (Ambion, AMS Biotechnology UK Ltd).

4.1 Controls

- Probes alone. This confirms the integrity of the RNA probes.
- tRNA instead of sample RNA. This is a good negative control. None of the probe should be protected and any bands in this lane indicate incomplete nuclease digestion.

- Preferably positive and negative cell line controls. It is useful to know exactly what size protected fragment to expect and to show that there is no cross-reactivity of the probe with total RNA.

4.2 Preparation of template DNA

The template DNA must be cloned into a suitable vector, such as Bluescript, which has the bacteriophage promoters T3 and T7 flanking the multiple cloning site.

The plasmid should then be linearized immediately downstream of the inserted template DNA, thus allowing the generation of a known size of anti-sense RNA transcript. (N.B. Ensure the transcript is anti-sense NOT sense.) The plasmid is often linearized within the insert to yield small probes (200–500 b.p.) for optimal use and to avoid the plasmid sequence. Inserts are often too long for preparation of a transcript suitable for protection. Purify this linearized template DNA on low melting point agarose, and dilute to 0.5 mg/ml for preparation of probe RNA.

Protocol 6. Preparation of probe

Equipment and reagents

NB. All reagents used should be RNase-free and made up with DEPC-treated water in baked glassware as appropriate.

- T7, T3, or SP6 RNA polymerases
- 5 × transcription buffer:
  For T3/T7 RNA polymerases: 200 mM Tris–HCl, pH 8.0, 40 mM MgCl$_2$, 10 mM spermidine, 250 mM NaCl
  For SP6: 200 mM Tris–HCl, pH 7.5, 30 mM MgCl$_2$, 10 mM spermidine
- 200 mM dithiothreitol (DTT)
- 3 NTP mix (use 4 mM each of ATP, UTP, and GTP)
- [α-^{32}P]CTP (Amersham cat. No. PB10162, 10 mCi/ml)[a]
- Water

- RNase inhibitor
- Template DNA
- Bacteria phage polymerase
- RNase-free DNase I
- Sephadex G-50 column (e.g. Boehringer-Mannheim)
- Eppendorf microcentrifuge tubes
- 5 × hybridization buffer: 200 mM Pipes, pH 6.4, 2 M NaCl, 5 mM EDTA (for working solution dilute 1/5 with deionized formamide; may be stored in aliquots at −20°C for one month)

Method

1. Mix together the following reagents:
   - 4 μl 5 × transcription buffer
   - 1 μl 200 mM DTT
   - 2 μl 3NTP mix
   - 5 μl (^{32}P) CTP
   - 5 μl water
   - 1 μl RNase inhibitor
   - 1 μl 0.5 mg/ml template DNA
   - 1 μl of the relevant bacteriophage polymerase (5 U)

43

Protocol 6. *Continued*

N.B. spermidine may cause precipitation of DNA, therefore the DNA should not be added until last.

2. Incubate for 15 min at 37°C.

3. Add 2 μl RNase-free DNase1 (10 μ) and incubate for a further 15 min at 37°C.

4. Remove unincorporated radioactivity by passing the reaction mixture through a G-50 spin column according to the manufacturer's instructions. (Typically: Allow spin column to drain, spin for 2.5 min at 2500 *g*. Transfer column to fresh Eppendorf microcentrifuge tube, load sample and spin at 2500 *g* for 4 min. The eluted probe is collected in the Eppendorf tube.)

5. Dilute probe with 500 μl of hybridization buffer and count 1 μl in a scintillation counter (Çerenkov).

[a] Cold CTP is sometimes added to ensure the full-length transcript is made.

Protocol 7. Hybridization

Equipment and reagents

- Sample RNAs
- DEPC-treated water
- Hybridization buffer (see *Protocol 6*)
- Probe RNA
- GAPDH probe
- 0.5 mg/ml template for GAPDH or actin control

Method

1. To 10 μg of sample RNA in 10 μl DEPC water, add 40 μl hybridization buffer containing $1-5 \times 10^5$ c.p.m. of probe RNA (and 1×10^5 c.p.m. of GAPDH probe as a loading control). (N.B. make sure probes are different sizes.)

2. Incubate for 5 min at 85°C to denature the RNA. Spin down any condensation.

3. Transfer to 45°C for hybridization overnight.

Protocol 8. Digestion of non-hybridized RNA with ribonuclease

Equipment and reagents

- Ribonuclease digestion buffer: 10 mM Tris–HCl, pH 7.5, 300 mM NaCl, 5 mM EDTA; to 10 ml of buffer add 20 ml ribonuclease T1 (500 000 U) and 20 mg/ml of ribonuclease A
- Proteinase K
- 20% SDS
- RNase T1 and A

Method

1. Add 350 μl of ribonuclease digestion buffer (containing 1/500 each of 20 μg/ml RNase A and 500 000 U/ml of RNase T1) to each tube.

2. Allow to stand at room temperature for 30 min. This digests any single-stranded RNA including unbound probe.

3. Inactivate the RNases by digestion with 2.5 μl of 20 μg/ml proteinase K in the presence of 10 μl 20% SDS.

Protocol 9. 'Cleaning up' the protected fragment

Equipment and reagents

- Tris-saturated phenol, pH 7.5–8
- Microcentrifuge and tubes
- Ethanol: 100%, 70%

Method

1. Extract once with 400 μl equilibrated phenol.

2. Spin 5 min in microcentrifuge and transfer 300 μl of the aqueous phase into a fresh tube.

3. Precipitate the RNA by adding 700 μl 100% ethanol and standing at −70°C for 30 min.

4. Pellet the RNA by centrifugation in a microcentrifuge for 10 min.

5. Wash the pellet thoroughly with 70% ethanol to remove the salt. (This is helped by warming to 37°C for 2–3 min.)

6. Centrifuge for 10 min in microcentrifuge to pellet the RNA and air dry the pellet. (The pellet is quite mobile at this stage.)

4.3 Analysis on sequencing gel

Protocol 10. Analysis

Equipment and reagents

- Gel loading buffer: 80% (v/v) deionized formamide, 1 mM EDTA, 0.1% Bromophenol blue, 0.1% xylene cyanol
- TBE buffer
- 10% APS
- Polyacrylamide/urea sequencing gel (Sequegel concentrate and diluent are convenient to use—National Diagnostics EC830 and EC840)
- TEMED

Protocol 10. *Continued*

Method

1. Prepare 6% polyacrylamide/urea sequencing gel as follows:
   for 150 ml gel:
   - 15 ml Sequegel concentrate
   - 99 ml Sequegel diluent
   - 15 ml 10 × TBE
   - 21 ml water
   - polymerize with 100 μl TEMED and 500 μl 10% APS.

2. Resuspend the pellets in 10 μl of gel loading buffer, heat to 95°C for 5 min, and cool rapidly on ice.

3. Run the samples on the sequencing gel at 100 W (50°C).

4. Dry and visualize the gel by autoradiography. This can be quantitated by image analysis or phospho-imaging. Relative mRNA expression may be determined by comparison with the internal GAPDH loading control.

5. *In situ* hybridization

The procedure can be divided into the following stages:

(a) Rapid fixation/snap freezing of sample (RNA degrades within minutes) and sectioning.

(b) Pre-hybridization treatments to optimize probe access to target mRNA sequences, and reduce non-specific interactions.

(c) Probe hybridization.

(d) Post-hybridization washes to remove unbound probe reducing background signal.

(e) Detection of bound probe.

For each tissue/cell sample outlined methods may have to be adjusted to optimize cytological detail, probe hybridization, and signal-to-noise ratios.

5.1 Sample handling

Tissue samples such as biopsy specimens should be snap frozen in liquid nitrogen immediately upon resection. Though we have found the use of frozen material more successful, wax embedded tissue can also be used provided it is rapidly fixed in formal saline (time of fixation is dependent on tissue type and sample size).

Peripheral blood mononuclear cells or cell suspensions should be cytospun immediately, but the slides can then be stored at −70°C following fixation in 4% paraformaldehyde at room temperature for 5 min and dehydration through an alcohol gradient.

5.2 Pre-hybridization

These steps are used to optimize probe access and reduce backgrounds.

Proteinase K treatment is included in the protocol for tissues but not for more fragile samples such as cytospin preparations. Controlled protease digestion partially removes cellular proteins to improve probe access.

The amount of proteinase K required has to be optimized for each tissue (the range may vary widely from 1 μg/ml to 50 μg/ml), overdigestion will result in the loss of morphology.

Treatment with 0.2 M HCl also provides an increase in signal though certain samples may lose morphological detail.

Acetylation with 0.25% acetic anhydride in triethanolamine reduces non-specific binding to the section and the slide by neutralizing electrostatic interactions.

5.3 Probe choice, labelling, and hybridization

Sensitivity of the *in situ* technique can be dictated to a large extent by the type of probe used. Anti-sense asymmetric probes have been used by many investigators to achieve good sensitivity along with low backgrounds.

Advantages of asymmetric RNA probes:

(a) They contain only the sequences complementary to the target mRNA whereas denatured double-stranded probes may reassociate during hybridization.

(b) Although DNA probes are more easily labelled by nick-translation or random priming these labelling methods do not offer the specific activity of RNA probes labelled by the *in vitro* transcription protocol outlined later.

(c) Single-stranded (unbound) probe can be removed by incorporating a ribonuclease A digestion step in the post-hybridization stages reducing background.

(d) The thermal stability of RNA/RNA hybrids is greater than that of RNA/DNA hybrids allowing increased hybridization and washing temperatures again serving to reduce non-specific binding.

The main disadvantage of using an RNA probe is the requirement to have the sequence of interest cloned into a suitable vector. Most of the large molecular biology companies now provide kits to enable easy cloning and *in vitro* transcription. The sensitivity of RNA probes to degradation requires that suitable precautions are taken.

The increasing availability of oligonucleotides may provide an alternative to RNA probes offering resistance to degradation, single-strandedness, and tailored probe-length. Use of overlapping oligo 'cocktails' may provide good overall specific activity. It is widely accepted that short probe fragments (100–200 nucleotides) yield higher signals. In the case of RNA probes this can be achieved using controlled alkaline hydrolysis as outlined in the protocol.

Digestion time in the alkaline digestion buffer is calculated using the following equation:

$$t = \frac{(L_o - L_f)}{kL_oL_f}$$

t = time in min; L_o = initial probe length in kb; L_f = desired probe length in kb; $k = 0.11$ (from Cox *et al.*, 7).

The label of choice in this system is isotopic. The choice of isotope is ^{32}S offering a compromise between low scatter (and therefore reasonable resolution) and relatively short exposure time. ^{33}P is also widely used requiring shorter exposure times although some reports suggest increased scatter.

Non-isotopic systems of labelling are improving all the time. Digoxigenin-based labelling systems are thought to be the more sensitive non-isotopic option. Boehringer-Mannheim supply easy to use DIG-labelling kits for preparing riboprobes.

5.4 Controls

The complexity of the technique demands rigorous controls: positive controls for the integrity of the probe and the integrity of the mRNA in the tissue and negative controls to verify that silver grain deposition in emulsion is due to probe localization rather than, for example, noise or emulsion contamination.

5.4.1 Positive controls

Suitable probe controls would include tissues or cells known to contain cells expressing target mRNA (see Northern analysis controls). We also have excellent tissue controls in the form of cell-lines transfected with the cytokine gene of interest established as tumours in nude mice. mRNA integrity in tissue can be verified by probing for an ubiquitously expressed message such as beta actin or glyceraldehyde phosphate dehydrogenase (GAPDH). Probes for polyadenylated sequences are likely to pick up partially degraded mRNA and because of the vast difference in levels of abundance relative to cytokine mRNA these would not be a relevant control.

5.4.2 Negative controls

Certain *in vitro* transcription vectors have paired RNA polymerase binding sites positioned on either side of the cloning site. These allow the generation of sense RNA probes from the insert. These will have comparable activity to the anti-sense probe yet will not bind to target sequences, thereby giving an idea of background levels. If these vectors are not available any irrelevant probe of similar activity may be used.

An idea of non-specific binding to tissue can be gained from pre-treating a small proportion of sections with ribonuclease A (see post-hybridization washes) in the pre-hybridization stages.

Protocol 11. Slide preparation for ISH

Equipment and reagents

- 400 ml glass staining troughs
- Slide racks
- Slides
- Coverslips
- 10% Decon-90
- Millipore water
- TESPA (Sigma)
- Acetone
- Repelcote

Method

1. Wash slides in hot 10% Decon 90 detergent overnight.
2. Rinse in running tap-water for several hours.
3. Stand in Millipore water for at least 1 h.
4. Bake overnight at 250°C.
5. To improve adherence of tissue sections slides are treated as follows:
   (a) Make 2% TESPA (Sigma) in acetone.
   (b) Incubate slides at room temperature for 10 sec.
   (c) Wash twice in acetone then water.
   (d) Bake dry.
6. Siliconize coverslips by dipping in Repelcote, dry dust-free, then bake.

5.5 Sample preparation

5.5.1 Solid tissue

Solid tissue (e.g. biopsy material) should be snap-frozen immediately in liquid nitrogen then stored at −90°C prior to sectioning.

Cut 5 μm cryostat sections on to TESPA-coated slides using a new blade for each specimen. If this is not practical the blade should be frequently swabbed with alcohol. Cut sections may be stored in slide boxes with silica gel for short periods (weeks) at −90°C. Wax embedded sections can be cut in a similar manner and run as for cryostat sections following standard de-waxing procedures.

M. Stuart Naylor et al.

5.5.2 Cells in suspension

Prepare cells in suspension, e.g. non-adherent cell lines or peripheral blood using a cytospin. Apply 100 μl at 1×10^6/ml at 500 r.p.m. for 3 min. The procedure is then the same as is used for cryostat sections but treatment with proteinase K is omitted.

Cells may also be grown on coverslips and run as for cryostat sections.

Protocol 12. Pre-hybridization

Equipment and reagents

- PBS
- Millipore distilled water
- 0.2 M HCl
- 4% paraformaldehyde—prepare fresh (dissolve at 80°C in fume hood)
- 5–20 μg/ml proteinase K made up in appropriate buffer (50 mM Tris, 5 mM EDTA pH 7.5)
- 0.1 M triethanolamine with 0.25% acetic anhydride added prior to use
- Alcohols—30%, 50%, 70%, 95%, 100%, in distilled water
- Rotary shaker

Method

All incubations at room temp. on a rotary shaker. Place slides in racks, incubate in the following, in suitable slide troughs, e.g. 400 ml dishes (see *Protocol 11*)

1.	PBS	5 min
2.	distilled water	5 min
3.	0.2 M HCl	20 min
4.	PBS	5 min
5.	4% PFA	15 min
6.	PBS	2 × 5 min
7.	5 μg/ml Proteinase K made up in appropriate buffer (50 mM Tris, 5 mM EDTA pH 7.5)	7.5 min
8.	PBS	5 min
9.	4% PFA	5 min
10.	distilled water dip	
11.	0.1 M Triethanolamine + 0.25% acetic anhydride	2 × 10 min
12.	PBS	5 min
13.	Dehydrate 30%, 50%, 70%, 95%, 100%	2 min in each
14.	Air-dry in a dust-free environment	

Protocol 13. Probe labelling

(most of the following components and *in vitro* transcription kits available from Promega or Stratagene)

Equipment and reagents

- 5 × transcription buffer (see *Protocol 6*)
- 100 mM DTT
- RNasin
- 10 mM of each of ATP, CTP, GTP
- ^{35}S-UTP, ^{33}P, or digoxigenin
- Relevant RNA polymerase (e.g. T7, T3, SP6)
- RQ1 DNase
- Ultrapure phenol
- 10 M ammonium acetate
- Absolute alcohol
- 10 mg/ml carrier RNA (either tRNA or rRNA)
- Dry ice/cardice

- Alkaline digestion buffer (40 mM NaHCO$_3$, 60 mM Na$_2$CO$_3$ pH 10.2, add DTT to 10 mM before use)
- 1 M sodium acetate
- 5% acetic acid
- Scintillation fluid
- 1.5 ml microcentrifuge tubes
- Vacuum desiccator
- Alkaline digestion buffer
- Dry block
- Phenol
- Scintillant (e.g. Picouflor)

Method

1. To a sterile ribonuclease-free 1.5 ml microcentrifuge tube add the following:

   (a) 5 × transcription buffer 4 µl

   (b) 100 mM DTT 2 µl

   (c) RNasin 0.8 µl

   (d) ATP, CTP, GTP 4 µl
   (mixed in a ratio of 1:1:1:1 with distilled water)

   (e) Linearized transcription vector template 1 µl (= 1 µg)

   (f) ^{35}S-UTP (> 1000 µCi/nmol) or ^{33}P-UTP 10 µl

   (g) Relevant polymerase 1 µl

2. Pipette up and down gently to mix. Incubate at 37°C for 1 h.

3. Add 2 units of RQ1 DNase, pipette up and down gently to mix.

4. Incubate at 37°C for 30 min.

5. Add 80 µl of distilled water, phenol extract, add 2 µl of 10 mg/ml carrier RNA then precipitate with 20 µl of 10 M ammonium acetate and 300 µl of absolute alcohol.

6. Put on dry ice for 10 min. Microcentrifuge for 10 min at 4°C. Carefully remove the supernate.

7. Dry the pellet in a vacuum desiccator (or invert microcentrifuge tube and dry in RNase-free environment at room temp.) and take up in 100 µl of alkaline digestion buffer.

Protocol 13. *Continued*

8. Incubate in a dry block or water bath at 60°C for 75 min (dependent on probe-length; see Section 5.3).

9. Stop this reaction with 10 μl 1M sodium acetate, 10 μl 5% acetic acid, and 2 μl 10 mg/ml carrier RNA

10. Phenol extract twice then precipitate as above, dry and resuspend in 40 μl of 10 mM DTT.

11. Count 1 μl in approximately 10 ml of suitable scintillant, e.g. Picoflor (expect approximately 2–4 × 10⁶ d.p.m./μl).

For DIG-labelling follow manufacturer's protocol.

Protocol 14. Hybridization

Equipment and reagents

(Storage temperature shown in brackets after each reagent)

- 1 M DTT (−20°C)
- Deionized formamide (add 50 ml of formamide to 5 g of a mixed bed resin, Bio-Rad AG 501-X8, stir gently at 4°C, filter, aliquot and store at −20°C)
- 100 × Denhardts (4°C)
- 1 M Tris pH 8 (room temperature)
- 5 M NaCl (room temperature)
- 0.5 M EDTA (room temperature)
- 10 mg/ml Poly A (−20°C)
- 10 mg/ml carrier RNA, rRNA or tRNA (−20°C)
- 20 mM cold S-UTP (−20°C) [may be omitted]
- 50% dextran sulphate (4°C)
- 1.5 ml microcentrifuge tubes
- Siliconized coverslips (see *Protocol 11*)
- Sealed slide box
- 5 × SSC

Method

1. Prepare the hybridization solution by adding the following (figures in number of μl) to 1.5 ml microcentrifuge tube. Select the required volume according to number of sections. 10–15 μl is required for each section.

Reagent	μl	μl	Final conc.
(a) 1 M DTT	5	10	10 mM
(b) Deionized formamide	300	600	60%
(c) 100 × Denhardts	5	10	1 ×
(d) 1 M Tris pH 8	5	10	10 mM
(e) 5 M NaCl	30	60	0.3 mM
(f) 0.5 M EDTA	5	10	1 mM
(g) 10 mg/ml Poly A	15	30	300 μg/μl
(h) 10 mg/ml carrier RNA	15	30	300 μg/μl
(i) 20 mM cold S-UTP	11	22	500 μM
(j) 50% dextran sulphate	100	200	10%
Total volume (μl)	491	982	

2. Add the probe to a concentration of 5×10^4 c.p.m./µl.

3. Heat in a dry block for 2 min at 80°C.

4. Apply 10–15 µl/section, cover with a siliconized coverslip.

5. Incubate overnight in a 50°C incubator in a sealed slide-box humidified with a tissue soaked in 50% formamide, 5 × SSC.

Protocol 15. Post-hybridization washes

Equipment and reagents

- 2-mercaptoethanol
- 5 × SSC (see *Protocol 14*)
- 4 × SSC
- 2 × SSC
- 0.1 × SSC
- Formamide
- Ribonuclease buffer (0.5 M NaCl, 10 mM Tris, 5 mM EDTA pH 8) 10 mg/ml ribonuclease A

- Alcohols—30%, 50%, 70%, 95%, 100%, in distilled water
- 0.1% gelatine
- Slide racks
- 400 ml slide dishes
- Waterbaths at 37°C, 50°C, and 65°C
- Autoradiograph equipment and facilities (*Protocol 16*)

Method

1. Slides should be transferred to racks and incubated in the following buffers in 400 ml slide dishes equilibrated in 37°C, 50°C, and 65°C waterbaths as directed. If possible slides should be gently agitated during the incubations.

Buffer	Temperature	Time
(a) 5 × SSC, 0.1% 2-mercaptoethanol (2-ME)	50°C	3 × 20 min
(b) 50% formamide, 2 × SSC, 0.1% 2-ME	65°C	30 min
(c) Ribonuclease Buffer (0.5M NaCl, 10mM Tris, 5mM EDTA pH 8)	37°C	2 × 10 min
(d) Ribonuclease A, 20 µg/ml in the above	37°C	30 min
(e) Wash in (c)	37°C	15 min
(f) As (b)		
(g) 2 × SSC	room temp.	15 min
(h) 0.1 × SSC	room temp.	15 min

2. Dehydrate 30%, 50%, 70%, 95%, 100% alcohol   2 min in each

3. Air-dry in a dust-free environment.

4. Dip slides in filtered 0.1% gelatine.

5. Air-dry in a dust-free environment.

6. Autoradiograph.

Protocol 16. Autoradiography

Equipment and reagents

- 0.1% gelatine
- Ilford K5 emulsion or equivalent
- Silica gel
- Kodak D-19 developer or equivalent
- 1% acetic acid
- 30% sodium thiosulphate (made fresh)

- Foil
- Toluidine blue
- 50°C waterbath
- Slide mailing box
- Light-proof box

Method

1. Pre-warm 10 ml of 0.1% gelatine at 50°C in a 50 ml measuring cylinder.
2. In a dark-room (suitable safelight may be used), top up to 20 ml with Ilford K5 emulsion.
3. Melt emulsion in a 50°C waterbath, rock gently (avoid bubbles) to mix.
4. Pour into a slide mailing box.
5. Dip slides with forceps, wipe backs then dry for 2 h in a light-proof box in a dark room.
6. Transfer slides to a slide box containing silica gel then wrap box in foil, put at room temperature overnight, then put to 4°C for 7–10 days.
7. In a dark room, at room temp., incubate slides in the following:
   (a) D-19 developer or equivalent for 2.5 min.
   (b) 1% acetic acid for 0.5 min.
   (c) 30% sodium thiosulphate (freshly made) for 5 min.
8. Transfer slides into distilled water then remove from the dark-room and wash in running distilled water tap for 60 min.
9. Counterstain with Toluidine blue or other suitable general stain.

5.6 Interpretation

Probe localization will be detected under brightfield microscopy as a peri-nuclear deposition of silver grains over the cells expressing the target message often giving a corona-like appearance (see *Figure 3*). If available, darkfield microscopy will accentuate this deposition but will also increase the background considerably under high magnification.

Artifactual positives are common and may be due to several inapparent reasons, e.g. contamination of emulsion with talc from gloves, precipitate in the gelatine, precipitation of components in the hybridization mix, bubbles in the emulsion. Negative controls should be screened thoroughly and employed widely.

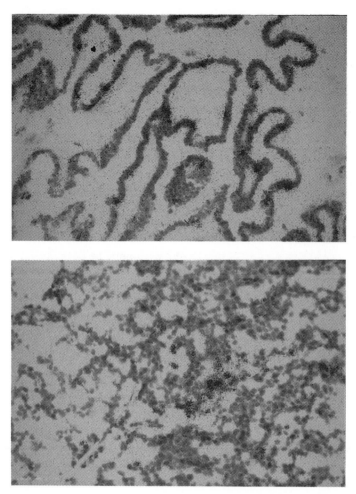

Figure 3. *In situ* hybridization of human ovarian cancer tissue sections with a riboprobe to MCP-1. (a) and (b) show the mRNA localization in two biopsies of ovarian cancer of differing malignancy. (Reproduced with kind permission from R. Negus Biological Therapy Laboratory ICRF.)

High backgrounds may be a result of many factors, for instance insufficient changes of washing buffer, insufficiently high temperatures, poor ribonuclease A digestion post-hybridization, or exposure to extraneous radiation during autoradiography.

References

1. Chomczynski, P. and Sacchi, N. (1987). *Analytical Biochemistry*, **162**, 156.
2. Puissant, C. and Houdebine, L.-M. (1990). *BioTechniques*, **8** (2), 148.

3. Feinberg, A. P. and Vogelstein, B. (1983). *Anal. Biochem.*, **132**, 6.
4. Feinberg, A. P. and Vogelstein, B. (1984). *Anal. Biochem.*, **137**, 266.
5. Church, G. M. and Gilbert, W. (1984). *Proc. Natl. Acad. Sci. USA*, **81**, 1991.
6. *Current Protocols in Molecular Biology*, **1** (unit 4.7.1), Supplement 24 CPMB, Wiley Interscience.
7. Cox, K. H., Deleon, D. V., Angerer, L. M., and Angerer, R. C. (1984). *Devel. Biol.*, **101**, 485.

4

Polymerase chain reaction to quantitate cytokine mRNA

CORNELIA PLATZER and THOMAS BLANKENSTEIN

1. Introduction

Cytokines display their function preferentially at local sites, occurring transiently and in low concentrations. For more effective intervention in pathophysiologic processes mediated by cytokines it is important to study their complex cross-regulatory network *in vivo*. Therefore, it is necessary to quantify cytokines in local sites of expression. Detection of the protein is hampered by usually low production of cytokines, by their short half-life *in vivo* and by the lack of specificity of certain antibodies for intracellular estimation. Consequently, analysis of mRNA has been widely used as substitute for studying variations in local cytokine levels. However, conventional methods of mRNA analysis such as 'Northern' and 'dot blot' hybridization, and nuclease protection mapping are not sensitive enough to detect mRNA in samples limited by low cell number or low copy-number per cell. Moreover, these methods permit only crude quantification. Thus, polymerase chain reaction (PCR) strategies have been developed that allow quantification of cytokine mRNA at a very sensitive level. Various quantitative RT-PCR procedures have been reported, most relying on the use of external standards and comparing products generated in separate reactions. However, because of the exponential nature of the PCR reaction small variations in the amplification efficiency may result in considerable changes of product yields. In addition, the amount of product can reach plateaux during later stages of the reaction due to consumption of PCR compounds or generation of inhibitors. These problems can be overcome by the use of internal controls in competitive PCR as has been employed by several groups (1–6).

The most stringent requirement for an internal DNA control standard is the content of the same primer template sequences as the target, because the amplification efficiency is primarily determined by the primer annealing reaction. Differences between intervening sequences of control and target DNA may influence denaturation or polymerase characteristics but this does not play any role when comparing mRNA in different samples. In competitive PCR, an internal control fragment constructed by cloning an irrelevant sequence flanked by the desired primer templates, competes with the target DNA for primer binding and amplification. It is possible to determine the amount of target cDNA by spiking different known molar quantities of the competitor DNA in a series of PCR reactions containing equal amounts of the target DNA. As the initial ratio of target-to-competitor DNA remains constant throughout the amplification, it is not necessary to obtain data during the exponential phase of the reaction. This is perhaps the greatest advantage of competitive PCR.

Several analytical approaches following PCR have been described. The most simple is to distinguish the products generated by the control and target by size in agarose gel electrophoresis and to estimate the amounts of PCR products by ethidium-bromide staining. Competitive quantification may start at the level of cDNA synthesis using internal control RNA. By quantifying the cDNA of a housekeeping gene, e.g. beta-actin, and subsequent equilibration of different cDNA samples to be compared according to their beta-actin cDNA content it is convenient to avoid the more complicated handling with control RNA.

The ability to perform quantitative analysis of mRNAs by competitive RT-PCR will undoubtedly expand the use and applications of PCR.

2. General remarks

- A single-step GuSCN phenol-chloroform extraction method is best for isolating total RNA, especially from tissue (see Chapters 3 and 5 for methods)
- Add carrier RNA (e.g. tRNA) during preparation from very low cell numbers
- Collect all samples to be compared as total RNA or as cell lysate in GuSCN at −80°C before beginning quantification
- Run agarose gel electrophoresis to check RNA quality by means of the integrity of the 18S and 28S rRNA and to roughly quantify total RNA
- Choose primer sequences of about 20 bp located on different exons if genomic sequence is available in order to recognize DNA contamination of the RNA preparation
- Primer pairs should amplify a fragment of about 200–700 bp
- If possible, perform RNA preparation, cDNA and the PCR reaction in another room to PCR product analysis to avoid contamination
- Use different pipettes for buffers, control fragment, RNA and cDNA, PCR product

Preparation, concentration and storage of some of the above mentioned standard reagents are according to ref. 7. The same holds true for standard techniques such as RNA preparation and DNA sequence analysis; oligo-nucleotide synthesis requires further equipment and is not shown here.

3. Cytokine RT-PCR

Carry out cDNA synthesis as recommended by the supplier of reverse transcriptase. Conditions for PCR amplification (e.g. cycle number, annealing temperature, hot start) have to be optimized for each primer pair. For instance, amount of Taq-polymerase and cycle number can be reduced up to 0.5 units/reaction tube and 25 cycles, respectively, if β-actin or another high abundant cDNA is amplified. Pipette both reaction mixes on ice.

Protocol 1. cDNA synthesis

Equipment and reagents

- Total RNA
- Eppendorf tubes
- 5 × reaction buffer: 250 mM Tris–HCl, pH 8.3 at room temperature, 375 mM KCl, 15 mM MgCl$_2$
- 0.1 M DTT
- 0.2 mg/ml oligo-dT or random hexamer primer
- Mixed dNTP stock: 2.5 mM dATP, dCTP, dGTP, and dTTP
- RNasin (usually 40 U/μl)
- M-MLV reverse transcriptase (RT) (usually 200 U/μl)
- DEPC-treated water
- Centrifuge with Eppendorf rotor and tubes

Method

1. Place 1 μg total RNA in an Eppendorf tube.
2. Add 11.5 μl reaction mix, consisting of 4 μl 5 × reaction buffer; 2 μl DTT; 2 μl oligo (dT) or random primer, 2 μl dNTP mix; 0.5 μl (20 units) RNasin; 1 μl RT (200 units).
3. Add DEPC-treated water to a final reaction volume of 20 μl.
4. Mix well, centrifuge briefly.
5. Incubate at room temp. for 10 min.
6. Incubate at 37°C for 60 min.
7. Heat-inactivate RT for 10 min at 70°C.
8. Store at −20°C.

Protocol 2. PCR reaction

Equipment and reagents

- cDNA (see *Protocol 1*)
- PCR tubes
- 10 × PCR reaction buffer: 500 mM KCl, 100 mM Tris–HCl, pH 8.3 at room temperature, 15 mM MgCl$_2$, 0.01% (w/v) gelatin or 1% Triton X-100
- NTP mix (see *Protocol 1*)
- 10 μM stock primers
- Taq-polymerase (usually 5 U/μl)
- Light mineral oil
- Thermocycler
- Loading buffer with bromophenol blue
- Agarose
- DNA molecular weight marker
- Restriction enzymes
- Ethidium bromide
- Submarine gel electrophoresis chamber
- Transilluminator for UV light

Protocol 2. *Continued*

Method

1. Place 1 µl of each cDNA in PCR tubes, except in a control tube (to exclude contamination).

2. Add 24 µl reaction mix, consisting of 2.5 µl 10 × reaction buffer; 2 µl NTP mix; 0.5 µl of each primer; 0.2 µl Taq-polymerase (1 unit); 18.3 µl H_2O (if hot start is necessary to increase amplification specificity, Taq-polymerase has to be added to each tube after overlaying and heating up to 95°C).

3. Overlay with mineral oil.

4. Subject to amplification in a thermocycler: 3 min 95°C; (30 sec. 95°C, 1 min 60°C, 1 min 72°C) × 30–35, depending on the abundance of the target cDNA and the amplification efficiency of the primer pair; 5 min 72°C.

5. Add 2 µl loading buffer to 8 µl of the PCR-product and analyse by agarose gel electrophoresis.

6. Confirm specificity of PCR product by 'Southern' hybridization with a DNA probe or by restriction analysis with enzymes indicative for the amplified sequence. This step is mandatory in the beginning and may be omitted once you have demonstrated specificity of PCR reaction.

4. Control fragment design and construction

Standards with single specificity can be easily obtained using PCR amplification of cDNA with specific primers subsequent to deletion or insertion of a DNA fragment of 50–100 bp. If the genomic sequence is known to have a small intron, the genomic DNA amplified with the same primers as the cDNA may serve as internal standard. Another more convenient approach for control fragment construction is described in *Protocol 3*.

When a number of different cytokine transcripts have to be quantified the use of a multispecific control fragment is desirable (*Protocol 4*).

Protocol 3. Construction of control fragments with single specificity

Equipment and reagents

- Genomic DNA
- 10 × reaction buffer (see *Protocol 2*)
- dNTP mix (see *Protocol 1*)
- Primers
- Taq-polymerase (see *Protocol 2*)

- Light mineral oil
- Thermocycler
- Agarose gel electrophoresis equipment and reagents (see *Protocol 2*)
- Gene clean kit (Bio 101 Inc.)

Method

1. Place 1 μg genomic DNA of an evolutionary distantly related species in a PCR tube.

2. Add 10 μl 10 × reaction buffer; 8 μl dNTP mix; 4 μl of each specific primer; 0.4 μl (2 units) Taq-polymerase.

3. Add H_2O to a final volume of 100 μl and overlay with oil.

4. Subject to amplification in a thermocycler: 5 min 95°C; (1 min 95°C, 2 min 37°C, 2 min 72°C) × 3; (1 min 95°C, 2 min 60°C, 2 min 72°C) × 35; 5 min 72°C.

5. Analyse 10 μl of the PCR product in agarose gel. Due to the low annealing temperature during the first 3 cycles several bands of different size should appear.

6. Perform preparative gel electrophoresis with the remaining 90 μl of PCR product.

7. Excise gel slice containing a DNA fragment of appropriate size.

8. Isolate the fragment from the gel (e.g. with Gene clean).

9. Dilute the fragment at least 1:1000 and repeat the procedure of amplification (except the 3 low-stringent annealing cycles) and purification.

10. To exclude the possibility that the fragment is flanked by one of the primers at both ends, perform amplification with each one of the primers.

An example for generations of control fragments by cross-species PCR is shown in *Figure* 1. It shows construction of an IL-1 specific control fragment obtained by mouse IL-1 primer and chicken genomic DNA (3).

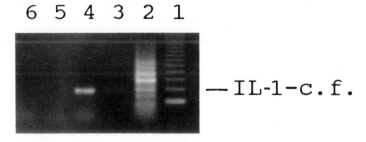

Figure 1. Analysis of amplification products obtained by cross-species PCR with mouse IL-1 primers and genomic chicken DNA. **Lane 1:** 1 μg 123 bp ladder; **lane 2:** equivalent of 0.2 μg of genomic chicken DNA amplified under low-stringency annealing conditions; **lane 3:** purified IL-1 competitor DNA fragment (IL-1 cf); **lanes 4–6:** PCR in the presence of 1 fg of competitor DNA fragment and either both primers (4), sense (5) or anti-sense (6) primer alone. Taken from Überla *et al.* (1991). *PCR Methods Applic.*, **1**, 136–9, with permission.

Protocol 4. Design and construction of a multispecific control fragment

Equipment and reagents

- GENMON computer program (GBF Braunschweig)
- DNA ligase and appropriate reaction buffer
- Agarose gel electrophoresis equipment and reagents
- Plasmid (e.g. pBluescript II KS+ from Stratagene)
- Sequence analysis equipment and reagents

Method

1. Design the location of the primers on the control DNA in such a tandem array of 5' and 3' sequences that the amplified fragment differs by approximately 100 bp in size from the cDNA amplification product (*Figure 2*). Only the oligonucleotides at the extreme ends of the synthetic DNA fragment should contain restriction sites for subsequent cloning into a plasmid.

2. Select 38–65mer overlapping complementary oligonucleotides comprising the sequences of all cytokine specific 5' and 3' primers with an appropriate computer program (e.g. GENMON from GBF Braunschweig) and synthesize them.

3. Hybridize the oligonucleotides and ligate them.

4. Visualize the ligation product of correct size by agarose gel electrophoresis and purify it.

5. Clone the control fragment into an appropriate plasmid and verify it by sequence analysis.

By the above protocol we have constructed mouse cytokine control plasmid pMCQ (*Figure 2*) and recently human cytokine control plasmids pHCQ1 and pHCQ2 (*Figure 3*) have been made by one of us (C. P.). The mouse and human primer sequences and PCR product sizes of cDNA and control fragment are listed in *Tables 1* and *2,* respectively (4, 6).

5. Quantitative PCR

In most cases determination of relative cytokine mRNA levels in two cDNA samples is envisaged. In order to correct for variations of cDNA amounts in different preparations, all samples to be compared have to be adjusted to contain identical cDNA concentrations. This can be based on cDNA content of a constitutively expressed gene, e.g. beta-actin. Subsequently, the calibrated samples can be quantified for cytokine cDNA levels. For the quantification of each cDNA, the appropriate amount of the control fragment to be

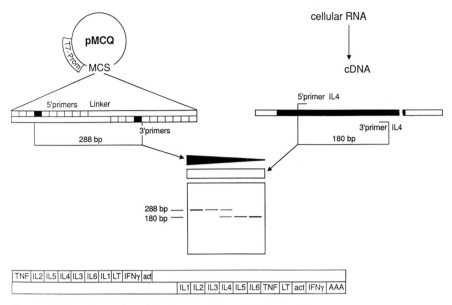

Figure 2. Experimental outline for cytokine mRNA quantification by competitive PCR with control plasmid pMCQ. The amount of PCR product is proportional to the amount of input DNA as demonstrated by co-amplification of constant amounts of target cDNA and serially diluted control fragment. At the bottom, the order of 5′ and 3′ primer-specific sequences in plasmid pMCQ is given. Taken from Platzer *et al.* (1992). *Eur. J. Immunol.*, **22**, 1179–84, with permission.

Table 1. Oligonucleotides used to amplify mouse target cDNA and control fragment (cf) DNA from pMCQ and lengths (bp) of their PCR products

	sense antisense	cDNA	cf
b-actin	TGGAATCCTGTGGCATCCATGAAAC TAAAACGCAGCTCAGTAACAGTCCG	348	264
IL–1a	CTCTAGAGCACCATGCTACAGAC TGGAATCCAGGGGAAACACTG	308	138
IL–2	TGATGGACCTACAGGAGCTCCTGAG GAGTCAAATCCAGAACATGCCGCAG	167	289
IL–3	GAAGTGGATCCTGAGGACAGATACG GACCATGGGCCATGAGGAACATTC	292	238
IL–4	CGAAGAACACCACAGAGAGTGAGCT GACTCATTCATGGTGCAGCTTATCG	180	288
IL–5	ATGACTGTGCCTCTGTGCCTGGAGC CTGTTTTTCCTGGAGTAAACTGGGG	242	338
IL–6	TGGAGTCACAGAAGGAGTGGCTAAG TCTGACCACAGTGAGGAATGTCCAC	154	288
TNF	GGCAGGTCTACTTTGGAGTCATTGC ACATTCGAGGCTCCAGTGAATTCGG	307	438
LT	TGGCTGGGAACAGGGGAAGGTTGAC CGTGCTTTCTTCTAGAACCCCTTGG	205	290
IFNg	AGCGGCTGACTGAACTCAGATTGTAG GTCACAGTTTTCAGCTGTATAGGG	243	315

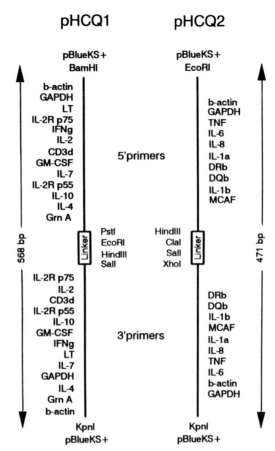

Figure 3. Control fragments for human cytokine quantification. The order of 5′ and 3′ primer-specific sequences in plasmids pHCQ1 and pHCQ2 is given. Taken from Platzer *et al.* (1994), *Transplantation* **58**, 264, with permission.

added to the PCR reactions has to be estimated. According to the competition of the two substrates for the primer the ratio of the two templates remains constant during the amplification. This is the case if the target to standard ratio is between 0.1 and 10. In our hands the above described multispecific control fragments (lengths 471–568 bp, concentration 10–25 ng/ μl) were diluted 10^{-3} to 10^{-5} to analyse the housekeeping gene expression, and 10^{-4} to 10^{-10} to estimate gene expression of different cytokines. In general, when determining absolute initial amounts of mRNAs by competitive PCR using standard DNA fragments, one must take into account the fact that the efficiency of reverse transcription is about 40–50%. Thus, calculations of the absolute amount indicate the minimum number of mRNA molecules present in a given sample. In order to determine relative mRNA

Table 2. Oligonucleotides used to amplify human target cDNA and control fragment DNA from pHCQ1 and 2 and lengths of their PCR products

	sense antisense	cDNA	cf
b-actin	CGGGAAATCGTGCGTGACAT GAACTTTGGGGGATGCTCGC	712	557/438 [a]
GAPDH	GCAGGGGGGAGCCAAAAGGG TGCCAGCCCCAGCGTCAAAG	567	477/438 [a]
IL–1a	CTCACGGCTGCTGCATTACA ACCTACGCCTGGTTTTCCAG	365	258
IL–1b	TGCCCGTCTTCCTGGGAGGG GGCTGGGGATTGGCCCTGAA	288	141
IL–2	CCTCAACTCCTGCCACAATG TTGCTGATTAAGTCCCTGGG	340	232
IL–2R p55	CCTGCCTCGTCACAACAACA AAAACGCAGGCAAGCACAAC	312	182
IL–2R p75	TCAAACAACATTCCACCCCA CAAATGGTCAGCAGCCCTCT	389	252
IL–4	GCTTCCCCCTCTGTTCTTCC TCTGGTTGGCTTCCTTCACA	371	289
IL–6	TAGCCGCCCCACACAGACAG GGCTGGCATTTGTGGTTGGG	408	358
IL–7	TTTTATTCCGTGCTGCTCGC GCCCTAATCCGTTTTGACCA	429	316
IL–8	GGGTCTGTTGTAGGGTTGCC TGTGGATCCTGGCTAGCAGA	403	298
IL–10	CTGAGAACCAAGACCCAGACATCAAGG CAATAAGGTTTCTCAAGGGGCTGG	351	196
IFNg	TCGTTTTGGGTTCTCTTGGC GCAGGCAGGACAACCATTAC	477	357
CD3d	CTGGACCTGGGAAAACGCATC GTACTGAGCATCATCTCGATC	309	233
GM–CSF	CTCGCCCAGCCCCAGCACGC GCAGCTCCCCGGCTTGGCCA	411	276
Grn A	TGAACAAAAGGTCCCAGGTC ATTCATTCCAATCACAGGGT	368	289
TNF	CTCTGGCCCAGGCAGTCAGA GGCGTTTGGGAAGGTTGGAT	519	358
LT	ACCCTCAAACCTGCTGCTCA AGAATGGAGGCAGAATGGGG	509	417
MCAF	ATGAAAGTCTCTGCCGCCCT TCTTCGGAGTTTGGGTTTGC	295	141
DRb	CCGGATCCTTCGTGTCCCCACAGCACG CTCCCCAACCCCGTAGTTGTGTCTGCA	271	151
DQb	CTCGGATCCGCATGTGCTACTTCACCAACG GAGCTGCAGGTAGTTGTGTCTGCACAC	234	151

[a] From control plasmids pHCQ1 and 2, respectively

levels between two cDNA probes it is not necessary to run PCR with many-fold dilutions of the competitor. It is convenient to add a constant amount of competitor to PCR reactions which contain in comparison the different cDNA samples calibrated before to equal concentrations. The increase or decrease of their product ratio from sample to sample directly reflects the initial relative abundance of the target mRNA within each sample.

Protocol 5. Adjustment of cDNA concentrations by competitive PCR between competitor DNA and a housekeeping gene

Equipment and reagents

- Standard DNA
- PCR tubes
- β-actin or GAPDH primers
- Thermocycler

- Agarose gel electrophoresis equipment and reagents
- CCD imaging system

Method

1. Prepare a series of 1:10 dilutions of the standard DNA.

2. Prepare 3 to 5 PCR tubes for each cDNA included in the comparison.

3. Give 1 μl undiluted cDNA in the tubes.

4. Place 1 μl of the different dilutions of the standard DNA.

5. Add PCR reaction mix with β-actin or GAPDH primers and run PCR for 25–27 cycles as described above.

6. Analyse PCR products by agarose gel electrophoresis.

7. The relative amounts of target and standard product in each sample are compared visually, by CCD imaging or incorporation of radioactivity during PCR.

8. The initial amounts of target and competitor are assumed to be equal in those reactions where the molar ratio of their products are judged to be equal (after correction for size differences).

9. All cDNAs which subsequently should be compared for cytokine cDNA amounts according to *Protocol 6* (see below) should be diluted to the level of the cDNA with the lowest housekeeping gene content.

10. Run a test-PCR with a single control fragment concentration which should give equal band intensities with all adjusted cDNAs.

Protocol 6. Quantification of cytokine cDNA

Equipment and reagents

- Adjusted cDNA from *Protocol 5*
- Standard DNA
- PCR reaction mix (see *Protocol 2*)
- Cytokine primers

- Thermocycler
- Agarose gel electrophoresis equipment and reagents
- CCD imaging system

Method

1. Place 1 µl of each adjusted cDNA into tubes 1–5.

2. Add 1 µl of different dilutions of the standard DNA to each cDNA.

3. Add PCR reaction mix with the respective cytokine primers and run PCR for 27–40 cycles as described above.

4. Analyse the PCR products by agarose gel electrophoresis.

5. Determine the ratio of amplified target to competitor products in each sample as described in *Protocol 5*.

6. Differences in the ratios indicate the relative differences in mRNA levels between the samples (semi-quantitative approach) or, using a dilution series of competitor, the log of the ratios is graphed as a function of the log of the known amount of competitor added to the PCR reaction (quantitative approach).

An example for cytokine quantification according to *Protocols 5* and *6* is shown in *Figure 4*. It shows determination of IL-4 mRNA levels in spleen cells derived from IL-4 transgene heterozygous versus homozygous mice. The cDNAs were first adjusted to equal concentrations by means of β-actin content and subsequently IL-4 mRNA levels were determined. The results show a two-fold increase of IL-4 mRNA in the homozygous compared to heterozygous animals (gene dosage effect) (4).

An alternative comparative method for quantitating individual cytokine mRNA is to use radioactive PCR amplification and comparison to a house-keeping gene. This method is briefly described in the next chapter.

Acknowledgements

This work was supported by the Deutsche Krebshilfe, Mildred Scheel Stifung, e.V., and the Deutsche Forschungsgemeinschaft.

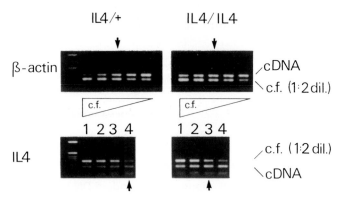

Figure 4. IL-4 transgenic homozygous animals contain twice as much IL-4 mRNA as heterozygous mice. **Top**: cDNA samples were adjusted to identical concentrations of β-actin cDNA by co-amplification of constant amounts of cDNA and two-fold dilutions of control fragment (cf). Arrows indicate the titration point when equal band intensities were obtained for target cDNA and control fragment. **Bottom**: identical amounts of cDNA were used for PCR with IL-4-specific primers in the presence of two-fold dilutions of control fragment. In the IL-4-transgenic homozygous mouse, the IL-4 cDNA successfully competes with twice as much control fragment as in the heterozygous mouse (arrows). Taken from Platzer *et al.* (1992). *Eur. J. Immunol.*, **22**, 1179–84, with permission.

References

1. Wang, A. M., Doyle, M. V., and Mark, D. F. (1989). *Proc. Natl. Acad. Sci. USA*, **86**, 9717.
2. Gilliland, G., Perrin, S., Blanchard, K., and Bunn, H. F. (1990). *Proc. Natl. Acad. Sci. USA*, **87**, 2725.
3. Überla, K., Platzer, C., Diamantstein, T., and Blankenstein, T. (1991). *PCR Methods Applic.*, **1**, 136.
4. Platzer, C., Richter, G., Überla, K., Müller, W., Blöcker, H., Diamantstein, T., and Blankenstein, Th. (1992). *Eur. J. Immunol.*, **22**, 1179.
5. Bouaboula, M., Legoux, P., Pésségué, B., Delpech, B., Dumont, X., Piechaczyk, M., Casselas, P., and Shire, D. (1992). *J. Biol. Chem.*, **267**, 21 830.
6. Platzer, C., Ode-Hakim, S., Reinke, P., Döcke, W.-D., Ewert, R., and Volk, H. D. (1994). *Transplantation*, **58**, 264.
7. Sambrook, J., Fritzsch, E. F., and Maniatis, T. (ed.) (1989). *Molecular cloning, a laboratory manual* (2nd edn). Cold Spring Harbor Laboratory Press, NY.

5

Comparative polymerase chain reaction

SYLVIE DE KOSSODO

1. Introduction

Comparative PCR may be used to compare the expression of cytokines in a series of diverse settings: for example, in different tissues or cell lines; in the same tissues or cell lines; before and after specific treatments; during kinetic studies; in uninfected versus infected organs etc. In each case the small amount of tissue, low cell-numbers or low copy-number would limit the utilization of Northern blots or the RNAse-protection assays described in Chapter 3. In contrast to the techniques described in Chapter 4, the method detailed here does not necessitate the construction of a multispecific control fragment. Comparison is achieved by subjecting samples to amplification, in sub-saturating conditions, of the constitutively expressed glyceraldehyde phosphate dehydrogenase (GAPDH) gene and the specific cytokine, either in parallel, or preferably, in the same tube, in the presence of a radiolabelled deoxynucleotide. The relative product amounts are quantitated by measuring the incorporated radioactivity of the specific bands from dried gels. This can be done by phosphor-imager analysis or, after exposure to film, by conventional densitometry. The values obtained by measuring the incorporated radioactivity in the cytokine amplification products are normalized to the values obtained in the GAPDH amplification products.

It is always useful to perform a titration curve by plotting the c.p.m. associated with each amplified fragment of the appropriate size to the dilution of cDNA. The values obtained for each cytokine and GAPDH are then used to generate a linear regression line. Also, a sample of cDNA can be subjected to sequential cycles of amplification. The c.p.m.s obtained for each amplified fragment are plotted against the number of cycles. The amounts of PCR-generated bands increase logarithmically up to a certain number of cycles, reaching a plateau thereafter. Under the conditions used one must ensure there is a linear relationship between the initial amount of cDNA and the PCR signal.

2. Methods

Protocol 1. RNA extraction

The RNA extraction method we use is based on a published method by Chomczynski and Sacchi (1). A similar method is described in Chapter 3.

Equipment and reagents

- Denaturing solution: 23.6 g guanidinium thiocyanate (prepared in autoclaved H_2O), 1.25 ml of 1 M Na-citrate pH 7 (autoclaved), 2.5 ml of 10% lauryl sarcosyl (heated 10 min at 65°C), 330 μl 2-mercaptoethanol, complete to 50 ml with autoclaved H_2O. This solution can be stored at least 3 months at room temperature.
- 2 M Na-acetate, pH 4 (autoclaved)
- Phenol (water or TE-saturated)
- Chloroform/isoamylalcohol mixture (49:1)
- Isopropanol or ethanol.
- Autoclaved Eppendorf tubes

Method

1. Homogenize the tissue (around 20–100 mg or 1 confluent plate) in 0.5 ml of denaturing solution. Should not be too gooey, if so, add more denaturing solution and adapt the rest of the amounts. Put in an autoclaved Eppendorf tube.

2. Add 50 μl of 2 M Na-acetate. Mix well by inverting the tube.

3. Add 0.5 ml phenol. Mix well by inversion.

4. Add 0.1 ml chloroform/isoamylalcohol. Mix vigorously by inversion (for 10 sec).

5. Leave on ice for 15 min.

6. Centrifuge at 10 000 g (or 14 000 r.p.m. in a microcentrifuge) for 20 min at 4°C.

7. Take supernatant into new Eppendorf and add same volume of isopropanol or 1 ml of ethanol. Leave a minimum of 2 h at −70°C.

8. Centrifuge at 10 000 g (14 000 r.p.m. in a microcentrifuge) for 15 min at 4°C.

9. Keep pellet, wash in 70% ethanol. Centrifuge at 10 000 g (14 000 r.p.m. in a microcentrifuge) for 15 min at 4°C.

10. Dry pellet.

11. Resuspend in 20–40 μl autoclaved H_2O. Use 2 μl for OD_{260} nm. Verify the integrity of the RNAs by electrophoresis as described before (Chapter 3, *Protocol 1*).

Protocol 2. DNAse treatment

The DNAse step is used to ensure that the specific PCR-generated fragments derive from the amplification of cDNA and not genomic DNA. This is not only important in the case where the primers chosen do not encompass an intron, but also to avoid competition between cDNA and genomic DNA templates during the amplification reaction.

Equipment and reagents

- Promega *in vitro* Transcription Kit: 5 × transcription buffer, RNAsin, 0.1 M dTT, RQ1 DNAse (from Promega f.ex.)
- 3 M Na-acetate pH 5.4 (autoclaved)
- Ethanol
- TE buffer (Tris 10 mM ph7–EDTA 1 mM pH 7) (autoclaved)
- Eppendorf tubes and microcentrifuge
- Autoclaved water

Method

1. In an Eppendorf tube add, in this precise order: 4 μl of 5 × transcription buffer, 1 μl 0.1 M dTT, 1 μl of RNAsin, X μl of RNA (between 1 and 20 μg of total RNA), 1 μl of RQ1 DNAse, H_2O to a final volume of 20 μl.
2. Leave at 37°C for 30 min.
3. Add 80 μl of TE. Mix well.
4. Add 100 μl of phenol saturated in TE. Mix well.
5. Centrifuge for 2 min at RT on a microcentrifuge.
6. Put supernatant in new tube. Add 100 μl of chloroform/isoamylalcohol. Mix well.
7. Centrifuge for 2 min at RT on a microcentrifuge.
8. Put supernatant in new tube. Add 10 μl of 3M Na-acetate pH 5.4. Mix well. Add 200 μl of ethanol. Mix well.
9. Leave at least 2 h at −70°C.
10. Centrifuge at 10 000 *g* (14 000 r.p.m. in a microcentrifuge) for 15 min at 4°C.
11. Keep pellet, wash in 70% ethanol. Centrifuge at 10 000 *g* (14 000 r.p.m. in a microcentrifuge) for 15 min at 4°C.
12. Dry pellet.
13. Resuspend in 12 μl autoclaved H_2O.

Protocol 3. Reverse transcription

It is important to compare the efficacy of oligo(dT) versus random primers when preparing cDNA. For example, if primers are situated in the 5′ end portion of the gene, and the gene is either large (above 2000 bp) or has a long 3′ untranslated region, it might prove best to prepare the cDNA with random primers.

Equipment and reagents

- cDNA Synthesis kit (Boehringer Mannheim AG, Switzerland): Solution 1 (buffer), Solution 2 (RNAsin), Solution 3 (deoxynucleotide mix), Solution 4 (oligo(dT)$_{15}$) *or* Solution 12 (Random primers), Solution 5 (AMV reverse transcriptase).

Method

1. To the 12 μl of RNA (generally 20 μg RNA) add, in this precise order:
   4 μl of Sol. 1
   1 μl of Sol. 2
   1 μl of Sol. 3
   1 μl of Sol. 4 *or* 12
   1 μl of Sol. 5

2. Leave at 42°C for 1 h.

3. Add 80 μl of H_2O (to have a 200 ng RNA per μl solution, this is not imperative).

4. Keep at −20°C until use.

Protocol 4. PCR reaction

This method employs radioactivity, and special care should be taken in all steps beyond the PCR reaction, especially while handling the gel and buffer.

Equipment and reagents

- 10 × Buffer: 250 mM TAPS-HCl pH 9.3, 500 mM KCl, 10 mM 2-mercaptoethanol, 125 mg/ml denatured salmon sperm DNA
- Deoxynucleotide mix: 200 mM of each dATP, dCTP, dTTP, and dGTP
- [α-^{32}P]dCTP (3000 Ci mmol^{-1}, Amersham)
- AmpliTaq (Perkin Elmer, Cetus)
- Ampli-primer mix: 0.2 mM of each 5′ and 3′ specific primers
- Mineral oil
- Thermocycler

Method

1. To the desired amount of cDNA to be amplified (e.g. 4 μl or 200 ng) add 5 μl of 10 × buffer, 4 μl of deoxynucleotide mix, 1 μl of ampli-

72

primer mix, 0.5 μl of ^{32}P-dCTP, 0.5 μl of ampliTaq (2.5 Units) and water to a final volume of 50 μl.

2. Overlay with 1 drop of mineral oil.

3. Amplify samples in a DNA thermal cycler. Conditions of amplification vary with each specific set of primers. A typical amplification reaction for GAPDH would be: 94°C for 5 min, 55°C for 2 min, and 72°C for 2 min for 1 cycle followed by 30 cycles at 94°C for 1 min, 55°C for 2 min, and 72°C for 2 min. When possible, amplify for GAPDH and the specific cytokine in the same tube. If this is not possible, work in parallel tubes. All samples should be assayed in duplicate.

Protocol 5. Quantification

Equipment and reagents

- Electrophoresis grade agarose
- 5 × TBE buffer: 54 g Trizma base, 27.5 g boric acid, and 4.65 g EDTA for 1 l
- Ethidium bromide at 10 mg/ml
- Loading dye: 0.25% xylene cyanol, 0.25% bromophenol blue, and 4% sucrose
- 3 MM chromatography paper (Whatman)
- Gel dryer
- Phosphor screen cassette and phosphor-imager
- UV transilluminator
- Saran wrap

Method

1. Prepare a 1.2% agarose gel in 1 × TBE, dissolve by heating in a microwave. Once it has cooled down to approximately 50°C, add 5 μl of ethidium bromide and pour into a gel tray.

2. Mix 15 μl of the amplification reaction with 3 μl of loading dye and load on to gel.

3. Migrate for 2 h at 80 V.

4. Visualize the PCR products on a UV transilluminator.

5. Place the gel on top of a 3 MM chromatography paper (Whatman) and cover it with Saran wrap. Dry the gel in a gel dryer (no heat) for 30 min.

6. Place the dried gel in a phosphor screen cassette (Molecular Dynamics) and expose for 24 h.

7. Place the screen in a Molecular Dynamics 400 A phosphor-imager and read the incorporated radioactivity in each of the specific PCR-generated fragments.

8. Normalize the values obtained by measuring the incorporated radio-activity in the specific cytokine fragments to the GAPDH fragments.

A typical result is shown in *Figure 1*.

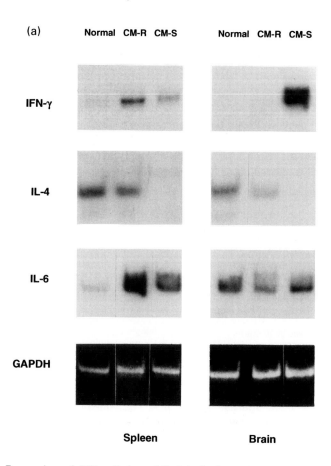

Figure 1. Expression of IFN-γ, IL-4, and IL-6 in brain and spleen of uninfected, and *Plasmodium berghei* infected cerebral malaria-resistant (CM-R) and cerebral malaria-susceptible (CM-S) mice. (a) The equivalent of 200 ng of total RNA was reverse-transcribed and amplified under the conditions specified. Shown is a representative gel of the organs from 1 mouse. Exposure times were variable (4 h to 1 day at room temp. 1) but consistent for each cytokine. The GAPDH-specific fragments are shown as visualized by ethidium bromide staining. (b) The radioactivity of the PCR products in brain was quantitated with a phosphor-imager, normalizing the values obtained (expressed as mean ± SD, 4 mice per group) to the values obtained by quantitating the GAPDH amplification products. Taken from ref. 2.

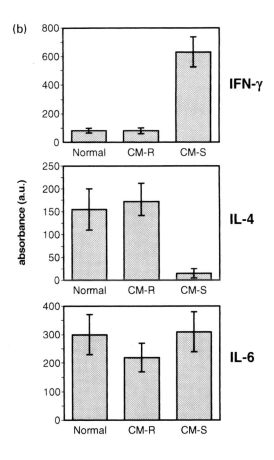

(b)

References

1. Chomczynski, P. and Sacchi, N. (1987). *Analytical Biochemistry*, **162**, 156.
2. De Kossodo, S. and Grau, G. E. (1993). *J. Immunol.*, **151**, 4811.

6

Receptor-binding studies

THOMAS M. MARIANO, LARA IZOTOVA, and
SIDNEY PESTKA

1. Introduction

This chapter will deal primarily with radiolabelling interferons, IFNs, the analysis of binding assays, and a procedure for covalently cross-linking labelled IFN to its receptor. Finally, some data on the cloning of the Hu- and Mu-IFN-γ receptors and a functional Hu-IFN-α/β receptor will be presented. Although most of the data presented are from our studies of the interaction of IFN-γ and its receptor, the procedures should be applicable to other ligand–receptor systems.

The interferons (IFNs) are a family of proteins which elicit a multitude of cellular responses including antiviral, antiproliferative, and immunoregulatory activities. There are four classes of human interferons that have been isolated and characterized (1–9). These are the leukocyte or interferon alpha (IFN-α), fibroblast or interferon beta (IFN-β), immune or interferon gamma (IFN-γ), and interferon omega (IFN-ω). Binding to a specific cell surface receptor is a necessary, but apparently not sufficient, condition for cellular activation (10–13). IFN-α, IFN-β, and IFN-ω share a common receptor which is distinct from the IFN-γ receptor (5, 14).

During the course of our studies of the IFN receptors, our group has developed a novel procedure for radiolabelling ligands to very high activities (15–18). The method is based on the fact that human (Hu), murine (Mu), rat (Ra), and even bovine (Bo) IFN-γs contain at least one recognition sequence for the cAMP-dependent protein kinase from bovine heart. Thus, this fortuitous circumstance has enabled us to phosphorylate several recombinant IFN-γs to very high activities by utilizing [γ-^{32}P]ATP and a commercially-available protein kinase. None of the phosphorylated proteins exhibited any significant loss of biological activity when assayed in a standard cytopathic effect inhibition assay for interferon activity.

It was subsequently demonstrated that serine is the only amino acid phosphorylated. Mu- and Ra-IFN-γ each have a single phosphorylation site (19) whereas Hu-IFN-γ has two phosphorylation sites (20). The phosphorylation sites reside at the carboxy-terminal end of the proteins. Because of its nearly-perfect identity with Hu-IFN-γ in the carboxy-terminal region, we have speculated that Bo-IFN-γ also has two sites of phosphorylation (19).

Recently, the labelling procedure has been extended to proteins which lack an intrinsic phosphorylation site. By employing oligonucleotide-directed mutagenesis, phosphorylation sites were introduced in Hu-IFN-αA (21, 22), Hu-IFN-αB2 (23), and Hu-IFN-αA/D(*Bgl*) (23). Several mutant proteins were isolated and each was capable of being phosphorylated with no loss of biological activity. The [^{32}P]Hu-IFN-αA-Pl and Hu-IFN-αB2-P proteins have been used extensively in our laboratory to study the Hu-IFN-α/β receptor.

Due to the very high labelling efficiency of the phosphorylation reaction (up to 10-fold higher than many iodination procedures) and high retention of biological activity, the ^{32}P-labelled IFNs have become invaluable reagents in our efforts to characterize and clone the IFN receptors. Binding studies can allow one to estimate the dissociation constant (K_d) of the ligand–receptor complex as well as the number of receptors per cell or per gram of protein in a preparation of membranes. From this type of information, a suitable source for purifying the receptor protein can be identified. We have also set up a translation system in which a binding assay is performed directly on *Xenopus laevis* oocytes 48–72 h after microinjection with mRNA (24). The oocyte binding assay has been used to evaluate various mRNAs for their relative content of IFN-γ receptor RNA as well as to test the binding capability of putative receptor clones whose RNA was transcribed *in vitro* from vectors driven by the strong SP6 or T7 phage promoters (25). If the bound [^{32}P]IFN-γ is then covalently cross-linked to the receptor, the molecular size of the receptor can be estimated after SDS-polyacrylamide gel electrophoresis of the covalent complex and visualization by autoradiography.

2. Methods

2.1 Cell culture

Grow tissue culture cells in appropriate media (Life Technologies or Sigma) containing 10% heat-inactivated fetal bovine serum and 50 μg/ml gentamicin sulfate at 37°C in 5% CO_2. For binding experiments, wash suspension cells once and then resuspend in fresh medium. For adherent cells, remove medium, wash cells once with phosphate-buffered saline lacking Ca^{2+} and Mg^{2+} (PBS), release with trypsin-EDTA, wash, and finally resuspend in fresh medium. (Some receptors may be trypsin sensitive—see Section 5.)

The biologic activity of the interferons can be assayed in a standard cytopathic effect inhibition assay (27) with cells and challenge virus appropriate to the type and species of interferon being examined.

2.2 Interferon radiolabelling

The conditions of the phosphorylation reaction can vary somewhat depending on the desired specific activity of the material. The following protocol usually yields high levels of labelling:

Protocol 1. Interferon radiolabelling

Equipment and reagents

- [γ-^{32}P]ATP (>6000 Ci/mmol, Amersham or New England Nuclear) dried in a Savant Speed-Vac Concentrator prior to use
- Catalytic subunit of bovine heart cAMP-dependent protein kinase (Sigma); prepared in 6 mg/ml (39 mM) dithiothreitol at 15 U/μl and stored in liquid nitrogen in small aliquots

- 1 × reaction buffer 20 mM Tris–HCl, pH 7.4, 100 mM NaCl, 12 mM MgCl$_2$
- Recombinant Mu-and Hu-IFN-γ (24, 26) stored in liquid nitrogen in small aliquots
- Bovine serum albumin
- 10 mM sodium pyrophosphate, pH 6.7

Method

1. Incubate about 0.5–1 μg of protein at 30°C for 60 min with 1 mCi of [γ-^{32}P]ATP and 15–30 units of the catalytic subunit of the cAMP-dependent protein kinase. The reaction volume of 60 μl should also contain 20 mM Tris–HCl, pH 7.4, 100 mM NaCl, 12 mM MgCl$_2$, and 0.65–1.3 mM dithiothreitol (depending on how much kinase is used).

2. Dilute the reaction mixture with 0.25–0.5 ml of a cold solution of bovine serum albumin (5 mg/ml) in 10 mM sodium pyrophosphate, pH 6.7. The sodium pyrophosphate is used to inhibit dephosphorylation.

3. Dialyze twice against 1 litre or three times against 250 ml of 10 mM sodium pyrophosphate, pH 6.7 at 4°C to remove the unincorporated [γ-^{32}P]ATP. Continue dialysis for at least 6 h before changing the solution. Store the labelled protein in small aliquots in liquid nitrogen.

Typically, the degree of phosphorylation has ranged from about 1000–4000 Ci/mmol (65–254 μCi/μg) for Mu-IFN-γ, 630–6000 Ci/mmol (36–340 μCi/μg) for Hu-IFN-γ, 890–2600 Ci/mmol (45–135 μCi/μg) for Hu-IFN-αA-P1, 1111 Ci/mmol (~57 μCi/μg) for Hu-IFN-αB2-P and 1028 Ci/mmol (~53 μCi/μg) for Hu-IFN-αA/D-P.

2.3 Binding of [^{32}P]Mu-IFN-γ to tissue culture cells

Protocol 2. Binding of [^{32}P]Mu-IFN-γ to tissue culture cells

Equipment and reagents

- [^{32}P]Mu-IFN-γ stored in liquid nitrogen in small aliquots
- Costar Spin-X filter unit
- 5–10% (w/v) sucrose in PBS

- Long 0.4 ml polypropylene microcentrifuge tubes
- Brinkmann model 5413 centrifuge with horizontal rotor

Protocol 2. *Continued*

Method

1. Prior to binding to cells, centrifuge the [^{32}P]Mu-IFN-γ at 15 000 g for 15 min or filter (e.g. Costar Spin-X filter unit) for 1–2 min at 4°C. This procedure removes protein aggregates which form from freezing and thawing the [^{32}P]Mu-IFN-γ solution.

2. Prepare cells as described in Section 2.1 and resuspend in medium at a concentration of about 5 × 10^6/ml.

3. Typically, place 240 μl of cells into the first tube in each of two sets of 1.5 ml polypropylene tubes. The remaining tubes should contain 120 μl of cells. The first set of tubes contains untreated cells whereas the second set of tubes contains cells which were treated with unlabelled Mu-IFN-γ at a concentration of 1 μg/ml for 5 min prior to making the aliquots. The tested cells should contain unlabelled ligand at a concentration about 100-fold greater than the K_d. In practice, at least a 100-fold excess over the concentration of the radioactive ligand is usually sufficient.

4. Add about 1–2.5 × 10^6 counts/min (c.p.m.) of [^{32}P]Mu-IFN-γ to the first tube of each set and make 2-fold serial dilutions. Gently resuspend the cells every 15 min.

5. Following incubation at 22–24°C for 70–90 min, layer an aliquot of 50 μl (in duplicate) from each tube over a 0.35 ml cushion of 5–10% (w/v) sucrose in PBS in a long 0.4 ml polypropylene microcentrifuge tube.

6. Centrifuge the tubes for 2 min at about 11 000 r.p.m. in a horizontal rotor (e.g. Brinkmann model 5413 centrifuge).

7. Freeze tubes in liquid nitrogen and cut off the tips containing the cell pellets with cutting pliers.

8. Determine the radioactivity of the tips (bound c.p.m.) and the rest of the tube (free c.p.m.) in a liquid scintillation counter. Define specific binding as the difference between binding in the absence (total binding) and binding in the presence (non-specific binding) of excess unlabelled IFN.

A typical plot of bound c.p.m. versus free c.p.m. is shown in *Figure 1*. Transformation of the data according to the method of Scatchard (28) should give rise to a straight line when *bound c.p.m./free c.p.m.* is plotted versus *bound c.p.m.* (*Figure 2*). By converting the *bound c.p.m.* on the abscissa into a molar quantity of IFN bound, the slope of the plot is then equal to $-(K_d)^{-1}$. By determining the intercept on the abscissa in c.p.m., converting it to the number of molecules of IFN from the specific radioactivity of the [^{32}P]Mu-

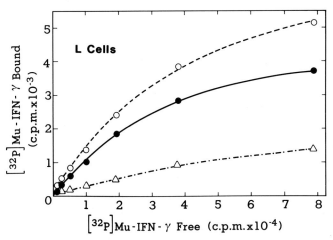

Figure 1. Concentration dependence of [^{32}P]Mu-IFN-γ binding to mouse L cells. Binding to L cells (an adherent cell line) was carried out as described in *Protocol 2* and bound c.p.m. were plotted versus free c.p.m. for total binding (○), non-specific binding (△), and the calculated specific binding (●, i.e., total minus non-specific binding). The data are taken from Langer *et al.* (18) with permission.

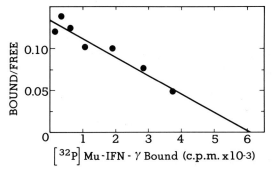

Figure 2. Scatchard analysis of binding data. The data for specific binding from *Figure 1* were plotted according to the method of Scatchard (28). The slope of the resulting straight line is related to the K_d of the ligand–receptor interaction and the intercept on the abscissa is related to the number of receptors per cell. For L cells, several experiments gave a range of 3.0–6.9 × 10^{-10} M for the K_d and 6.0–16 × 10^3 receptors per cell. The data are taken from Langer *et al.* (18) with permission.

IFN-γ, and dividing by the number of cells in the 50 μl sample analysed, one can estimate the number of IFN receptors per cell.

2.4 Binding of [^{32}P]Mu-IFN-γ to *X. laevis* oocytes

The procedures for microinjecting RNA into frog oocytes as well as the binding assay have already been described in detail (24). Only the protocol for the binding assay will be repeated here for the sake of completeness.

Protocol 3. Binding of [^{32}P]Mu-IFN-γ to *X. laevis* oocytes

Equipment and reagents

- Large *X. laevis* females (NASCO)
- 5 ml polypropylene tubes
- 0.5 × Leibovitz L-15 medium, pH 7.6
- Binding solution: 0.5 × Leibovitz L-15 medium (Life Technologies), 1 mg/ml bovine serum albumin, 5 × 10^5 c.p.m. [^{32}P]Mu-IFN-γ

- Unlabelled Mu-IFN-γ
- Rotary shaker
- 1.5 ml polypropylene tubes
- Bray's solution (National Diagnostics)
- Ultrasonic processor (model W-375, Heat Systems Inc.)
- RNA

Method

1. Incubate injected oocytes at 18°C for about 48 h. Place 10 oocytes (in duplicate or triplicate) in each 5 ml polypropylene tube in 200 µl of binding solution. Another tube contains 10 oocytes, 200 µl of the same radioactive binding solution, and, in addition, 0.1 µg of unlabelled Mu-IFN-γ as competitor. Clarify the binding solutions by centrifugation or filtration prior to use.

2. Incubate the oocytes at room temperature on a rotary shaker for 90 min.

3. Wash the oocytes three times with 3 ml of ice-cold 0.5 × L-15 medium. Washing consists of adding the medium, allowing the oocytes to settle to the bottom of the tube, and then carefully aspirating the supernatant.

4. Transfer the oocytes to 1.5 ml polypropylene tubes and add 1 ml of Bray's solution (National Diagnostics).

5. Disrupt the oocytes in an ultrasonic processor (Model W-375, Heat Systems, Inc.) and measure the associated radioactivity in a liquid scintillation counter (see an example of results in *Table 1*).

2.5 Covalent cross-linking of [^{32}P]Mu-IFN-γ to cells and oocytes

Protocol 4. Covalent cross-linking of [^{32}P]Mu-IFN-γ to cells and oocytes

Equipment and reagents

- Large *X. Laevis* females (NASCO)
- [^{32}P]Mu-IFN-γ (see *Protocol 2*)
- Unlabelled Mu-IFN-γ
- 0.5 × Leibovitz L-15, pH 7.6
- RNA
- Rotary shaker
- Microcentrifuge
- PBS

- Disuccinimidyl suberate (DSS, Pierce) prepared fresh at 18.3 mg/ml (50 mM) in dimethylsulfoxide
- 1 M Tris–HCl, pH 7.5
- Triton X-100
- EDTA
- SDS-polyacrylamide gel electrophoresis system
- Autoradiography system

Method

1. Resuspend tissue culture cells in medium at a density of 5×10^6/ml. Add about 10^6 c.p.m. of [^{32}P]Mu-IFN-γ, with or without 1 μg of unlabelled Mu-IFN-γ as competitor, to 0.3 ml of the cell suspension. For cross-linking to the receptor expressed in oocytes, inject with RNA, and, after about 48 h, place 50 oocytes in 800 μl of $0.5 \times$ Leibovitz's L-15 medium to which is added about 6×10^5 c.p.m. of [^{32}P]Mu-IFN-γ, with or without 0.6 μg of unlabelled Mu-IFN-γ.

2. Allow binding to proceed for 90 min at room temperature. Gently resuspend cells every 15 min; incubate oocytes on a rotary shaker.

3. Centrifuge cells at 15 000 g for 20 sec in a microcentrifuge, wash twice with 1 ml of cold PBS, and finally resuspend in 500 μl of PBS. Wash oocytes three times with cold $0.5 \times$ L-15 medium, three times with cold PBS, and finally resuspend in 800 μl of PBS.

4. Initiate covalent cross-linking by the addition of DSS to a final concentration of 0.5 mM (5 μl of 50 mM DSS per 500 μl of PBS).

5. After 20 min on ice, quench the reactions for 5 min on ice by the addition of Tris–HCl, pH 7.5, to a final concentration of 20 mM (10 μl of 1 M Tris–HCl, pH 7.5, per 500 μl of PBS).

6. Pellet cells and extract with 120 μl of PBS containing 0.5% (v/v) Triton X-100 and 5 mM EDTA. Allow oocytes to settle, carefully remove the supernatant, and extract the oocytes with 240 μl of the Triton X-100 solution.

7. After 20 min on ice, sediment insoluble material at 15 000 g for 10 min at 4°C.

8. Analyse the supernatants by SDS-polyacrylamide gel electrophoresis (29). Dry gel and subject to autoradiography with an intensifying screen.

The cross-linked [^{32}P]Mu-IFN-γ-receptor complex (*Figure 3*) is seen as a broad band of about 105–115 kDa. The monomeric and dimeric forms of [^{32}P]Mu-IFN-γ can also be seen if a gradient gel system is used. It is noteworthy that the IFN-γ receptors are heavily glycosylated with about 25–30% of the mass of the receptor consisting of carbohydrate (30, 31). The cross-linking data presented show that the receptor protein synthesized in oocytes is glycosylated to an extent similar to that seen on intact cells. However, it should also be noted that the glycosylation process in different cell types may vary somewhat and this could give rise to slightly different banding patterns when the cross-linked ligand-receptor complex is analysed on gels.

Table 1. Binding of [^{32}P]Mu-IFN-γ to cells and oocytes

Cell type	[^{32}P]Mu-IFN-γ bound (c.p.m.)		Ratio
	+	−	(−/+)
ABPL4 cell	1608	57 720	35.9
HL60 cells	1536	984	0.6
Injected oocytes	157	3000	19.0
Uninjected oocytes	285	312	1.1
Buffer-injected oocytes	178	211	1.2

The data in the table are expressed as c.p.m./1.5 × 10^6 cells or c.p.m./10 oocytes. Cells or oocytes injected with ABPL4 mRNA (25 ng/oocyte) were incubated with [^{32}P]Mu-IFN-γ with (+) or without (−) unlabelled competitor Mu-IFN-γ as described in *Protocols 2* and *3*. The results of a typical experiment are given for mouse ABPL4 and human HL60 cells which were used as positive and negative controls, respectively, for the binding reaction. Values for a typical experiment with uninjected and buffer-injected oocytes are also shown. All values for the oocytes represent the average of duplicate or triplicate determinations. The ratio (−/+) represents the c.p.m. bound in the absence to that in the presence of unlabelled competitor Mu-IFN-γ. The data were taken from Kumar *et al.* (24) with permission.

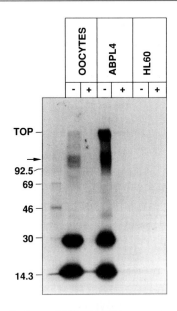

Figure 3. Covalent cross-linking of [^{32}P]Mu-IFN-γ to receptors on cells and oocytes. [^{32}P]Mu-IFN-γ was cross-linked to murine ABPL4 cells (suspension cells), human HL60 cells (suspension cells), or *X. laevis* oocytes injected with mRNA (25 ng/oocyte) from ABPL4 cells as described in *Protocol 4*. The detergent extracts were analysed on a 7.5 to 15% linear SDS-polyacrylamide gel. The binding reactions were carried out in the absence (−) or presence (+) of excess unlabelled Mu-IFN-γ. Positions of ^{14}C-labelled molecular weight markers are indicated on the far left side of the figure. Molecular masses are expressed in kDa. The position of the broad 105–115 kDa [^{32}P]Mu-IFN-γ–receptor complex is marked with an arrow. Since the wells for the oocyte (−) and ABPL4 cell (−) contained about the same c.p.m., it is evident that the cross-linking reaction is more efficient on the ABPL4 cells than on oocytes. Human HL60 cells served as a negative control. The figure is reproduced from Kumar *et al.* (24) with permission.

3. Cloning of the Mu-IFN-γ receptor (Mu-IFN-γR)

3.1 Introduction

By examining the binding and cross-linking patterns of ^{32}P-labelled Hu-IFN-γ to panels of mouse-human and hamster-human somatic cell hybrids, and Mu-IFN-γ to hamster-mouse hybrids, the Hu- and Mu-IFN-γR genes were localized to human chromosome 6q (16) and mouse chromosome 10 (13), respectively. During these studies, it was observed that the human and mouse homologues for the *myb* proto-oncogene are localized to the same chromosomes as those encoding the IFN-γR genes (32, 33). Since the IFN-γR and *myb* genes were genetically linked, we tested several mouse plasmacytoid lymphosarcoma cell lines for their ability to bind [^{32}P]Mu-IFN-γ. These cell lines, established from BALB/c mice injected with Abelson virus, exhibit enhanced transcription of *myb* sequences. We felt that, if the linkage between the two genes was close enough, the Mu-IFN-γR gene might also experience increased transcription and, hence, increased translation of Mu-IFN-γR. One cell line, designated ABPL4, did express five- to ten-fold more receptors than any other mouse cell we had previously tested.

The ABPL4 cells were therefore used as a source of Mu-IFN-γR mRNA to construct a cDNA library. We then proceeded to use the oocyte assay system to clone the Mu-IFN-γR from the ABPL4 library by hybrid selection (34), hybrid-arrested translation (35), and direct expression of RNA transcripts prepared *in vitro* from pools of the cDNA library. As the experiments were progressing, the isolation of a cDNA clone for the Hu-IFN-γR was reported (36). We therefore isolated a Hu-IFN-γR cDNA clone by using the polymerase chain reaction (PCR). The human clone was then used as a probe to screen the ABPL4 library and identify cDNAs for the Mu-IFN-γR (25).

3.2 Procedures for isolating the Mu-IFN-γR cDNA

The procedures for isolating the Mu-IFN-γR cDNA have already been described in detail (25) and will only be highlighted here:

(a) RNA was isolated from human Raji or mouse ABPL4 cells by a guanidine isothiocyanate extraction method (37) and polyadenylated mRNA was purified by two cycles of chromatography on oligo(dT)-cellulose (38).

(b) The cDNA encoding the Hu-IFN-γR was isolated by PCR and cloned into pCK1, a modified pGEM vector (25), and was used to screen an ABPL4 cDNA library prepared as described in the following sections.

(c) The ABPL4 polyadenylated mRNA was fractionated on a 5–20% sucrose gradient as described (39). Fractions of 1 ml were collected and the RNA was precipitated with ethanol three times, washed, dried, and dissolved in sterile water.

(d) The RNA from each fraction was microinjected into oocytes as reported elsewhere (24) and binding with [^{32}P]Mu-IFN-γ was examined about 48 h later as described in *Protocol 3*.

(e) The fraction exhibiting the highest receptor activity was used to prepare a cDNA library in the λGEM2 vector (Promega).

(f) Initially, 10 positive clones were identified that hybridized strongly to the Hu-IFN-γR probe (plasmid pHuIFN-γR8). These clones were then hybridized to DNA fragments which represented the binding, transmembrane, and cytoplasmic domains of the Hu-IFN-γR. Three clones that hybridized strongly to all three human probes were further characterized.

(g) Various fragments from the isolated clones were then subcloned into M13mp18 and M13mp19 vectors.

(h) Both strands of the largest cDNA (λMuIFN-γR4) were sequenced by the dideoxy chain termination method with Sequenase 2.0. *Figure 4* shows a comparison between the Mu- and Hu-IFN-γR sequences at the protein level.

(i) The insert of λMuIFN-γR4 was subcloned into vector pCK1 (25). The resulting subclone, pMuIFN-γR36, was linearized and *in vitro* transcription was carried out with the translation cap analogue 7-methyl-G(5′)ppp(5′)G to increase translation of the RNA after microinjection into frog oocytes. The binding of [^{32}P]Mu-IFN-γ to oocytes injected with sense or antisense Mu-IFN-γR RNA is summarized in *Table 2*.

Cross-linking of [^{32}P]Mu-IFN-γ to oocytes injected with the receptor RNA indicated that the size of the major cross-linked band is about 105–115 kDa, the same as that seen by cross-linking to intact ABPL4 cells.

4. Cloning functional type I and II interferon receptors

Functional receptors for the type I (IFN-α, IFN-β and IFN-ω) and type II (IFN-γ) interferon receptors have recently been obtained. As noted above, the IFN-γ receptor was required for binding IFN-γ to cells (16), but alone was not sufficient for activity. Another component we designated an accessory factor (AF-1) encoded on human chromosome 21 was required (10–12). The gene (40, 41) and cDNA (42) for human AF-1 were cloned. AF-1 is required for signal transduction, but not for IFN-γ binding (40–42); it is a member of the class 2 cytokine receptor family of transmembrane proteins (42). A clone for mouse AF-1 has also been identified (43). Thus, the IFN-γ receptor consists of at least two components: the ligand binding chain (Hu-IFN-γR) and the accessory chain, AF-1. It is possible that more than one accessory chain exists in the case of the human IFN-γR (41).

The isolation of the gene for AF-1 was undertaken through a new tech-

HOMOLOGY BETWEEN MOUSE AND HUMAN INTERFERON GAMMA RECEPTORS

```
  1  MGPQAAAGRMILLVVLMLSAKVGSGALTSTEDPEPPSVPVPTNVLIKSYNLNPVVCWEYQ   60
        || ||  |    |  | |    || | | |||| |||| | ||| || | ||||
  1  ........MALLFLLPLVMQGVSRAEMGTADLGPSSVPTPTNVTIESYNMNPIVYWEYQ    51

 61  NMSQTPIFTVQVKVYS...GSWTDSCTNISDHCCNIYGQIMYPDVSAWARVKAKVGQKES  117
     | | | ||| || |     | | | ||| | |||        | | ||| |||||||
 52  IMPQVPVFTVEVKNYGVKNSEWIDACINISHHYCNISDHVGDPSNSLWVRVKARVGQKES  111

118  DYARSKEFLMCLKGKVGPPGLEIRRKKEEQLSVLVFHPEVVVNGESQGTMFGDGSTCYTF  177
     || | ||  |   || ||| | || || |    ||| ||| |    |         |||
112  AYAKSEEFAVCRDGKIGPPKLDI.RKEEKQIMIDIFHPSVFVNGDEQEVDYDPETTCYIR  170

178  DYTVYVEHNRSGEILHTKHTVEKEECNETLCELNISVSTLDSRYCISVDGISSFWQVRTE  237
     | ||| |   ||    |  ||     | | | ||| || ||| || |  |    | | ||
171  VYNVYVRMN.GSEIQYKILTQKEDDCDEIQCQLAIPVSSLNSQYCVSAEGVLHVWGVTTE  229

238  KSKDVCIPPFHDDRKDSIWILVVAPLTVFTVVILVFAYWYTKK.NSFKRKSIMLPKSLLS  296
     ||| |||     |    | || |||   | ||    |||  || |   | ||| ||||| |
230  KSKEVCITIFNSSIKGSLWIPVVAALLLFLVLSLVFICFYIKKINPLKEKSIILPKSLIS  289

297  VVKSATLETKPESKY.SLVTPHQPAVLESETVICEEPLSTVTAP......DSPEAAEQEE  349
     || |||||||||||| ||    || | | | ||||||| | |              | |
290  VVRSATLETKPESKYVSLITSYQPFSLEKE.VVCEEPLSPATVPGMHTEDNPGKVEHTEE  348

350  LSKETKALEAGGSTSAMTPDSPPTPTQRRSFSLLSSNQSGPCS..LTAYHSRNGSDSGL.  406
     ||  |       | |   || | | | ||||||| |    | |      |||||| | |
349  LSSITEVVTTEENIPDVVPGSHLTPIERESSSPLSSNQSEPGSIALNSYHSRNCSESDHS  408

407  .........VGSGSSISDLESLPNNNSETKMAEHDPPPVRKAPMASGYDKPHMLVDVLV  456
              || || |  ||| |  | |||      | ||| |||||| ||| || ||
409  RNGFDTDSSCLESHSSLSDSEFPPNNKGEIKTEGQELITVIKAPTSFGYDKPHVLVDLLV  468

457  DVGGKESLMGYRLTGEAQELS   477
     |  |||||| ||| |   | |
469  DDSGKESLIGYRPTEDSKEFS   489
```

Figure 4. Comparison between the murine and human IFN-γ receptor sequences at the protein level. The first and second sequences represent the murine and human amino acid sequences, respectively. There is 49.5% homology in the extracellular domain (murine residues 28–254), 52% homology in the transmembrane domain (residues 255–277), and 54% homology in the cytoplasmic domain (residues 278–477) between the murine and human proteins. The residues enclosed within the darkened boxes at the beginning and middle of the sequences represent the putative signal peptide and transmembrane domains, respectively. The figure is reproduced from Kumar *et al.* (24) with permission.

nology for screening large segments of the human genome in yeast artificial chromosomes (YACs) into which mammalian selectable markers were introduced (40–42, 44, 45). As this was successful in localizing the gene for AF-1, we applied this same technology to identify a functional type I receptor. A YAC was identified that contained genes encoding the type I receptor

Table 2. Binding of [^{32}P]Mu-IFN-γ to oocytes injected with RNA transcribed *in vitro* from the Mu-IFN-γ receptor cDNA

RNA	Concentration of unlabelled Mu- or Hu-IFN-γ (competitor)	[^{32}P]Mu-IFN-γ bound (c.p.m.)	Ratio (−/+)
Sense RNA	None	3180	
(pMuIFN-γR36)(0.05 ng/oocyte)	(3.1 × 10^{-9} M) (Mu)	647	4.9
	(3.1 × 10^{-8} M) (Mu)	243	13.1
	(1.6 × 10^{-7} M) (Mu)	176	18.0
	(3.1 × 10^{-7} M) (Mu)	122	26.0
	(3.4 × 10^{-7} M) (Hu)	3130	1.0
Sense RNA	None	18080	
(pMuIFN-γR36)(10 ng/oocyte)	(3.1 × 10^{-9} M) (Mu)	14550	1.2
	(3.1 × 10^{-8} M) (Mu)	4470	4.0
	(1.6 × 10^{-7} M) (Mu)	2384	7.6
	(3.1 × 10^{-7} M) (Mu)	1850	9.8
	(3.4 × 10^{-7} M) (Hu)	19580	0.9
Anti-sense RNA	None	106	
(pMuIFN-γR36)(10 ng/oocyte)	(3.1 × 10^{-7} M) (Mu)	120	0.9
None	None	115	
	(3.1 × 10^{-7} M) (Mu)	112	1.0

The data in the table are expressed as c.p.m./10 oocytes. Oocytes injected with RNA transcribed *in vitro* from the murine receptor cDNA clone (pMuIFN-γR36) were incubated with [^{32}P]Mu-IFN-γ without (−) unlabelled competitor or with increasing amounts of unlabelled (+) competitor as given in parentheses. Details are given in *Protocol 3*. All values represent the average of duplicate determinations. The ratio (−/+) represents the c.p.m. of [^{32}P]Mu-IFN-γ bound in the absence to that in the presence of unlabelled competitor. The data are reproduced from Kumar *et al.* (25) with permission.

complex (46). Furthermore, by deleting one known gene, Hu-IFN-αR1 (47), within the YAC, we demonstrated that this gene was an essential component of the type I interferon receptor complex (48). There are additional genes on this YAC clone (46, 48) that contribute to binding of type I interferons and to functional activation of the receptor.

5. Concluding remarks

The binding and/or cross-linking procedures outlined here may require modification when applied to other ligand–receptor systems. In our experience, we have not detected any significant differences in the cross-linked bands from suspension and adherent cell lines. This indicates that the IFN-γR is relatively insensitive to the trypsin treatment used to release adherent cells from the tissue culture substrates. If other receptors are found to be degraded by treatment with trypsin, it may be necessary to do the binding and cross-linking procedures directly in the tissue culture plates or one can release adherent cells with PBS containing 5 mM EDTA and proceed as described for trypsin-released cells. For binding studies, the bound radioactivity can be determined after lysing the cells in tissue culture dishes with 1 N NaOH. Similarly, in a cross-linking experiment, the detergent extraction would be done in the plate and the supernatant after centrifugation would be analysed by SDS-polyacrylamide gel electrophoresis.

The phosphorylation of interferons as described in this chapter enables proteins to be radiolabelled to very high specific activities. Although Mu- and Hu-IFN-γ have intrinsic site(s) for phosphorylation, genetic engineering procedures were used to extend the labelling method to other proteins (21–23). The ^{32}P-labelled IFNs described in this chapter allowed us to set up a very sensitive expression and detection system in *X. laevis* oocytes. The assay system was also used to evaluate different tissues and cell lines for their content of receptor RNA as a prelude to cloning the IFN-γ receptor. These procedures are being used to define the activity and function of cloned receptors in heterologous cells. Furthermore, the ^{32}P-labelled ligands have been used to study binding to purified soluble receptors in a convenient and sensitive dot blot assay (49–51).

Acknowledgements

The work presented here was supported in part by United States Public Health Services Grant CA 46465-06 (awarded to S. Pestka) from the National Cancer Institute.

References

1. Pestka, S. and Baron, S. (1981). In *Methods in enzymology* (ed. S. Pestka), Vol. **78**, pp. 3–14. Academic Press, London.
2. Stewart, W. E., II (1979). *The interferon system*. Springer-Verlag, New York.
3. Pestka, S. (1983). *Arch. Biochem. Biophys.*, **221**, 1.
4. Langer, J. A. and Pestka, S. (1984). *J. Invest. Dermatol.*, **83**, 128s.
5. Pestka, S., Langer, J. A., Zoon, K. C., and Samuel, C. E. (1987). In *Annual review biochemistry* (ed. C. E. Richardson, P. D. Boyer, I. B. David, and A. Meister), Vol. **56**, pp. 727–7. Annual Reviews, Inc., Palo Alto, California.
6. Baron, S., Dianzani, F., Stanton, G. J., and Fleischmann, W. R. Jr. (ed.) (1987). *The interferon system: a current review to 1987*. University of Texas Press, Austin.
7. Hauptmann, R. and Swetly, P. (1985). *Nucleic Acids Res.*, **13**, 4739.
8. Feinstein, S. I., Mory, Y., Chernajovsky, Y., Maroteaux, L., Nir, U., Lavie, V., and Revel, M. (1985). *Mol. Cell. Biol.*, **5**, 510.
9. Capon, D. J., Shepard, H. M., and Goeddel, D. V. (1985). *Mol. Cell. Biol.*, **5**, 768.
10. Jung, V., Rashidbaigi, A., Jones, C., Tischfield, J. A., Shows, T. B., and Pestka, S. (1987). *Proc. Natl. Acad. Sci. USA*, **84**, 415.
11. Jung, V., Jones, C., Rashidbaigi, A., Geyer, D. D., Morse, H. G., Wright, R. B., and Pestka, S. (1988). *Somatic Cell Mol. Genet.*, **14**, 583.

12. Jung, V., Jones, C., Kumar, C. S., Stefanos, S., O'Connell, S., and Pestka, S. (1990). *J. Biol. Chem.*, **265**, 1827.
13. Mariano, T. M., Kozak, C. A., Langer, J. A., and Pestka, S. (1987). *J. Biol. Chem.*, **262**, 5812.
14. Flores, I., Mariano, T. M., and Pestka, S. (1991). *J. Biol. Chem.*, **266**, 19875.
15. Rashidbaigi, A., Kung, H.-F., and Pestka, S. (1985). *J. Biol. Chem.*, **260**, 8514.
16. Rashidbaigi, A., Langer, J. A., Jung, V., Jones, C., Morse, H. G., Tischfield, J. A., Trill, J. J., Kung, H.-F., and Pestka, S. (1986). *Proc. Natl. Acad. Sci. USA*, **83**, 384.
17. Kung, H.-F. and Bekesi, E. (1986). In *Methods in enzymology* (ed. S. Pestka), Vol. **119**, pp. 296–301. Academic Press, London.
18. Langer, J. A., Rashidbaigi, A., and Pestka, S. (1986). *J. Biol. Chem.*, **261**, 9801.
19. Fields, R., Mariano, T. M., Stein, S., and Pestka, S. (1988). *J. Interferon Res.*, **8**, 549.
20. Arakawa, T., Parker, C. G., and Lai, P.-H. (1986). *Biochem. Biophys. Res. Commun.*, **136**, 679.
21. Li, B.-L., Langer, J. A., Schwartz, B., and Pestka, S. (1989). *Proc. Natl. Acad. Sci. USA*, **86**, 558.
22. Zhao, X.-X., Li, B.-L., Langer, J. A., Van Riper, G., and Pestka, S. (1989). *Anal. Biochem.*, **178**, 342.
23. Wang, P., Izotova, L., Mariano, T. M., Donnelly, R. J., and Pestka, S. (1994). *J. Interferon Res.*, **14**, 41.
24. Kumar, C. S., Mariano, T. M., Noe, M., Deshpande, A. K., Rose, P. M., and Pestka, S. (1988). *J. Biol. Chem.*, **263**, 13493.
25. Kumar, C. S., Muthukumaran, G., Frost, L. J., Noe, M., Ahn, Y. H., Mariano, T. M., and Pestka, S. (1989). *J. Biol. Chem.*, **264**, 17939.
26. Kung, H.-F., Pan, Y.-C. E., Moschera, J., Tsai, K., Bekesi, E., Chang, M., Sugino, H., and Honda, S. (1986). In *Methods in enzymology* (ed. S. Pestka), Vol. **119**, pp. 204–10. Academic Press, London.
27. Familletti, P. C., Rubinstein, S., and Pestka, S. (1981). In *Methods in enzymology* (ed. S. Pestka), Vol. **78**, pp. 387–94. Academic Press, London.
28. Scatchard, G. (1949). *Ann. N.Y. Acad. Sci.*, **51**, 660.
29. Laemmli, U. S. (1970). *Nature*, **227**, 680.
30. Pestka, S., Rashidbaigi, A., Langer, J. A., Mariano, T. M., Kung, H.-F., Jones, C., and Tischfield, J. A. (1986). In *UCLA Symp. Mol. Cell. Biol., New Series* (ed. R. M. Friedman, T. Merigan, and T. Sreevalsan), Vol. **50**, pp. 259–68. Alan R. Liss, Inc., New York.
31. Calderon, J., Sheehan, K. C. F., Chance, C., Thomas, M. L., and Schreiber, R. D. (1988). *Proc. Natl. Acad. Sci. USA*, **85**, 4837.
32. Dalla-Favera, R., Franchini, G. F., Martinotti, S., Wong-Staal, F., Gallo, R. C., and Croce, C. M. (1982). *Proc. Natl. Acad. Sci. USA*, **79**, 4714.
33. Sakaguchi, A. Y., Lalley, P. A., Zabel, B. U., Ellis, R., Scolnick, E., and Naylor, S. L. (1984). *Cytogenet. Cell Genet.*, **37**, 573.
34. McCandliss, R., Sloma, A., and Pestka, S. (1981). In *Methods in enzymology* (ed. S. Pestka), Vol. **79**, pp. 618–22. Academic Press, London.
35. Paterson, B. M., Roberts, B. E., and Kuff, E. L. (1977). *Proc. Natl. Acad. Sci. USA*, **74**, 4370.
36. Aguet, M., Dembric, Z., and Merlin, G. (1988). *Cell*, **55**, 273.

37. Chirgwin, J. M., Przybla, A. A., MacDonald, R. J., and Rutter, W. J. (1979). *Biochemistry*, **18**, 5294.
38. Aviv, H. and Leder, P. (1972). *Proc. Natl. Acad. Sci. USA*, **69**, 1408.
39. McCandliss, R., Sloma, A., and Pestka, S. (1981). In *Methods in enzymology* (ed. S. Pestka), Vol. **79**, pp. 51–9. Academic Press, London.
40. Soh, J., Donnelly, R. J., Mariano, T. M., Cook, J. R., Schwartz, B., and Pestka, S. (1993). *Proc. Natl. Acad. Sci. USA*, **90**, 8737.
41. Cook, J. R., Emanuel, S. L., Donnelly, R. J., Soh, J., Mariano, T. M., Schwartz, B., Rhee, S., and Pestka, S. (1994). *J. Biol. Chem.*, **269**, 7013.
42. Soh, J., Donnelly, R. J., Kotenko, S., Mariano, T. M., Cook, J. R., Wang, N., Emanuel, S. L., Schwartz, B., Miki, T., and Pestka, S. (1994). *Cell*, **76**, 793.
43. Hemmi, S., Böhni, R., Stark, G., Di Marco, F., and Aguet, M. (1994). *Cell*, **76**, 803.
44. Soh, J., Mariano, T. M., Bradshaw, G., Donnelly, R. J., and Pestka, S. (1994). *DNA Cell Biol.*, **13**, 301.
45. Cook, J. R., Emanuel, S. L., and Pestka, S. (1993). *Gene Analysis: Techniques Applic.*, **10**, 109.
46. Soh, J., Mariano, T. M., Lim, J.-K., Izotova, L., Mirochnitchenko, O., Schwartz, B., Langer, J., and Pestka, S. (1994). *J. Biol. Chem.*, **269**, 18, 102.
47. Uzé, G., Lutfalla, G., and Gresser, I. (1990). *Cell*, **60**, 225.
48. Cleary, C. M., Donnelly, R. J., Soh, J., Mariano, T. M., and Pestka, S. (1994). *J. Biol. Chem.*, **269**, 18, 747.
49. Stefanos, S., Ahn, Y. H., and Pestka, S. (1989). *J. Interferon Res.*, **9**, 719.
50. Stefanos, S. and Pestka, S. (1990). *J. Biol. Regul. Homeosta. Agents*, **4**, 57.
51. Puvanakrishnan, R. and Langer, J. A. (1990). *J. Interferon Res.*, **10**, 299.

Measurement of cytokine induction of PC-specific phospholipase C and sphingomyelinases

STEFAN SCHÜTZE and MARTIN KRÖNKE

Cytokines play a vital role as mediators of intercellular communication within the immune system. One of the hallmarks of cytokine function is the marked redundancy of biological activities. The overlapping biological functions of TNF/IL-1 and GM-CSF/IL-3 are just two examples. The redundancy of cytokine activities may be based on the utilization of shared key components of intracellular signalling pathways. The principal architecture of major signalling pathways used by cytokines consists of series of proteins including G-proteins, second messenger generating enzymes, protein kinases, and phosphatases. Receptor-mediated hydrolysis of cellular phospholipids by phospholipases of distinct specificities (PLA_2, PLC, PLD, SMase) is a ubiquitous event of central importance in cellular signalling. Activation of these enzymes results in generation of early second messenger molecules which further transduce receptor signals to cytoplasmic target proteins. Since multiple intracellular targets exist for a given second messenger, the cytokine signal may branch off to more complex and diversified signals.

In this chapter, we focus on the phospholipase C and sphingomyelin system, generating two distinct early lipid second messenger molecules, 1,2-diacylglycerol (DAG) and ceramide, respectively.

Evidence has accumulated in recent years (see (1) for review) that many cytokines (IL-1, TNF, IL-3, CSF-1, PDGF, EGF, HGF, EPO, and IFNs) after binding to their cell surface receptors elicit intracellular responses by activating phosphatidylcholine-specific phospholipase C (PC-PLC). The functional significance of PC-PLC can be explained by the generation of the lipid second messenger diacylglycerol (DAG), a potent activator of protein kinase C (see (2) for review) and acidic sphingomyelinase (3–5). Contrary to the widespread occurrence of PC-PLC, the physiological role of this phospholipase for cell growth and differentiation is just beginning to be understood.

In general, several mechanisms may be responsible for agonist-induced formation of diacylglycerol (DAG) (see *Figure 1*). One extensively studied

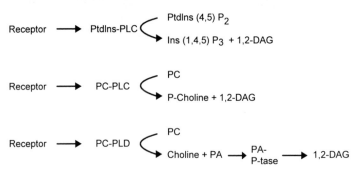

Figure 1. Pathways of diacylglycerol generation.

class of phospholipases C act on phosphatidylinositol-4,5-bisphosphate (PIP_2) to generate DAG and inositol-1,4,5-trisphosphate (IP_3). Another distinct phospholipase C species acts exclusively on phosphatidylcholine (PC), the most abundant phospholipid species of mammalian membranes, which results in the production of DAG and phosphorylcholine (Pchol). A further pathway to generate DAG involves PC cleavage by a phospholipase D (PLD) generating phosphatidic acid (PA), which can be subsequently converted to DAG by PA-phosphohydrolase.

Since the second messenger molecule DAG generated by these different pathways is apparently linked to different biological functions (see (1, 6) for review), it is important to discriminate between the phospholipases involved.

In addition to the lipid second messenger molecule DAG, the sphingolipid-derived second messenger ceramide has been recently implied in various important pathways of TNF and IL-1 (see (7, 8) for review). Ceramide can be generated from sphingomyelin (SM) hydrolysis (*Figure 2*) by two types of TNF-responsive SMases with neutral and acidic pH optima, respectively (see (9, 10) for review).

The metabolic active cleavage product of SMase, ceramide, is recognized as a novel multifunctional lipid second messenger, capable of inducing various signalling systems.

Acidic, endosomal SMase has been implicated in transducing TNF signals to the cell nucleus via activation of the nuclear transcription factor NF-κB (5). In addition to acidic SMase, a distinct, plasma membrane SMase with a neutral pH-optimum is stimulated by TNF (11–13) or IL-1 (14). Neutral SMase-derived ceramide triggers the activation of a protein kinase (11, 15) which couple the TNF and IL-1 signals to the MAP-kinase cascade. A further ceramide target is a cytosolic phosphatase, whose role in TNF signal transduction remains to elucidated (16).

In this chapter methods are described for the detection of agonist-induced activation of PC-PLC and SMases and production of the lipid second messengers 1,2-DAG and ceramide, respectively.

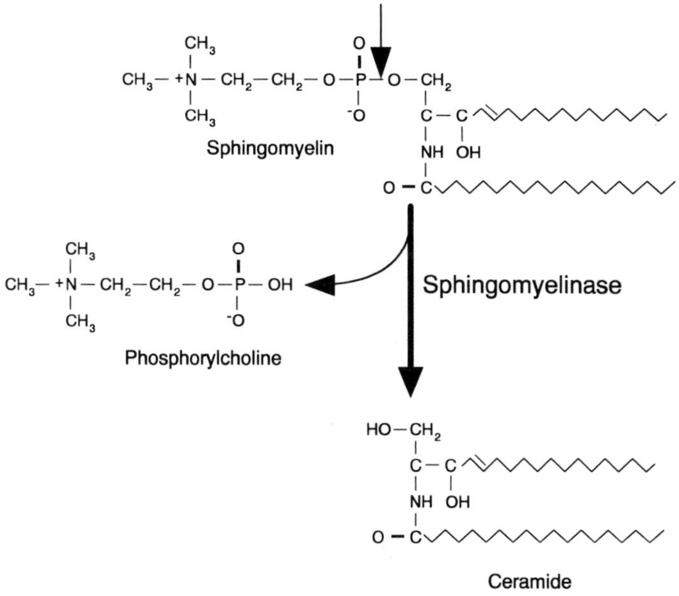

Figure 2. Generation of ceramide from sphingomyelin cleavage.

1. Phosphatidylcholine-specific phospholipase C

In general, cytokine activation of PC-PLC in whole cells can be monitored by estimating increments in the total amounts of the cleavage products, 1,2-diacylglycerol (DAG) and phosphorylcholine (Pchol) and by estimating the concomitant substrate degradation, i.e. the reduction of phospatidylcholine (PC). However, several problems have to be addressed: first, DAG-production can also result from cleavage of other phospholipid sources (PIP_2, PE) or from dephosphorylation of phosphatidic acid (PA) (*Figure 1*); second, the other cleavage product, Pchol, can also be produced by sphingomyelinase-(SMase-) mediated cleavage of sphingomyelin (SM) (*Figure 2*) or by phosphorylation of choline by choline-kinase; third, attempts of measuring changes in total mass of PC do not make sense, because PC is the major phospholipid component of cell membranes and the amount of PC cleaved in response to cytokines will be relatively minor compared to the overall concentration.

1.1 Detection of 1,2-DAG production from PC in metabolically labelled cells

Several methods have been successfully used to document changes in DAG levels in haematopoietic cells: first, by total mass analysis of dried organic extracts performed either by capillary gas chromatography (17), or by high performance liquid chromatography (HPLC) (18). DAG mass levels can also be estimated by phosphorylation of neutral lipid extracts employing DAG-kinase from *E. coli* and $[\gamma^{32}P]ATP$ (19). This assay system is commercially available (Amersham). This method, however, suffers from lack of internal controls to estimate effects of the preparation procedure on the yield of DAG isolated. A second method is based on metabolically labelling of the PC-pools of the cells using as precursor [3H]glycerol (20), [3H]myristic acid (21), 1-[3H]alkyl-2-lyso-glycero-3-phosphatidylcholine (22) or 1-[14C]palmitoyl-2-lyso-3-phosphatidylcholine (23). Radiolabelled DAG is subsequently analysed after extraction of neutral lipids by high performance thin layer chromatography (HPTLC). This method allows a considerable number of samples to be processed and the total material applied on TLC can be controlled (cholesterol or triglycerols as internal standards for neutral lipids extracted). In the following section, DAG-analysis will be described in cells labelled with 1-[14C]palmitoyl-2-lyso-3-phosphatidylcholine.

This approach is based on a rather selective labelling of PC and on the view that agonist-induced PC-cleavage is believed to occur on specific, hormone-sensitive pools (24).

Protocol 1. Metabolic cell labelling for detection of 1,2-DAG and cell stimulation

Equipment and reagents

- Hank's balanced salt solution (HBSS)
- Serum-free growth medium
- Bovine serum albumin
- 1-[14C]palmitoyl-2-lyso-3-phosphatidylcho-line (56.8 mCi/mmol, Amersham)
- Methanol
- Dry ice
- Microcentrifuge

Method

N.B. Growth factors present in fetal calf serum will influence basal DAG-levels. We therefore recommend to serum-starve the cells for 4 h prior to stimulation with agonists.

1. Wash cultured cells at least three times in Hank's balanced salt solution (HBSS) and recultivate in serum-free growth medium, supplemented with 2% bovine serum albumin (BSA) for 4 h.

2. After 2 h culture in serum-free medium, add 1-[14C]palmitoyl-2-lyso-3-phosphatidylcholine 1 µCi/ml.

3. After 2 h of labelling (total time 4 h serum-free culture), wash cells with HBSS and resuspend to 10^7 cells/ml for subsequent agonist stimulation.

4. Treat aliquots of 0.5 ml cell suspensions (1×10^7 cells/ml) with the

---

agonist of choice for various periods of time at 37°C. Treatments are stopped by immersion of the sample tubes in methanol/dry ice for 10 sec followed by centrifugation at 4°C in a microcentrifuge.

5. Discard supernatants and resuspend cell pellets in 1 ml cold methanol.

1.1.1 Lipid extraction

Two procedures are described which either allow for separation of neutral lipids, phospholipids and water soluble components according to the method of Bligh and Dyer (25) (*Protocol 2*), or a more rapid procedure, which separates only an organic phase from the aqueous phase without separation of neutral and phospholipids (Section 1.1.2).

Protocol 2. Lipid extraction

Equipment and reagents

- Methanol
- Chloroform
- 10 ml glass tubes
- Waterbath sonicator
- Vortex mixer
- Hexane
- Nitrogen gas

Method

1. Transfer samples that have been resuspended in 1 ml methanol to 10 ml glass-tubes containing 1 ml H_2O and 1.5 ml methanol. Sonicate tubes for 5 min in a waterbath sonicator.

2. Add 1.25 ml chloroform, briefly vortex and centrifuge for 10 min at 6000 *g*.

3. Save supernatants in new glass vials. Resuspend pellets in 1 ml H_2O, 2.5 ml methanol, 1.25 ml chloroform. Repeat sonication for 5 min and centrifugation for 5 min, 4000 *g*.

4. Add supernatant to first supernatant, add 2.5 ml H_2O and 2.5 ml chloroform to the combined supernatants, vortex, and centrifuge for phase-separation for 5 min at 4000 *g*.

5. Transfer and save the lower organic phase in new glass vials. Add 4 ml chloroform to the residual aqueous phase, mix, centrifuge for phase separation as above, and combine the lower, organic phase with the first.

6. Dry down the chloroform-phases under nitrogen.

7. Resuspend the dried samples for separation of neutral lipids and phospholipids in 4 ml hexane/3 ml methanol, mix, centrifuge as above for phase separation. The methanol phase contains phospholipids, the hexane phase contains neutral lipids.

Protocol 2. *Continued*

8. Transfer and safe-lower methanol phase into new glass vials, add 3 ml methanol and 2 ml hexane to the first residual hexane phase, mix, centrifuge for phase separation.

9. Transfer lower methanol phase to the first from step 8, repeat methanol washing of the hexane phase two more times.

10. Dry down separated methanol and hexane fractions under nitrogen.

11. Resuspend phospholipids from the methanol fraction and neutral lipids from the hexane fraction in 50 μl of chloroform/methanol (9:1) for TLC-analysis.

1.1.2 Rapid lipid extraction

In some cases, a separation of neutral and phospholipids by hexane/methanol extraction prior to TLC-analysis of DAG or phospholipids, respectively, can be omitted, because the solvent systems adopted for TLC-analysis of DAG and PL will also separate the two lipid fractions. If suitable for your purpose, follow steps 1–6 of *Protocol 2* and take the dried chloroform phase containing total lipids for TLC-separation analysis.

Protocol 3. Thin-layer chromatography for detection of 1,2-DAG

Equipment and reagents

- HPTLC silica gel 60 plates (Merck)
- Methanol
- Chloroform
- Benzene/ethylacetate (6.5:3.5) or other appropriate solvent system (see step 3)
- Standards: 1,2-DAG, 1,3-DAG, MAG, TG, cholesterol

- Kodak XAR-films
- Laser densitometry or liquid scintillation counting system
- Iodine vapour, charring densitometry system (see *Protocol 6*), or Coomassie blue

Method

High-performance thin-layer chromatography has been used extensively to characterize lipid metabolites.

1. Pre-run the plates in a solvent system composed of methanol/chloroform (1:1).

2. Dry the plates at 80°C for 30 min.

3. Pre-equilibrate the TLC-chambers for 1 h at room temperature with the solvent system to be used for lipid separation. For detection of 1,2-DAG, separate neutral lipids using a solvent system of benzene/ethylacetate (6.5/3.5).

4. Resuspend the dried samples in 50 μl chloroform/methanol (9:1) and spot the samples on the HPTLC plate in strips of 5–8 mm for each

sample about 2 cm from the base of the plate. Allow to dry. Load samples containing equal amounts of radioactivity. Use 1,2-DAG, 1,3-DAG, MAG, TG and cholesterol as standards.

5. Place TLC-plate into the chamber and allow the plates to develop at room temperature until the front is within 1 cm from the top.

6. Dry the plates, and expose to Kodak XAR-films at room temperature.

7. Autoradiographs can be analysed by two-dimensional scanning using laser-densitometry. Alternatively, the corresponding spots can be scraped off the plates and radioactivity is determined in a scintillation counter. A typical distribution of neutral lipids is given in *Figure 3A*.

Lipid standards can be visualized either by exposure to iodine vapour, by charring densitometry (26) (see *Protocol 6*), or by Coomassie-staining (27) (see Section 2.1.4).

1.2 Analysis of phosphatidylcholine (PC)

In contrast to labelling at equilibrium with ^{14}C-choline for 48 h, the rather selective labelling with lyso-PC for 2 h of agonist-sensitive PC-pools will more readily allow the detection of cytokine-induced changes in PC (23).

1.2.1 Metabolic labelling, cell stimulation, and extraction of PC

Agonist-induced changes in membrane PC-content can be monitored after short-term labelling with lyso-PC as described for detection of 1,2-DAG (*Protocol 1*).

Follow *Protocol 1* for radioactive labelling and cell stimulation. Instead of 1-[^{14}C]palmitoyl-2-lyso-3-phosphatidylcholine, also L-lyso-3-phosphatidyl [methyl[14]C]choline (1 μCi/ml, 22 mCi/mmol) can be used, revealing similar results (23). Extract phospholipids according to *Protocol 2*. The dried methanol phase is used as source for PC-analysis by TLC. Alternatively, follow *Protocol 3* and use the whole organic phase for TLC-analysis.

1.2.2 Thin-layer chromatography for detection of PC

After preparation of the silica 60 TLC-plates and equilibration of the chambers with the solvent according to *Protocol 3*, resolve the dried samples in chloroform/methanol (9:1), and spot on the TLC-plates. Separate phospholipids containing PC in a solvent system containing ($CHCl_3/CH_3OH/CH_3COOH/H_2O$) (100:60:20:5). Use lyso-PC, PC, PE, PI, PA and SM as standards and visualize by iodine vapour, Coomassie staining or charring densitometry (see *Protocol 6* and Section 2.1.4). Examine radiolabelled phospholipids by autoradiography followed by 2-D laser scanning. Spots can also be scraped-off followed by liquid scintillation counting. A typical distribution of phospholipids labelled with 1-[^{14}C]palmitoyl-2-lyso-3-phosphatidylcholine is presented in *Figure 3B*.

1.3 Detection of phosphorylcholine (Pchol)

The second cleavage product of PC-PLC, phosphorylcholine can be detected after radiolabelling of cells with either [methyl^{14}C]choline or lyso-phosphatidyl-[methyl^{14}C]choline. Cells should also be kept serum starved for 4 h prior to stimulation with agonists.

1.3.1 Metabolic labelling for detection of Pchol

For detection of water-soluble choline metabolites, label cells with [methyl^{14}C] choline (1 μCi/ml, spec. act. 56.4 mCi/mmol) for 48 h. This will result in at-equilibrium labelling. Alternatively cells can be labelled for 2 h with L-lyso-3-phosphatidyl-[methyl^{14}C]choline (spec. act. 22 mCi/ml).

Protocol 4. Pchol extraction

Choline metabolites are prepared from the **aqueous phase** obtained following *Protocol 2*

Method

1. Follow step 1–4, *Protocol 2.*

2. Transfer the upper, aqueous phase to new vial.

3. Freeze-dry samples.

Protocol 5. Thin-layer chromatography for detection of Pchol

Equipment and reagents

- Methanol
- TLC plates and chambers (see *Protocol 3*)
- Solvent system: CH$_3$OH/0.5% NaCl/NH$_4$OH (100:100:2)

- Standards: phosphorylcholine, glycero-phosphorylcholine, acetylcholine, choline
- Autoradiography system

Method

1. Resolve the freeze-dried samples in 50 μl methanol.

2. Prepare TLC-plates and TLC-chambers as described in *Protocol 3.*

3. Separate water-soluble components, containing Pchol on silica 60 TLC-plates using the solvent system CH$_3$OH/0.5% NaCl/NH$_4$OH (100:100:2). Phosphorylcholine, glycerophosphorylcholine, acetylcholine, and choline are used as standards.

4. Evaluation of radiolabelled choline metabolites by autoradiography and visualization of the standards can be performed as described in *Protocol 3.*

1.4 Estimation of combined phospholipase D (PLD)/ phosphatidic acid-phosphatase (PA-Ptase) activities in DAG-formation

PC-PLD activity can be monitored directly by estimating changes in its cleavage product phosphatidic acid (PA), or by determining the formation of labelled phosphatidylethanol (Peth), which is expected from PLD's transphosphatidylation activity in the presence of ethanol (28). A possible involvement of a combined PLD and PA-Ptase activity on DAG-formation can be tested by use of a PA-phosphohydrolase inhibitor, propranolol (29).

1.4.1 Estimation of phosphatidic acid (PA)

1. For detection of PA, label cells with 1-[^{14}C]palmitoyl-2-lyso-3-phosphatidylcholine (1 μCi/ml (spec. act. 56.8 mCi/mmol, Amersham) for 2 h as described in *Protocol 1* for detection of DAG (Section 1.1.1).
2. Extract phospholipids according to Section 1.1.2, following either *Protocol 2*, using the dried methanol phase or according to Section 1.1.2, using the whole organic phase for PA-analysis by TLC, respectively.

1.4.2 Thin-layer chromatography for detection of PA

After preparation of the silica 60 TLC-plates using aceton as solvent, TLC-chambers are equilibrated with the solvent according to *Protocol 3*. The dried samples are resolved in chloroform/methanol (9:1), spot on the TLC-plates, and phospholipids containing PA are separated in a solvent system containing ($CHCl_3/CH_3OH/CH_3COOH$) (65:15:5). PA, PC, PI, PE, SM and lyso-PC are used as standards and visualized by iodine vapour, Coomassie staining or charring densitometry (see *Protocol 6* and Section 2.1.4). Radiolabelled phospholipids are examined by autoradiography followed by 2-D laser scanning or liquid scintillation counting of scraped-off spots.

1.4.3 Transphosphatidylation in the presence of ethanol

To detect phosphatidylethanol (Peth), cells are labelled with 1-[^{14}C]palmitoyl-2-lyso-3-phosphatidylcholine (1 μCi/ml (spec. act. 56.8 mCi/mmol, Amersham) for 2 h as described for the detection of PA (Section 1.4.1). Prior to agonist stimulation, cells are pre-incubated for 2 min with 1% ethanol. Phospholipid extraction was performed as described for PA (Section 1.4.1). For separation of Peth from PA and other phospholipids, TLC is performed in a solvent system consisting of the **organic phase** of ethylacetate/acetic acid/H_2O (110:20:110).

1.4.4 Effect of the PA-phosphatase inhibitor propranolol on agonist-induced DAG-production

Cell are labelled with 1-[^{14}C]palmitoyl-2-lyso-3-phosphatidylcholine for 2 h as described in Section 1.4.1 and pretreated with 200 μM propranolol for 15 min prior to agonist-stimulation. The concentration of propranolol used for inhibition of PA-phosphatase may vary depending on the cell-line employed. The propranolol concentration needed has thus to be determined by stimulating the cells with 100 nM PMA for 30 min, which results in PKC-mediated activation of PLD (30). For estimating the effect of propranolol treatment on DAG-production, follow the procedures described in Section 1.1.

1.5 Inhibitors of PC-PLC

To evaluate biological functions of PC-PLC in cytokine signalling, a rather selective PC-PLC inhibitor, the xanthogenate D609 has been employed, originally obtained from Merz AG, Frankfurt, FRG and Drs Sauer and Amtmann, Heidelberg (5, 31, 32). This compound does not block PI-PLC, PC-PLD, PLA$_2$, SMase, PKC, Tyr-PK, PI-kinase (5, 33). Effective dosage may depend on the cell type used and the D609 charge obtained from the producer. Usually cells were treated with 100 µg/ml D609 for 30 min prior to agonist stimulation.

D609 is now commercially available from Biomol Research Laboratories Inc., PA and from Molecular Probes Inc., OR. Depending on the length of D609 treatment, nonspecific side-effects cannot be excluded.

2. Sphingomyelinases

Sphingolipids constitute an important class of membrane lipid molecules. Sphingomyelin (SM) breakdown products have emerged in recent years as biological active molecules exhibiting various functions within the cell (for review see 8, 34, 35). SM hydrolysis catalysed by sphingomyelinases results in generation of ceramide (see *Figure 2*), a potent neutral lipid second messenger molecule which is implied in the activation of a plasma membrane protein kinase (11, 36, 37), a cytosolic protein phosphatase (16) activation of the nuclear transcription factor NF-κB (5, 38, 39), downregulation of the *c-myc* proto-oncogene (40) and stimulation of the cyclo-oxygenase gene (41). The biological responses linked to SMase activation includes growth inhibition, induction of monocytic differentiation, apoptosis in tumour cell-lines, and proliferation of fibroblasts (see (8) for review). Ceramide is further metabolized by ceramidase to sphingosine, a physiological inhibitor of protein kinase C (34). The cytokines TNF and IL-1 stimulate at least two different SMases, a neutral, plasma membrane bound (N-SMase) (11–14) and an acidic, endosomal SMase (A-SMase) (5). The pleiotropic effect of ceramide in mediating various different biological effects may be based on the different localization and regulation of the two SMases (43). In addition to the membrane-bound N-and A-SMase, a cytosolic N-SMase was recently identified (42).

Activation of the SM-hydrolysis pathway in whole cells can be monitored either by measuring reduction in SM content in radiolabelled cells, or production of the cleavage products ceramide and Pchol. Ceramide can be detected from labelled cells by TLC or by mass analysis using DAG-kinase from *E. coli* (19). Pchol is measured as described for analysis of PC-PLC action (Section 1.3). Pchol can be produced by various different pathways *in vivo*, involving either PC-PLC, SMase, or choline kinase. We, therefore, use this cleavage product as read-out for SMase-activity only in a micellar assay system with defined, exogenous substrates (see Section 2.2).

2.1 Detection of SM hydrolysis in metabolically labelled cells

Hydrolysis of SM in whole cells can be measured either after radiolabelling of the SM pools, using [methyl-^{14}C]choline for labelling the choline moiety of the sphingolipid, or by analysis of total SM mass by charring densitometry.

2.1.1 Cell culture and metabolic labelling for analysis of SM

For detection of SM, cells are metabolically labelled for 48 h with [methyl-^{14}C]choline (1 μCi/ml, spec. act. 56.4 mCi/mol) in culture medium containing FCS. Prior to agonist-stimulation, cells are washed serum-free with HBSS and kept serum-starved for 4 h in medium supplemented with 2% BSA as described in *Protocol 1*. Stimulation of cells is performed as described in *Protocol 1*, steps 4 and 5.

2.1.2 SM extraction

Cell pellets (step 5 in *Protocol 1*) are resuspended in cold methanol and SM is extracted with the methanol (phospholipid) fraction described in Section 1.1.1, following *Protocol 2*. Alternatively, SM can also be resolved from total lipids from the chloroform fraction, see Section 1.1.2.

2.1.3 Thin-layer chromatography for detection of SM

Dried lipids are resolved in chloroform/methanol (9:1) and spot on to silica 60 TLC plates, which have been pre-treated with chloroform/methanol as described in *Protocol 3*. For separation of SM from other phospholipids, the solvent system ($CHCl_3/CH_3OH/CH_3COOH/H_2O$) (100:60:20:5) is employed. SM, PC, PA, PI, PE, and lyso-PC are used as standards and visualized by iodine vapour, Coomassie staining or charring densitometry (see *Protocol 6* and Section 2.1.4). Radiolabelled phospholipids are examined by autoradiography followed by 2-D laser scanning or by scraping off the spots and liquid scintillation counting. A typical resolution of ^{14}C-SM is shown in *Figure 3D*.

Protocol 6. Charring densitometry

Equipment and reagents

- 10% copper sulfate in aqueous phosphoric acid
- Oven

- Liquid scintillation counting *or* 2-dimensional laser densitometry system (Molecular Dynamics Personal Densitometer)

Method

After TLC-separation, phospholipids can be visualized and quantitated by charring of the plates with cupric reagent (26).

1. Remove TLC-plates after separation of samples from the TLC chamber and dry the plates for 10 min at 180°C.

2. Let the plates cool down to room temp. and expose for 15 sec to a solution of 10% copper sulfate in 8% aqueous phosphoric acid.

3. Transfer the plates back to the heater and dry for 2 min at 110°C.

4. Charring is performed at 175°C for approximately 10 min.

5. For quantitation of SM the charred bands are scraped off the plates and radioactivity is determined by liquid scintillation counting. Alternatively, non-radioactive SM can be analysed by scanning the charred TLC plates by 2 dimensional laser densitometry (Molecular Dynamics Personal Densitometer).

2.1.4 Coomassie staining

After separation by TLC, lipids can be visualized by staining the plates in a solution of 35% methanol, 100 mM NaCl, and 0.03% Coomassie brilliant blue R250 (27). Plates are destained in 35% methanol, 100 mM NaCl.

2.2 Detection of ceramide

The neutral lipid cleavage product of SMases, ceramide, can either be measured by total mass analysis, employing charring densitometry of TLC-plates as described in *Protocol 6* for analysis of SM, or by total mass determinations using DAG-kinase (12, 19). For this approach, the DAG-kinase assay system from Amersham Co. may be used. The third method is based on determination of radiolabelled ceramide isolated from metabolic labelled cells.

2.2.1 Metabolic labelling

For detection of ceramide, the SM pool of serum-starved cells are metabolically labelled with 1-[^{14}C]palmitoyl-2-lyso-3-phosphatidylcholine (1 μCi/ml, spec. act. 56.8 mCi/mmol) for 2 h as described for detection of DAG (*Protocol 1*). Stimulation of cells is performed as described in this section.

2.2.2 Ceramide extraction and TLC-analysis

Cell pellets are resuspended in cold methanol and neutral lipids are extracted. Ceramides are recovered in the hexane fraction following the lipid extraction, *Protocol 2*. Alternatively, ceramides can also be recovered from the total lipid (chloroform) fraction, see Section 1.1.2.

For separation of ceramide from other neutral lipids, dried samples are resolved in chloroform/methanol (9:1) and spotted on to silica 60 TLC plates, which have been pretreated as described in *Protocol 3*. For detection of ceramide, the solvent system ($CHCl_3/CH_3OH/7N$ NH_4OH/H_2O) (85:15:0.5:0.5) is used. Ceramide, DAG, MAG, TG are used as standards and visualized by iodine vapour, Coomassie staining or charring densitometry (see *Protocol 6* and Section 2.1.4). Radiolabelled ceramide is examined by autoradiography followed by 2-D laser scanning. A typical resolution of ceramide is shown in *Figure 3C*.

The solvent systems used for separation of the various lipid metabolites is summarized in *Table 1*.

2.3 Estimation of SMase activity *in vitro*

Agonist-stimulated SMase activities can also be estimated by *in vitro* enzymatic micellar assays of detergent-solubilized SMase preparations. With this method, a selective estimation of neutral and acidic SMases activities can be performed. Acidic SMase is a rather stable enzyme and insensitive to proteolysis. Thus acidic SMase-preparations can even be frozen down and thawed for retaining full activity. Neutral SMase, in contrast is highly susceptible to proteolyic inactivation. Therefore, the preparation of this enzyme has to be performed in the presence of various inhibitors, and enzymatic assays should be performed on the same day.

2.3.1 Cell stimulation

Serum-starve cells for 4 h prior to agonist-stimulation. No radioactive labelling is required.

1. Stimulate aliquots of 0.5 ml cell suspensions (1×10^7 cells/ml) with the agonist of choice for various periods of time at 37°C.

Table 1. Solvent systems for lipid analysis on TLC silica gel 60

Lipid metabolite	Solvent system	Ratio
1,2-diacylglycerol	benzene:ethylacetate	65:35
Ceramide	$CHCl_3:CH_3OH:7N\ NH_4OH:H_2O$	85:15:0.5:0.5
Phosphorylcholine	$CH_3OH:0.5\%\ NaCl:NH_4OH$	100:100:2
Phosphatidylcholine		
Phosphatidic acid	$CHCl_3:CH_3OH:CH_3COOH:H_2O$	100:60:20:5
Sphingomyelin		
Phosphatidylethanol	ethylacetate:$CH_3COOH:H_2O$	110:20:110

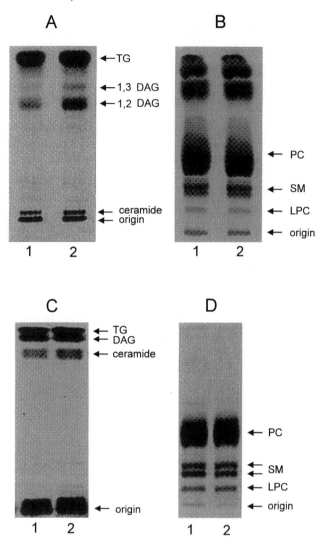

Figure 3. Resolution of ^{14}C-labelled diacylglycerol (A), phosphatidylcholine (B), ceramide (C) and sphingomyelin (D). U937 cells were labelled either for 2 h with 1-[^{14}C]palmitoyl-2-lyso-3-phosphatidylcholine (A, B, C) or for 24 h with [methyl-^{14}C]choline (D). Aliquots were left untreated (**lanes 1**) or were stimulated with 100 ng/ml TNF; (**lanes 2**) for 2 min (A,B,C) or 5 min (D). The lipid metabolites were isolated according to sections 1.1 (A); 1.2 (B); 2.2 (C) or 2.1 (D) and developed on TLC using the solvent systems summarized in *Table 1* for the respective components.

2. Stop treatments by immersion of the sample tubes in methanol/dry ice for 10 sec. Follow by centrifugation at 4°C in a microcentrifuge.

Protocol 7. Extraction of neutral SMase

Equipment and reagents

- Buffer: 20 mM Hepes, pH 7.4, 10 mM MgCl$_2$, 2 mM EDTA, 5 mM DTT, 0.1 mM Na$_3$VO, 0.1 mM Na$_2$MoO$_4$, 30 mM *p*-nitrophenylphosphate, 10 mM α-glycerophosphate, 1 mM
- PMSF, 10 μM leupeptin, 10 μM pepstatin, 750 μM ATP, 0.2% Triton X-100
- 18-gauge needle and syringe

Method

1. Dissolve cell pellets in buffer.

2. After 5 min at 5°C, homogenize cells by repeated squeezing through an 18-gauge needle.

3. Remove cell debris and nuclei by low-speed centrifugation (800 *g*) for 5 min.

4. Use the supernatants containing Triton X-100-solubilized enzymes for *in vitro* assay of agonist-stimulated N-SMase activities.

Protocol 8. Assay for neutral SMase activity

Equipment and reagents

- Reaction buffer: 20 mM Hepes, pH 7.4, 1 mM MgCl$_2$, 2.25 μl (1 nmol) [*N*-methyl-^{14}C] sphingomyelin (0.2 μCi/ml, 56.6 mCi/mmol)
- Chloroform/methanol (2:1, v/v)
- Vortex mixer
- Microcentrifuge and tubes
- Liquid scintillation counting system

To warrant linear reactions, the amount of ^{14}C-SM hydrolysed should not exceed 10% of the total amount of radioactive SM added to the assay.

Method

1. Add 50 μg of protein to a reaction buffer in a micro-tube in a total volume of 50 μl.

2. Incubate samples for 2 h at 37°C.

3. Stop reactions by adding 800 μl chloroform/methanol (2:1), vortex-mix and add 250 μl H$_2$O, vortex-mix and centrifuge 2 min full speed in a microcentrifuge for phase separation.

4. 200 μl of the aqueous upper phase of each sample are analysed for ^{14}C-labelled Pchol by liquid scintillation counting.

Protocol 9. Extraction of acidic SMase

Equipment and reagents

- 0.2% Triton X-100 or 0.05% Nonidet P-40
- 18-gauge needle and syringe

Method

1. Resuspend cell pellets in 200 μl 0.2% Triton X-100. Alternatively, 0.05% Nonidet P-40 (NP-40) can be used as detergent (42).

2. After 5 min at 5°C, homogenize cells by repeated squeezing of cells through an 18-gauge needle.

3. Remove cell debris and nuclei by low-speed centrifugation (800 *g*) for 5 min.

4. Use the supernatants, containing Triton X-100-solubilized enzymes for *in vitro* assay of agonist-stimulated A-SMase activities.

Protocol 10. Assay for acidic SMase

Equipment and reagents

- Reaction buffer: 250 mM sodium acetate, pH 5.0, 1 mM EDTA, 2.25 μl (1 nmol) [*N*-methyl-^{14}C]sphingomyelin (0.2 μCi/ml, 56.6 mCi/mmol)
- Chloroform/methanol (2:1, v/v)
- Vortex mixer
- Microcentrifuge and tubes
- Liquid scintillation counting system

Method

1. Add 50 μg of protein to the reaction buffer in a microcentrifuge tube in a total volume of 50 μl.

2. Incubate samples for 2 h at 37°C.

3. Stop reactions by adding 800 μl chloroform/methanol (2:1), vortex-mix and add 250 μl H$_2$O, vortex-mix and centrifuge 2 min full-speed in a microcentrifuge for phase separation.

4. Analyse 200 μl of the aqueous upper phase of each sample for ^{14}C-labelled Pchol by liquid scintillation counting.

Acknowledgements

The work of our laboratory was supported by the Deutsche Forschungsgemeinschaft (DFG) and the Wilhelm Sander Stiftung.

References

1. Schütze, S. and Krönke, M. (1994). In *Signal-activated phospholipases* (ed. M. Liscovitch), pp. 101–24. R. G. Landes, Biomedical Publishers, USA.
2. Nishizuka, Y. (1992). *Science*, **258**, 607.
3. Quintern, L. E., Weitz, G., Nehrkorn, H., Tager, J. M., Schram, A. W., and Sandhoff, K. (1987). *Biochim. Biophys. Acta*, **922**, 323.
4. Kolesnick, R. N. (1987). *J. Biol. Chem.*, **262**, 16759.
5. Schütze, S., Potthoff, K., Machleidt, T., Berkovic, D., Wiegmann, K., and Krönke, M. (1992). *Cell*, **71**, 765.
6. Liscovitch, M. (1992). *Trends. Biol. Sci.*, **17**, 393.
7. Kolesnick, R. N. (1992). *Trends. Cell. Biol.*, **2**, 232.
8. Hannun, Y. A. (1994). *J. Biol. Chem.*, **269**(5), 3125.
9. Chatterjee, S. (1993). *Adv. Lipid Res.*, **26**, 25.
10. Spence, M. W. (1993). *Adv. Lipid Res.*, **26**, 3.
11. Mathias, S., Dressler, K. A., and Kolesnick, R. N. (1991). *Proc. Natl. Acad. Sci. USA*, **88**, 10009.
12. Dressler, K. A., Mathias, S., and Kolesnick, R. N. (1992). *Science*, **255**, 1715.
13. Chatterjee, S. (1993). *J. Biol. Chem.*, **268**, 3401.
14. Mathias, S., Younes, A., Kan, C. C., Orlow, I., Joseph, C., and Kolesnick, R. N. (1993). *Science*, **259**, 519.
15. Raines, M. A., Kolesnick, R. N., and Golde, D. W. (1993). *J. Biol. Chem.*, **268**, 14572.
16. Dobrowsky, R. T. and Hannun, Y. A. (1993). *Adv. Lipid Res.*, **25**, 91.
17. Pessin, M. S., Baldassare, J. J., and Raben, D. M. (1990). *J. Biol. Chem.*, **265**, 7959.
18. Pettitt, T. R., Zaqqa, M., and Wakelam, J. O. (1994). *Biochem. J.*, **298**, 655.
19. Preiss, J., Loomis, C. R., Bishop, W. R., Stein, R., Niedel, J. E., and Bell, R. M. (1986). *J. Biol. Chem.*, **261**, 8597.
20. Rao, P. and Mufson, R. A. (1994). *Cancer Res.*, **54**, 777.
21. Ha, K. S. and Exton, J. H. (1993). *J. Biol. Chem.*, **268**, 10534.
22. Augert, G., Bocckino, S. B., Blackmore, P. F., and Exton, J. H. (1989). *J. Biol. Chem.*, **264**, 21689.
23. Schütze, S., Berkovic, D., Tomsing, O., Unger, C., and Krönke, M. (1991). *J. Exp. Med.*, **174**, 975.
24. Rana, R. S., Mertz, R. J., Kowluru, A., Dixon, J. F., Hokin, L. E., and MacDonald, M. J. (1985). *J. Biol. Chem.*, **260**, 7861.
25. Bligh, V. and Dyer, W. J. (1959). *Can. Biochem. J. Physiol.*, **37**, 911.
26. Rustenbeck, I. and Lenzen, S. (1990). *J. Chromatogr.*, **525**, 85.
27. Nakamura, K. and Handa, S. (1984). *Anal. Biochem.*, **142**, 406.
28. Huang, C. F. and Cabot, M. C. (1990). *J. Biol. Chem.*, **265**, 17468.
29. Billah, M. M., Eckel, S., Mullmann, T. J., Egan, R. W., and Siegel, M. I. (1989). *J. Biol. Chem.*, **264**, 17069.
30. Billah, M. M. (1993). *Curr. Opin. Immunol.*, **5**, 114.
31. Sauer, G., Amtmann, E., Melber, K., Knapp, A., Muller, K., Hummel, K., and Scherm, A. (1984). *Proc. Natl. Acad. Sci. USA*, **81**, 3263.
32. Muller Decker, K. (1989). *Biochem. Biophys. Res. Commun.*, **162**, 198.

33. Cai, H., Erhardt, P., Troppmair, J., Diaz-Meco, M. T., Sithanandam, G., Rapp, U. R., Moscat, J., and Cooper, G. M. (1993). *Mol. Cell Biol.*, **13:12**, 7645.
34. Hannun, Y. A. and Bell, R. M. (1993). *Adv. Lipid Res.*, **25**, 27.
35. Ballou, L. R. (1992). *Immunol. Today*, **13**, 339.
36. Joseph, C. K., Byun, H. S., Bittman, R., and Kolesnick, R. N. (1993). *J. Biol. Chem.*, **268**, 20002.
37. Liu, J., Mathias, S., Yang, Z., and Kolesnick, R. N. (1994). *J. Biol. Chem.*, **269(4)**, 3047.
38. Yang, Z., Costanzo, M., Golde, D. W., and Kolesnick, R. N. (1993). *J. Biol. Chem.* **268:27**, 20520.
39. Machleidt, T., Wiegmann, K., Henkel, T., Schütze, S., Baeuerle, P., and Krönke, M. (1994). *J. Biol. Chem.*, **269**, 13760.
40. Kim, M., Linardic, C. M., Obeid, L. M., and Hannun, Y. A. (1991). *J. Biol. Chem.*, **266**, 484.
41. Ballou, L. R., Chao, C. P., Holness, M. A., Barker, S. C., and Raghow, R. (1992). *J. Biol. Chem.*, **267**, 20044.
42. Okazaki, T., Bielawska, A., Domae, N., Bell, R. M., and Hannun, Y. A. (1994). *J. Biol. Chem.*, **269**, 4070.
43. Wiegmann, K., Shütze, S., Machleidt, T., Witte, D., and Krönke, M. (1994). *Cell*, **78**, 1005.

8

Two-dimensional gel electrophoresis for purification, sequencing, and identification of cytokine-induced proteins in normal and malignant cells

DIANA M. SMITH, HUU M. TRAN, and LOIS B. EPSTEIN

1. Introduction

Two-dimensional polyacrylamide gel electrophoresis (2D PAGE) permits high resolution and purification of individual proteins from complex mixtures of proteins. Proteins are separated first according to their charge, employing isoelectric focusing (IEF), and then in the second dimension, according to their molecular mass. 2D PAGE is unequalled in its ability to simultaneously resolve hundreds of proteins from cellular extracts and at the same time provide data on their isoelectric point (pI), molecular mass (M_r), electrophoretic pattern, and relative abundance (1–3). Thus, a large amount of global information concerning the regulatory effects of cytokines on the expression of multiple individual proteins can be obtained in one experiment.

2D PAGE, computer-based analysis of gels and the more recent advances in protein microsequencing technologies, coupled with the capability of searching protein databases for homologous proteins, permits investigators working in the field of cytokine research to answer basic and complex questions about the actions of these important regulatory molecules. For example, it is possible:

(a) To characterize proteins whose synthesis is suppressed or enhanced by individual cytokines, thus identifying specific pathways of the action of a single cytokine or combinations of cytokines. Specific information concerning the pI/M_r, electrophoretic pattern, and relative abundance of proteins affected by cytokine treatment can be obtained.

(b) To compare the effects of a given cytokine on different cell types.

(c) To identify individual proteins induced by cytokines through a particular type of receptor using agents that either mimic or block specific receptor–cytokine interactions.

(d) To compare the effects of a given cytokine in the presence of different physical or external environments (i.e. heat, ionizing radiation, hormones, chemotherapeutic agents, antibiotics, infectious agents).

(e) To define at the molecular level the effects of cytokines on ontogeny, growth or differentiation.

(f) To define the effects of cytokines on protein modification, i.e. glycosylation or phosphorylation.

(g) To isolate sufficient quantities of highly-purified cytokine-regulated proteins that are compatible with subsequent microsequencing. This technology can be used to determine the partial amino acid sequences of such proteins which, when matched to that of known proteins in protein databases, can then be used to establish the identity of most proteins.

(h) To develop protein databases on individual cell types examined.

2. Preparation of cell lysates

2.1 Samples for analytical gels

To determine if the synthesis of proteins in the cells under study is regulated by treatment with a given cytokine, combinations of cytokines, or cytokines combined with other agents, it is usually necessary to label the proteins metabolically with radiolabelled amino acids. This permits the preparation of autoradiograms from the 2D gels which provide a sensitive method for the quantitation of the cytokine-induced effects on individual proteins. In our laboratory we routinely label proteins with 15 ^{14}C-labelled amino acids. This allows for a more uniform incorporation of radioactivity into all *de novo* synthesized proteins as compared to labelling proteins with ^{35}S-labelled amino acids. Furthermore, lysates can be stored for an indefinite period of time prior to the running of gels.

Protocol 1. Preparation of cell lysates for analytical gels

Equipment and reagents
- Tissue culture medium +10% (v/v) serum +10 µg/ml gentamicin sulfate
- Labelling medium: medium (<2% usual contents of amino acids) + 10% dialysed FCS + 10 µg/ml gentamicin sulfate
- ^{14}C-labelled amino acid mixture[a] (>1.85 GBq, CFB.25, Amersham, Illinois, USA)
- Cytokines
- PBS + 0.1 g/litre CaCl$_2$ + 0.1 g/litre MgCl$_2$
- Sample buffer 1[b]: 0.3% (w/v) SDS, 200 mM DTT, 28 mM tris-HCl, 22 mM Tris-base
- Sample buffer 2[c]: 476 mM Tris–HCl, 24 mM Tris-base 50 mM MgCl$_2$, 1 mg/ml DNAase I, (DPFF Corp., Worthington Biochemical Corp., New Jersey, USA), >0.25 mg/ml RNAase A (RASE, Worthington Biochemical)
- 10% trichloroacetic acid (TCA)
- Tissue culture trays (12-well, 22 mm in diameter)
- Pipette tips
- 2 ml screw-cap microcentrifuge tubes
- Wet ice

- Dry ice
- Glass filters
- Scintillation cocktail and vials
- Lyophilizer
- Microcentrifuge
- Vortex

- Waterbath (65°C)
- Freezer (−70°C)
- Vacuum device for TCA precipitation of proteins
- β-scintillation counter

Method

1. Grow cells in usual tissue culture medium in 12-well tissue culture trays until about 80–90% confluent. Remove medium and replace with 0.8–1.0 ml medium containing labelled amino acids, cytokines and/or other agents. Omit cytokines from control wells. Set up two wells per treatment or control.

2. Incubate for desired amount of time (usually 6–48 h) at 37°C. Following treatment, chill trays on wet ice. Proceed with lysis of cells, keeping trays on ice.

3. Suction-off medium and wash cells in wells rapidly three times with 1.0 ml cold PBS.

4. Add 100 μl hot sample buffer 1 (heated to 65°C in a waterbath) to cells in well and tilt tray to spread buffer over entire well. Wait 1 min, then add 10 μl sample buffer 2.

5. Scrape cells off bottom of well with pipette tip and pipette solution up and down several times. Solution should change from very viscous to fluid.

6. Transfer lysate to a 2 ml screw-cap microcentrifuge tube. Lysates from the two wells of identical treatments can be combined in one tube.

7. Incubate the microcentrifuge tubes containing lysates for 5 min at 65°C in a waterbath, vortex samples well, then chill on ice.

8. Pellet cellular debris by centrifugation (2000 *g*, 5 min) and transfer supernatant fluid to a new microcentrifuge tube.

9. Mix samples, remove a small quantity of each and determine the amount of radioactivity incorporated into *de novo* synthesized proteins. Precipitate proteins by spotting 5 μl of lysates on glass filters that have been pre-wet with 10% TCA. After 2 min, apply vacuum and wash filters with additional 10% TCA. Allow filters to dry, place in scintillation vials with scintillation cocktail and count in β-scintillation counter.

10. Snap freeze lysates in an alcohol-dry ice bath, and store at −70°C.

[a] Dry down labelled amino acids on lyophilizer and resuspend in tissue culture labelling medium containing 10% serum. If dialysed serum is used and medium contains <2% of the cold amino acids that will compete with radiolabelled amino acids, labelling can be done with 50–100 μCi/ml. Otherwise label with 500–1000 μCi/ml.
[b] Aliquot and store at −70°C. This buffer is used to lyse cells and solubilize proteins.
[c] Mix Tris buffers, $MgCl_2$ and water, chill on ice and then add enzymes. Aliquot in 100 μl quantities and store at −70°C. This buffer digests DNA and RNA in the cell lysate samples. DNA and RNA can interfere with the electrophoretic pattern of the proteins in the 2D gels.

2.2 Samples for preparative gels

If identification of individual cytokine-regulated proteins is the ultimate goal, it is necessary to prepare larger cell lysates which contain a higher concentration of proteins, and run preparative-scale 2D gels. The quantities of the individual proteins isolated on one or more preparative gel should be sufficient for protein identification by either immunoblot or microsequence analysis.

Protocol 2. Preparation of cell lysates for preparative gels

Equipment and reagents

- Tissue culture medium + 10% (v/v) serum + 10 μg/ml gentamicin sulfate
- Cytokines
- PBS + 0.1 g/litre CaCl$_2$ + 0.1 g/litre MgCl$_2$
- Sample buffer 1: see *Protocol 1*
- Sample buffer 2: see *Protocol 1*
- Sample buffer 3[a]: 9 M urea, 4% (w/v) *n*-octyl glucoside, 100 mM DTT, 5.5% (v/v) ampholytes (3–10 pH, 2/D optimized, Millipore, Massachusetts, USA)
- Acetone (HPLC grade)
- Bio-Rad Protein Assay Kit II, Bradford method (4), (Bio-Rad, California, USA)

- Tissue culture dishes (150 mm in diameter)
- Cell scrapers
- 40–50 ml screw-cap centrifuge tubes
- Pipettes
- Wet ice
- Dry ice
- Vortex
- Waterbath (65°C)
- Freezer (−20°C)
- High-speed centrifuge
- Vacuum source

Method

1. Grow cells in 10 or more tissue culture dishes until about 80–90% confluent. Remove medium and replace with 30–40 ml medium containing cytokines and/or other agents.

2. Incubate for desired amount of time (6–48 h) at 37°C. Following treatment, chill dishes on ice. Then proceed with lysis of cells, keeping dishes on ice throughout entire procedure.

3. Suction-off medium and wash cells rapidly three times with 10 ml cold PBS per dish.

4. Add 0.8–2.0 ml hot (65°C) sample buffer 1 to cells in each dish and swirl to spread buffer over entire dish. Wait 1 min, then add one-tenth the volume sample buffer 2 (80–200 μl).

5. Scrape cells off bottom of well with a cell scraper. Solution should become very fluid.

6. Transfer lysates to 50 ml screw-cap centrifuge tubes, combining lysates of identical treatments.

7. Incubate lysates for 15 min at 65°C in a waterbath, vortex samples well, cool to room temperature and add 4 volumes of acetone. Mix well.

---

8. Precipitate proteins by storing lysates in acetone overnight in freezer at −20°C.

9. Pellet proteins by centrifugation (2600 *g*, 15 min), carefully remove the supernatants and resuspend protein pellets to one-fourth the volume of the original lysate in sample buffer 3. Note: Do not let protein-acetone pellets dry out. To resuspend protein pellets, vortex pellets in residual acetone and add sample buffer 3 a little at a time with mixing.

10. Determine the protein content of concentrated lysates using a modification of the method of Bradford (4), in which 5 μl of each lysate is added to 800 μl of water and 200 μl of Bio-Rad protein assay dye reagent. Add 5 μl of sample buffer 3 to 800 μl of appropriately diluted protein standards and 200 μl of the dye when preparing the standard curve. Add 5 μl of 0.1 N HCl to all samples and standards to neutralize the alkaline pH of urea present in sample buffer 3. Continue protein assay protocol as recommended by manufacturer.

11. Snap freeze lysates in an alcohol-dry ice bath, and store lysates at −70°C.

---

[a] Mix urea, 20 ml water, *n*-octyl glucoside, DTT and heat to 37°C to dissolve urea. Add ampholytes, bring volume to 50 ml with water and filter through a 0.45 μm filter. Aliquot in 0.5–1.0 ml quantities and store at −70°C. The presence of urea prevents the proteins from precipitating in aggregates.

3. Isolation of proteins from cell lysates by 2D PAGE

3.1 Resolving proteins on 2D analytical gels

For many years, our laboratory prepared 2D gels by the original procedure of O'Farrell (1). Subsequently, we have used larger format gels, developed in the laboratory of Dr James Garrels (2) and adapted for small individual research laboratories by several companies specializing in biotechnology instrumentation. These gels offer distinct advantages, as the electrophoretic patterns of proteins are highly reproducible and hundreds of proteins can be resolved in a given gel loaded with a whole cell lysate. At the invitation of Millipore Corporation, we tested their Investigator™ 2D Electrophoresis System for running large 22 × 22 cm format 2D gels and adapted it for use in our laboratory to run both analytical and preparative gels.

Protocol 3. Analytical 2D PAGE of radiolabelled whole cell lysates according to the method of Patton *et al.* (5) with minor modifications

Equipment and reagents

- Radiolabelled cell lysates (see *Protocol 1*)
- First-dimension IEF gel: 8.5 M urea, 2% (w/v) *n*-octyl glucoside, 4.5% (v/v) acrylamide/bis (30:0.65, Duracryl™[a], Millipore), 2.75% (v/v) ampholytes, 0.067% (w/v) ammonium persulfate (APS)
- IEF cathode buffer: 100 mM NaOH
- IEF anode buffer: 100 mM H_3PO_4
- IEF sample overlay buffer: 0.5 M urea, 0.2% (w/v) *n*-octyl glucoside, 0.1% (v/v) ampholytes, 50 mM DTT
- IEF gel equilibration buffer: 300 mM Trisbase, 75 mM Tris–HCl, 0.3% (w/v) SDS, 50 mM DTT, 0.02% (w/v) Bromophenol blue
- Second-dimension slab gels: 7.5–15% (v/v) acrylamide/bis (30:0.65, Duracryl™, Millipore), 375 mM Tris (Trizma[b] Pre-set crystals, pH 8.8, Sigma Chemical Co., Missouri, USA), 0.1% (w/v) SDS, 0.025% (w/v) APS, 0.05% (v/v) TEMED
- Second-dimension running buffer: 25 mM Tris (Trizma, Pre-set Crystals, pH 8.8, Sigma Chemical Co.), 192 mM glycine, 0.1% (w/v) SDS

- IEF standards (carbamylated creatine kinase markers, BDH pH calibration kit for IEF, Hoefer Scientific Instruments, California, USA)
- Rainbow™ Protein Molecular Weight Markers, ^{14}C-labelled, high mixture (Amersham Corp.)
- Equipment for running large format 2D PAGE gels (Investigator™ 2D Electrophoresis System, Millipore, or equivalent)
- Glycerin (Fisher Scientific, New Jersey, USA)
- Film
- Filter paper
- Pipettes
- Laboratory glassware
- Gel dryers
- Film cassettes
- Film processor
- System for computer analysis of gels (optional)
- [14]C Unique Standards[c] for autoradiography (ARC-146 (PL), American Radiolabeled Chemicals, Inc., Missouri, USA)

Method

1. Prepare first dimension IEF gel solution and cast IEF tube gels approximately 1 mm in diameter and 18 cm long. Stand tubes up in casting vial containing the IEF gel solution and slowly layer water on top of gel solution allowing tubes to fill with gel solution by capillary action. Allow at least 60 min for gels to polymerize.

2. Prepare the second dimension SDS slab gel solution and cast gels approximately 1 mm thick, 22 cm long, and 22 cm wide using slab gel casting chamber. Use a concentration of acrylamide in the separating gel that optimally resolves the proteins of interest (usually 10–12.5% for whole cell lysates) and cast the separating gels a day before using to allow for complete cross-linking of acrylamide monomers.

3. Remove IEF tube gels and load into IEF electrophoresis chamber. Debubble tube gels and layer 10 μl IEF sample overlay buffer on top of gels.

4. Prefocus gels for 1–2 h to a maximum of 1000 V while applying a maximum current of 110 mA/gel.

5. Dilute radiolabelled lysates to 10 000–90 000 d.p.m./μl in sample

116

buffer 3 (see *Protocol 2*), one volume of lysate to five or more volumes of buffer. Load 400 000–9 000 000 d.p.m. of the desired lysates on top of first-dimension tube gels in a total volume of 10–30 µl.

6. Focus gels for 17 h at 1000 V and then for an additional 30 min at 1500 V.

7. Assemble SDS slab gels in second-dimension running tank.

8. Chill IEF gels in tubes 5 min on ice to facilitate handling. Extrude IEF gels into equilibration buffer. Let gels sit in buffer for 1–2 min, then load IEF gels on top of SDS slab gels.

9. Electrophorese slab gels for 5–6 h at 14 W/gel. **Note:** When running gels at this high power, significant heat can build up in the upper buffer chamber. Start run with chilled buffer (4°C) and replace warm buffer with fresh chilled buffer when temperature increases to 27°C. Alternatively, after proteins have entered the stacking gels, run gels 12–24 h at lower power (500 mW/gel) and then 1–2 h at 16 W/gel at the end of the run to enhance protein resolution. Terminate run when dye front is 1–2 cm from the bottom of the gels.

10. Remove gels from plates and rinse gels briefly (1 or 2 min) in 2% (v/v) glycerin in water. Place gels on filter paper and dry gels on a gel dryer.

11. Place a [14]C Unique Standard along side the gels with film in the film cassette. Expose to film for desired period of time (usually 5–60 days).

12. Optional: Run IEF standards along with the protein lysates in the first dimension tube gels. After running both dimensions and silver stain-ing the gels (5), match specific proteins to the marker spots, identify pH gradient and assign pI values to proteins across the gel. Load molecular weight markers in wells on top of the SDS slab gels along the sides of the IEF tube gels and run the second dimension gels as in step 9 above. Match specific proteins to markers, identify the gradient of relative molecular masses (M_r) and assign M_r values to proteins throughout the gel.

13. As an alternative to step 12 above, assign pI and M_r values by computer analysis of autoradiograms of gels using values of known marker proteins.

---

[a] This preparation of acrylamide/bis produces gels with increased rigidity and makes for easier handling.
[b] The resolution and electrophoretic pattern of proteins in gels are more consistent when pre-pH Trizma crystals are used.
[c] These reusable standards are used to quantitate the amount of radioactivity incorporated into the individual proteins and allow for the quantitative analysis of proteins in computer-matched gels analysed by various software.

3.2 Resolving proteins on preparative 2D gels

Preparative 2D gels are run to isolate relatively large amounts of purified cytokine-regulated proteins from whole cell lysates for subsequent protein identification by either immunoblot or microsequence protocols. The methods are similar to those described for analytical gels above and modified by our laboratory to allow for a greater protein load.

Protocol 4. Preparative 2D PAGE of cell lysates by the method of Millipore with modifications (6)

Equipment and reagents

- Lysates for preparative gels (see *Protocol 2*)
- Radiolabelled cell lysates (see *Protocol 1*)
- First dimension IEF gels: 8.5 M urea, 2% (w/v) *n*-octyl glucoside, 4.5% (v/v) acrylamide/bis (30:0.65, Duracryl™, Millipore), 5.5% (v/v) ampholytes, 0.04% (w/v) APS, 0.12% (v/v) TEMED
- IEF cathode buffer: (see *Protocol 3*)
- IEF anode buffer: (see *Protocol 3*)
- IEF gel equilibration buffer: (see *Protocol 3*)
- Second-dimension slab gels: (see *Protocol 3*)
- Second-dimension stacking gels (pH 6.8): 4–7.5% (v/v) acrylamide/bis (30:0.65, Duracryl™, Millipore), 122 mM Tris–HCl, 4 mM Tris-base, 0.1% (w/v) SDS, 0.05% (w/v) APS, 0.1% (v/v) TEMED
- Second-dimension running buffer: 25 mM Tris (Trizma, Pre-set crystals, pH 8.8, Sigma Chemical Co.), 192 mM glycine, 0.1% (w/v) SDS, 0.1 mM sodium thioglycolate[a]
- CAPS transfer buffer (pH 11): 0.005% (w/v)

- SDS, 10 mM CAPS (3-[cyclohexylamino]-1-propanesulfonic acid, Sigma Chemical Co.)
- Coomassie blue G-250 Stain[b]: 17% (w/v) $(NH_4)_2SO_4$ 34% (v/v) methanol, 0.5% (v/v) acetic acid, 0.1% (w/v) Coomassie blue G-250 (Serva, New York, USA)
- Equipment for running large format 2D PAGE preparative gels (Investigator™ 2D Electrophoresis system, Millipore, or equivalent)
- Equipment for transferring large format 2D PAGE gels to PVDF membranes (Transphor® II Electrophoresis Unit, Hoefer Scientific Instruments, or equivalent)
- Orbital platform shaker
- PVDF vinyl membranes (Immobilon-P, Millipore)
- Filter paper
- Pipettes
- Large flat dishes for staining gels
- Laboratory glassware

Method

1. Prepare IEF gel solution and cast 3 mm thick and 18 cm long IEF tube gels as in *Protocol 3*, step 1. Use Millipore 3–10 pH/2D optimized ampholytes. The pH gradient generated in preparative gels using this preparation of ampholytes is more continuous and consistent when compared to gels cast with other ampholyte preparations.

2. Cast 1 mm thick, 22 cm wide and 19 cm long SDS slab gels as in *Protocol 3*, step 2, allowing for a 3 cm stacking gel. The day the gels are run, cast stacking gels consisting of one-half the concentration of acrylamide in the separating gel.

3. Load tube gels into IEF chamber, overlay each with 50 µl sample overlay buffer and pre-focus IEF gels to 750 V over a period of 4–5 h while applying a maximum current of 600 mA/gel.

4. Dilute lysates in sample buffer 3 and load each IEF gel with 300–1200 µg

---

protein in a volume of 100–300 μl. To isolate proteins for subsequent microsequencing, load the maximum amount of protein on gels that gives good spot resolution (800–1200 μg). Load 300–800 μg protein on gels when the identification of proteins by 2D immunoblot analysis is intended. To facilitate the identification of individual proteins on immunoblots, load gels with a mixture of radiolabelled and cold lysate (see *Protocol 4*, step 5).

5. Focus IEF gels for 17.5 h at 1000 V and then for 30 min at 1500 V.

6. Remove tubes from chamber and chill on ice for 15 min. Extrude IEF gels into 25 ml IEF equilibration buffer. Equilibrate twice in buffer for 15 min at slow speed on an orbital shaker, using a separate tube for each gel.

7. Place equilibrated IEF gels in alignment guides on top of slab gels and electrophorese as indicated in *Protocol 3*, step 10.

8. Carefully remove gels from between plates. If proteins are to be excised for subsequent microsequence analysis, stain preparative gels in Coomassie blue G-250 stain 18–24 h on an orbital shaker, one gel per dish, as gels are fragile. Destain in water, changing water often (20 min apart). Wrap gels in plastic wrap and store at 4°C prior to excising proteins of interest for microsequence analysis.

9. Blot proteins in gels to PVDF membranes (Immobilon-P) at 1.3 mA/cm^2 for 3 h in CAPS transfer buffer if proteins are to be identified by immunoblot analysis. Prepare autoradiograms of the blot. After the blot is probed with antibody, overlay the film on top of the blot to locate the protein(s) which are detected by specific antibody staining.

[a] Sodium thioglycolate is added to scavenge free radicals or oxidants trapped in the gel matrix which can modify proteins and complicate sequence analysis.
[b] The colloidal nature of this stain provides high protein-specific staining and low background-staining.

4. Identification of proteins by immunoblot analysis

2D PAGE immunoblot analysis is useful for confirming or establishing the identity of cellular proteins and may provide additional information concerning their regulation by cytokines. Several examples in which 2D PAGE immunoblot analysis can be useful are as follows:

- To confirm the identity of a protein established previously by microsequencing and protein data base homology searches.

- To locate the exact location of a tentatively identified protein in 2D gels, based on similarities (i.e. pI/M_r, regulation by cytokines) to a protein from another cell type isolated by 2D PAGE and previously identified by microsequencing techniques.

- To determine the effects of cytokine treatment on the amount of a known protein. 2D immunoblots have an advantage over 1D immunoblots when the protein of interest has a M_r similar to another cellular protein of great abundance (e.g. actin or tubulin).

- To detect multiple forms of a given protein (e.g. phosphorylated forms), which may possibly result from cytokine treatment.

Protocol 5. Identification of proteins by immunoblot analysis by the method of Lopez *et al.* (7) with minor modifications

Equipment and reagents

- PVDF blot of 2D gel (see *Protocol 4*)
- Blocking buffer: PBS, 5% (w/v) non-fat dry milk, 0.5% (v/v) fish gelatin 0.1% (v/v) Tween-20, 0.04% (w/v) sodium azide
- Wash buffer: PBS, 0.5% (w/v) non-dry milk, 0.5% (v/v) fish gelatin, 0.1% (v/v) Tween-20, 0.04% (w/v) sodium azide
- Primary antibodies: antiserum or monoclonal antibody (IgG) raised in either rabbits, goats, or mice

- Secondary antibody[a] conjugated to colloidal gold. AuroProbe BLplus (Amersham Corp.)
- Silver enhancement reagent: IntenSE™ BL Silver enhancement kit (Amersham Corp.)
- Methanol, double-distilled water
- Pipettes
- Large flat dishes (size of blots)
- Heat-sealable pouches
- Orbital platform shaker
- Pouch heat-sealer

Method

1. Pre-wet dried PVDF membranes in methanol for 5 sec. Rinse in water.

2. Place blot in a large flat dish, cover with blocking buffer, place on orbital shaker and incubate 18 h (or longer) at room temperature.

3. Wash blot twice for 60 min at room temperature in wash buffer on shaker in a large dish.

4. Place blot in sealing pouch, add primary antibody diluted in 30–40 ml wash buffer, seal pouch, place on orbital shaker and incubate 18 h (or longer) at room temperature.

5. Remove blot from pouch and wash blot twice as indicated above (step 3).

6. Place blot in new sealing pouch, add secondary antibody gold-conjugate diluted 1:200 in 30–40 ml wash buffer, seal pouch, place on orbital shaker and incubate 18–24 h at room temperature. Proteins staining with specific antibody usually appear as faint red spots.

7. Remove blot from pouch, place in flat dish and wash blot twice for 1 min in water.

8. Enhance the signal on blot with IntenSE™ BL following manufacturer's instruction.

---

[a] The secondary antibody conjugate must be directed toward the species of the animal or cells that produced the primary antibody.

5. Microsequencing of proteins isolated by preparative 2D PAGE

Some proteins isolated by preparative 2D PAGE can be microsequenced by automated Edman degradation following electroblotting to PVDF membranes. Previously, we have been successful in sequencing several cytokine-regulated proteins from blots of preparative 2D gels by this method (8, 9). However, the majority of mammalian proteins are modified at their N-termini and therefore are not readily sequenceable by automated Edman degradation (10). This problem can be overcome by either chemical or protease digestion of the intact molecule, separation of individual peptides and subsequent microsequencing of internal peptides. Methods that cleave proteins into several large peptides are often preferred because they permit SDS PAGE separation of the individual peptides. Although our laboratory currently sequences cellular proteins purified by preparative 2D PAGE using both Edman degradation and mass spectrometry, we will limit our protocols to those used with automated Edman microsequencing, as these technologies are more widely available to most individual scientists and require less technical expertise to analyse the data and deduce a partial amino acid sequence. Further discussion of the mass spectrometric approach to sequencing from preparative 2D gels is detailed in previous publications (6, 11).

Protocol 6. In-gel CNBr[a] cleavage of 2D PAGE isolated proteins and peptide separation by the method of Schagger and von Jagow (12)[b]

Equipment and reagents

- 2D PAGE gel pieces containing 50–100 pmol protein of interest (see *Protocol 4*). Note: protein can be obtained from multiple 2D gels.
- 2D PAGE gel pieces that do not contain any protein
- CNBr cocktail: 70% (v/v) formic acid, 50 mg/ml CNBr, 0.01% (w/v) SDS
- 1 M Tris (pH 8.0)
- Equilibration buffer: 0.5 M Tris (pH 9.0), 10% (v/v) glycerol, 10 mM DTT, 0.001% Bromophenol blue
- CAPS transfer buffer (pH 11): (see *Protocol 4*)

- Coomassie blue R250 Stain: 40% (v/v) methanol 1% (v/v) acetic acid, 0.1% (w/v) Coomassie blue R250 (Boehringer Mannheim Corp.)
- Destain: 50% (v/v) methanol
- Equipment for running vertical SDS slab gels (Sturdier® SE 410, Hoefer Scientific Instruments, or equivalent)
- 1.5 mm thick Tris-tricine-SDS gel: 16 cm separating gel (16.5% T/6% C), 3–4 cm spacer gel (10% T/3% C), 4–5 cm stacking gel (4% T/3% C)

Protocol 6. *Continued*

- Tris-tricine SDS PAGE running buffers: Anode (lower chamber)–200 mM Tris (pH 8.9); Cathode (upper chamber)–100 mM Tris (pH 8.25), 100 mM tricine, 0.1% (w/v) SDS
- Protein molecular weight standards, low range (GibcoBRL, New York, USA, or equivalent)
- Equipment for transferring large PAGE gels to PVDF membranes (Transphor® II Electrophoresis Unit, Hoefer Scientific Instruments, or equivalent)
- Scalpels
- Argon gas
- Pipette tips
- Screw-cap microcentrifuge tubes (air tight)
- Aluminium foil
- PVDF membranes (Immobilon-PSQ, Millipore)
- Large flat dishes for staining blots
- Speed-vac
- Shaking waterbath or nutator in incubator

A. *In-gel CNBr digestion of proteins*

1. Excise gel pieces containing the protein(s) of interest from gels with a scalpel and place gel pieces in a screw-cap microcentrifuge tube, one tube for each different protein.

2. Place tubes in speed-vac and reduce the size of the gel pieces to approximately half of their original volume (5–10 min). Do not allow gel pieces to dry completely, as protein recovery will be greatly reduced.

3. Add CNBr cocktail to partially dehydrated gel pieces in tubes, 200–300 µl or enough to cover and swell gel pieces to their original volume.

4. Gas tubes with an inert gas (e.g. argon), tighten caps of tubes, wrap tubes in foil, place tubes in water bath or on nutator and incubate 18–24 h at 37°C with constant agitation.

5. Place tubes in speed-vac and remove CNBr and formic acid from gel pieces under vacuum (5–10 min). Do not let gel plugs dry out.

6. Wash gel pieces three times for 5 min with 0.5 ml Tris buffer.

B. *Separation of peptides by Tris-tricine-SDS PAGE*

1. Cast a 1.5 mm thick discontinuous Tris-tricine-SDS gel consisting of a separating gel, spacer gel and stacking gel according to the method of Schagger and von Jagow (13). Use a wide-tooth comb when casting the stacking gel to create wells wide enough to accommodate gel pieces (approximately 8 mm wide).

2. Equilibrate the washed gel pieces after CNBr cleavage 5 min in equilibration buffer. Equilibrate the extra gel pieces (protein-free) as well.

3. Place equilibrated gel pieces in the wells of the stacking gel. Add extra gel pieces (protein-free) to any additional wells including those containing molecular weight standards. This will minimize distortions in the electrophoretic gradient.

4. Separate peptides by electrophoresis at 30 V for 1 h, and then at 120 V for 18–24 h or until peptides are resolved and dye front reaches the bottom of the gel.

5. Blot gel to PVDF membrane at 1.3 mA/cm^2 for 2.5 h in CAPS transfer buffer.

6. Wash blot in double-distilled water 15 min, then stain 45–60 sec in Coomassie blue R250 stain, and destain until peptide bands are resolved.

7. Excise peptide bands and load into an automated Edman microsequencer.

[a] CNBr cleaves proteins at the C-termini of methionine residues.
[b] This method permits the resolution of peptides approximately 1–100 kDa.

6. Analysis of the gels

6.1 Visual approach to analysis

The first phase in analysis of 2D gels is to determine the proteins of interest by visually comparing the autoradiograms of control and cytokine-treated gels and recording those proteins that are induced or suppressed. Although time intensive, this approach focuses the initial analysis, allows the investigator to obtain qualitative data and to become familiar with the protein pattern of the gels.

6.2 Computer based analysis

Dr Jim Garrels (13) developed some of the original software for computer-based analysis of 2D gels. In an earlier edition of this chapter we described our experience with the PDQUEST system employed by our laboratory at the time which permitted spot detection, quantitation of protein spots on autoradiograms, gel matching, data output and limited database management. There has been a dramatic improvement in the level, speed and ease at which gels can be computer analysed and in particular, the capacity to build and access information from protein databases. Recently, our research interests have concentrated more on the identification of key cytokine-regulated proteins isolated by 2D PAGE using immunoblot and microsequence analysis. Therefore, we refer the reader to a recent review by Patton (14) that describes the biological and gel considerations for good analytical image analysis, as well as recent state-of-the-art advances in the field.

7. Analytical 2D PAGE applications

7.1 Analytical gels for detecting alterations in gene/ protein expression

On autoradiograms of our large analytical 2D gels we routinely detect 500–1500 protein spots. These spots range in abundance from >10000 p.p.m.

(1%) to <20 ppm of total labelled protein. A protein that is present at 20 p.p.m. (20 000 molecules/cell) is a rare protein, but many important regulatory proteins, such as transcription factors and cytokine receptors, may exist at even lower levels—(500 copies/cell), and are therefore difficult to detect on 2D gels of total cell lysates. This approach can be compared with that of screening cDNA libraries by differential hybridization with cDNA probes from control or treated cells, where it has been possible (in practice) to identify induced or suppressed cDNAs corresponding to mRNAs of ~0.01% (100 p.p.m.), or greater. By contrast, we have been able to detect on 2D gels induced or suppressed proteins at 10–50 p.p.m. (15, Harris and Epstein, unpublished data).

7.2 Work from our laboratory employing analytical 2D gels

The use of 2D gel technology has permitted our laboratory to detect those proteins induced by interferon alpha (IFN-α) and gamma (IFN-γ) in human fibroblasts (16–18) and ovarian carcinoma cells (19). Furthermore, it has facilitated our definition of the molecular weight, isoelectric point and extent of induction of proteins induced in fibroblasts, ACHN human renal cell carcinoma, A549 lung carcinoma and A375 melanoma cells by one or more of the following cytokines: IFN-α, IFN-γ, interleukin-1 (IL-1), and tumour necrosis factor (TNF) (15, 20, 21 and Chen, Smith and Epstein, unpublished data). We have defined those proteins synergistically induced in ME-180 cervical carcinoma cells (21, 22) and A375 melanoma cells (20, 21) by TNF and IFN-γ. These studies revealed both differences and similarities in cytokine action on a given cell type, indicating that diverse molecular mechanisms are associated with the synergistic antiproliferative effects of TNF and IFN-γ. In other studies (23), we combined specific agonist and blocking antibodies to either 55 kDa or 75 kDa TNF receptors with 2D gel analysis to determine which receptor is responsible for mediating the induction of individual melanoma proteins. We found that TNF enhanced synthesis of plasminogen activator-2, manganese superoxide dismutase and protein 28/5.6 is selectively triggered by the 55 kDa TNF-receptor.

8. Preparative 2D PAGE applications

8.1 Preparative gels for detecting alterations in protein expression

Immunoblots of 2D gels are particularly useful for detecting structurally-related proteins or multiple forms of the same protein (3). For example, phosphorylated species can often be detected as satellite protein spots with pIs that are slightly more acidic than non-phosphorylated forms. Specific antibodies can be used which recognize proteins that are phosphorylated on certain amino acid residues such as tyrosine, to evaluate protein function. Carbamylated species usually run as a series of protein spots with increasing p*I*s. Glycosylated species often have slightly higher molecular masses than their parent molecules. The detection of these modifications can give valuable information concerning the nature of various cellular proteins, how they mediate cellular function and how they may be altered by cytokine treatment.

8.2 Work from our laboratory employing preparative 2D gels and microsequencing techniques

We have used preparative 2D PAGE for the isolation, purification and initial characterization of cellular proteins prior to microsequencing. Our microsequencing studies employing automated Edman degradation techniques are being conducted with the assistance of Dr Ralph Reid at the UCSF Biomolecular Research Centre. From blots of 2D preparative gels of human fibroblast lysates we obtained an N-terminal partial amino acid sequence for two cytokine-regulated proteins: manganese superoxide dismutase, a protein that is induced by TNF and IFN-γ (8), and leucine amino peptidase (9), a protein that is induced by IFN-γ. Because most mammalian proteins contain N-terminal modifications that limit Edman sequencing of intact proteins (10), we have developed the strategy for identification of proteins based on internal peptide sequences described above in *Protocol 6*. In one application (24), a 28 kDa, 6.2 p*I* melanoma protein was isolated from multiple 2D preparative gels, concentrated into one gel plug by additional electrophoresis and subjected to *in situ* CNBr cleavage. The resulting peptides were separated by 1D PAGE and blotted to a PVDF membrane prior to sequencing by automated Edman degradation. Sequence was obtained for the first 19 amino acids of an 18 kDa peptide fragment and matched that of amino acids 76–94 of the human high-mobility-group-1 protein (HMG-1), a chromosome-associated protein that preferentially binds single-stranded DNA and regulates DNA topology, thereby facilitating replication, transcription, repair and structural packaging of DNA (25). CNBr digestion and electrophoresis of an adjacent protein isolated from 2D gels of similar molecular mass but greater p*I* (28 kDa, 6.4 p*I*) produced an electrophoretic peptide map identical to that of the 28 kDa protein with sequence homology to HMG-1, indicating that these two proteins are structurally related. 2D immunoblot analysis confirmed the identity of these two proteins as members of the human HMG-1, 2 chromosomal protein family and showed that the relative amounts of these proteins are upregulated by IFN-γ and TNF in combination. The ability of cytokines, such as TNF and IFN-γ to induce HMG-1 suggests a biochemical pathway by which these cytokines may play an indirect role in regulating the expression of cellular genes.

Our microsequencing studies employing mass spectrometry are being done in collaboration with Drs A. Burlingame, S. Hall and Mr K. Clauser of the

UCSF Facility for Mass Spectrometry and Dr F. Masiarz of Chiron Corp. Although it is beyond the scope of this chapter to describe in detail the technology and our studies to date, we would like to stress the importance of preparative 2D PAGE in providing relatively large amounts of highly purified cellular proteins for this purpose. In one study (6) we combined preparative 2D PAGE, tandem mass spectrometry and Edman degradation for the isolation and identification, by partial sequencing, of 34 kDa, 6.4 p*I* human melanoma protein, lipocortin I. The protein was isolated from whole cell lysates on several 2D preparative gels. Tryptic peptides were separated by HPLC and fractions containing individual peptides were then subjected to liquid secondary ion mass spectrometry (LSIMS). Several HPLC peptide fractions contained molecular ions that matched the predicted mono-isotopic ion masses corresponding to protonated lipocortin I tryptic peptides. Mass spectrometric high-energy collision-induced dissociation (CID) analysis revealed the sequence and acetylation of the N-terminal tryptic peptide and an acrylamide-modified cysteine in another tryptic peptide. In other studies (11) we have employed matrix-assisted laser ionization desorption time-of-flight mass spectrometry to obtain molecular masses of tryptic peptides and tentative identification of several proteins isolated by preparative 2D PAGE. Thus, we can obtain knowledge concerning both the primary structure and covalent modifications of proteins isolated from 2D gels directly using mass spectrometry which is applicable to a broad range of biological problems, including the identification of cytokine-modified proteins and the nature of such modifications.

9. Concluding remarks

The power of 2D gels for analysing at the molecular level the effects of cytokines or other agents on cell physiology is enormous. No other technique can provide as much global information about the effects on gene/protein expression in a single experiment. However, 2D PAGE technology is time and labour intensive. The running of the gels and their analysis requires dexterity and patience, as the amount of data generated in each experiment is massive. The availability of pre-cast gels and systems for running multiple gels has reduced some of the time and effort in running the gels. Recent developments in laser scanning and manufacture of computers with increased speed and memory have alleviated some of the time spent on analysis.

Preparative scale 2D gel electrophoresis is one of the most powerful, highest resolution biochemical techniques available for the purification of proteins prior to identification by amino acid sequencing. Furthermore, preparative 2D PAGE permits the simultaneous purification of many proteins of diverse structure and function.

Acknowledgements

The work from our laboratory presented in this chapter was supported by NIH grants CA27903, CA44446, and AG08938.

References

1. O'Farrell, P. H. (1975). *J. Biol. Chem.*, **250**, 4007.
2. Garrels, J. (1979). *J. Biol. Chem.*, **254**, 7961.
3. Celis, J. E. and Bravo, E. (ed.) (1984). *Two-dimensional gel electrophoresis of proteins*. Academic Press, New York.
4. Bradford, M. M. (1976). *Anal. Biochem.*, **72**, 248.
5. Patton, W. F., Pluskal, M. G., Skea, W. M. *et al.* (1990). *Biotechniques*, **8**, 518.
6. Hall, S. C., Smith, D. M., Masiarz, F. R. *et al.* (1993). *Proc. Natl. Acad. Sci. USA*, **90**, 1927.
7. Lopez, M. F. and Patton, W. F. (1991). In *2D PAGE'91* (ed. M. J. Dunn), pp. 313–77. Department of Cardiothoracic Surgery, National Heart and Lung Institute (UK), London.
8. Harris, C. A., Derbin, K. S., Hunte-McDonough, B. *et al.* (1991). *J. Immunol.*, **147**, 149.
9. Harris, C. A., Hunte, B., Krauss, M. R., *et al.* (1992). *J. Biol. Chem.*, **267**, 6865.
10. Driessen, H. P. C., de Jong, W. W., Tesser, G. I., *et al.* (1984). *CRC Crit. Rev. Biochem.*, **18**, 281.
11. Clauser, K. R., Hall, S. C., Smith, D. M., Webb, W. W., Andrews, L. E., Tran, H. M., Epstein, L. B. and, Burlingame, A. L. (1995). *Proc. Natl. Acad. Sci. USA*, **92**, in press.
12. Schagger, H. and von Jagow, G. (1987). *Anal. Biochem.*, **166**, 368.
13. Garrels, J. I., Farrar, J. T., and Burwell, C. B. IV. (1984). In *Two-dimensional gel electrophoresis of proteins: methods and applications* (ed. J. E. Celis and R. Bravo), pp. 38–91. Academic Press, New York.
14. Patton, W. F. (1994). *J. Chromatography*, in press.
15. Beresini, M. H., Lempert, M. J., and Epstein, L. B. (1988). *J. Immunol.*, **140**, 485.
16. Weil, J., Epstein, C. J., and Epstein L. B. *et al.* (1983). *Nature*, **301**, 437.
17. Weil, J., Epstein, C. J., and Epstein, L. B., *et al.* (1983). *Antiviral Res.*, **3**, 303.
18. Weil, J., Epstein, C. J., and Epstein, L. B. (1983/1984). *Nat. Immun. Cell Growth Regul.*, **3**, 51.
19. Epstein, L. B., Hebert, S. J., and Lempert, M. J. (1983). In *The biology of the interferon system 1983* (ed. E. DeMaeyer and H. Schellekens), pp. 231–8. Elsevier, Amsterdam.
20. Smith, D. M. and Epstein, L. B. (1992). In *Tumor necrosis factor: structure-function relationship and clinical application* (ed. T. Osawa and B. Bonavida), pp. 173–82. Karger, Basel.
21. Smith, D. M. and Epstein, L. B. (1991). In *2D PAGE'91* (ed. M. J. Dunn), pp. 145–9. Department of Cardiothoracic Surgery, National Heart and Lung Institute (UK), London.

22. Beresini, M. H., Sugarman, B. J., Shepard, H. M. *et al.* (1990). *Electrophoresis*, **11**, 232.
23. Smith, D. M., Tran, H. M., Soo, V. W., *et al.* (1994). *J. Biol. Chem.*, **269**, 9898.
24. Smith, D. M., Tran, H. M., and Epstein, L. B. (1994). *Proc. American Assoc. for Cancer Research*, **35**, 489.
25. Bustin, M., Lehn, D. A., and Landsman, D. (1990). *Biochim. et Biophys. Acta*, **1049**, 231.

9

Antiviral activity of cytokines

JOHN A. LEWIS

1. Introduction

A variety of biological materials exhibit the ability to induce an antiviral state in cells. To date the best known antiviral agents are the interferons (IFNs), a family of proteins and glycoproteins which are synthesized by virus infected cells or by cells treated with various chemicals and are capable of inhibiting cell-growth as well as inhibiting virus replication and modulating immune responses. Certain other cytokines have been reported to show antiviral properties (1, 2) but in general these effects have not been investigated in depth and the mechanisms underlying such responses are not understood. Assays for specific cytokines are usually of two kinds: bioassays which measure the potency of a preparation are generally based on some specific response of cells to the agent of interest while enzyme-linked immuno-absorbent assay (ELISA) procedures offer a simple and usually sensitive means of measuring the mass of the substance. This chapter describes general procedures for assaying antiviral responses based on our experience with IFNs. Further procedures are described in Chapter 21 and a companion volume (3). These methods can be readily adapted for other cytokines with suspected or demonstrated antiviral activity.

1.1 Choice of an assay system

The principle of any assay for antiviral responses is to treat a selected cultured cell-line with the agent of interest, challenge the cells with an appropriate virus and estimate the degree of virus replication. Several considerations apply to choosing the assay system: the cells should be easy to grow and able to withstand manipulations; the virus to be used should be as safe as possible and of course able to infect the chosen cells; the measurement of virus infection should be as reproducible and as convenient as possible. In general the simplest approach is to use a cytopathic virus with cells that grow in monolayer culture and use a vital dye to stain surviving cells. This technique can be made quantitative by eluting the dye and measuring its optical density. For non-adherent cells a virus yield reduction assay involving a secondary assay step is used.

IFNs are active on nearly all somatic cells with the exception of mutant strains. They induce antiviral mechanisms which are active against almost all classes of virus ranging from large DNA viruses to retroviruses. Nonetheless some viruses are more susceptible than others; DNA viruses are generally more resistant to the effects of IFNs than are picornaviruses. These properties may not apply to other 'antiviral' cytokines and it may be necessary to try different cell-lines and viruses to find the appropriate combination. For example tumour necrosis factor (TNF) has been reported to be active on human Hep2 cells but not on HeLa cells (2). In our own experience Hep2 cells were not protected against vesicular stomatitis virus (VSV) by human TNF and this may reflect variations in the susceptibility of different strains of the same cell-line. We have also observed variations in responses of human WISH and other cells to human IFN-γ. IFNs exhibit varying degrees of species-specific response with human IFNs being inactive on mouse cells and vice versa. However, human IFN is active on bovine cells so species specificity is not an absolute parameter. Other cytokines may not exhibit species-specificity; for example human TNF functions on mouse cells.

Since little is known about the mechanisms by which cytokines other than IFNs exert their effects it is important to beware of artefactual responses. In particular it is probably a good idea to ensure that observed antiviral effects are not due to the production of endogenous IFN. Many cells produce low levels of interferon-β which cannot be detected by conventional assay procedures and the addition of other agents may enhance this production leading to antiviral effects in a secondary manner. In addition synergistic effects may occur between the cytokine of interest and endogenous IFN (1, 4, 5). One approach to test this would be to include specific antibodies capable of neutralizing either type I IFNs-α and -β (type II IFN-γ is only produced by certain T-lymphocytes) or the cytokine under study (see Chapter 19). A limitation here may be that extracellular antibodies will not completely prevent IFNs from interacting with their receptors in an autocrine situation. Another approach is to use highly sensitive procedures such as the polymerase chain reaction to detect and quantitate mRNAs, see Chapters 4 and 5 (4).

1.1.1 Experimental considerations

For those not accustomed to handling animal viruses it is advisable to consult a friendly virologist for tips on the safe and legal use of these agents. In different countries the use of various types of virus is regulated and it is important to ensure that you comply with local regulations. For instance, the use of VSV is restricted in the UK to a few licensed laboratories. Generally, the viruses most commonly used to measure antiviral responses are VSV, Mengo or encephalomyocarditis virus (EMCV), Semliki Forest virus (SFV) and Sindbis virus. These infectious agents pose a relatively low risk to humans athough care should always be taken to prevent accidental exposure. VSV may produce symptoms similar to the common cold and there is some evidence that EMCV may play a role in the aetiology of certain forms of diabetes. The factors determining the use of a virus are not only the risk of human infection; VSV replicates readily in many animals and limitations on its use reflect concern over potential danger to the agricultural industry. It is prudent to assume that all animal viruses are dangerous and work accordingly. In the USA the type of manipulation described here with VSV, EMCV and Semliki Forest virus can be performed under Biosafety Level 2 conditions:

- Set up a limited area for use of the viruses and restrict access of workers not involved in the experiments.
- Those using the viruses should be carefully trained and informed of the potential hazards to themselves, their colleagues and the environment.
- Protective clothing and gloves should be worn at all times.
- Vertical (not horizontal) laminar flow hoods rated for biohazardous agents (class II or better) should be used with the window closed to the recommended level; the functioning of the cabinet should be inspected regularly.
- Provide suitable disposal containers with autoclavable bags for waste material.
- All contaminated materials should be either autoclaved or treated with 20% Chlorox.
- Work surfaces should be swabbed daily and after each use with 20% Chlorox followed by water and then irradiated with ultra-violet light which inactivates most viruses.
- Use mechanical pipetting devices and never mouth pipette.
- Avoid forming an aerosol of contaminated solutions (e.g. by blowing out the last drops from a pipette).
- After completing the experiments wash hands thoroughly with soap and water.

In most countries Government agencies responsible for regulating research with hazardous materials publish guidelines for the safe use of particular viruses (e.g. HHS publication (NIH)–88–8395) (7).

2. Assaying for antiviral effects

Three procedures are described here. The first is a simple and reliable cytopathic effect assay which allows the detection of antiviral effects in a semi-quantitative way. A second procedure, which is a little more tedious, permits better quantitation of the antiviral cytokine. A third procedure which can be used for both monolayer and non-attached cells probably is capable of providing the most accurate quantitative data in the hands of an experienced worker but is also the most tedious and time-consuming.

2.1 Multi-well assay procedure

In this assay cells growing in 24-well cluster dishes are exposed to the cytokine, infected with virus and stained with a vital dye to determine the degree of protection against virus replication.

Protocol 1. Multi-well assay

Equipment and reagents

- Normal growth medium
- Cell culture
- 24-well culture dishes (Falcon 3047)
- Cytokine
- Sterile tubes (Falcon 2006)
- Pasteur pipettes
- Vortex mixer
- Suction system
- Virus[b] (2.5×10^7 p.f.u./ml)
- PBS: 80 g NaCl, 2 g KCl, 11.5 g Na_2HPO_4 (anhydrous), 0.1 g phenol red, to 1000 ml with water, adjust pH to 7.3

- Inverted microscope
- 20% Chlorox
- Methyl violet stain: 0.5% methyl violet 2B (Matheson, Coleman, and Bell), 0.9% (w/v) NaCl, 2% (w/v) formaldehyde, 50% (w/v) ethanol
- Shaking platform
- Dye elution solution: 0.5 M NaCl, 50% (w/v) ethanol
- Automated ELISA plate scanner

Method

1. Seed 3×10^5 cells in 1.0 ml growth medium in 24-well culture dishes (e.g. Falcon 3047) and grow to confluency overnight.

2. Prepare serial dilutions[a] of the cytokine in normal growth medium. Pipette 2.4 ml of growth medium into 5 sterile tubes (e.g. Falcon 2006). Prepare the highest concentration of cytokine to be tested by dilution into 3 ml of growth medium. Transfer 0.6 ml to the first of the diluent tubes, mix thoroughly by pipetting up and down or on a vortex shaker and, with a new pipette, transfer 0.6 ml to the next tube in the series. Continue until all the tubes have received cytokine. This provides a 5-fold serial dilution step with enough dilution for duplicate 1 ml amounts.

3. Remove medium from monolayers with a Pasteur pipette connected to a suction system. Be careful not to touch the monolayer or use too strong a vacuum or some cells will be lost giving a circular, clear space where the tip of the pipette was applied.

4. Add the cytokine dilutions (0.5 to 1 ml) to different wells in duplicate. As controls, add normal growth medium to 2 wells. Incubate cells for 16–24 h.

5. Remove medium and add 0.2 ml of virus[b] (2.5×10^7 p.f.u./ml) diluted in phosphate-buffered saline (PBS) to one set of experimental wells (multiplicity of infection [m.o.i.] = 10 p.f.u./cell). Add virus to a control

---

well which received only growth medium (virus control). Add 0.2 ml of PBS to the other control (cell control) and to the second set of experimental wells. Incubate for 1 h at 37°C.

6. Remove virus inoculum with a Pasteur pipette and carefully discard into a container. Replace with 1 ml of normal growth medium. Incubate cells at 37°C for 1–2 days. Check cultures in an inverted microscope for cytopathic effects.[c] When over 90% of the cells in the virus control show a cytopathic effect proceed to step 7.

7. Stain monolayers by carefully shaking the medium into a pan containing diluted Chlorox (20%) being careful not to splash the infected liquid. Rinse with PBS to remove debris.[d] Add 0.3 to 0.5 ml of methyl violet stain solution and incubate for 5–10 min on a shaking platform at room temperature. Shake the stain out into a sink and rinse with several changes of tap H_2O (ambient temperature; if the H_2O is too cold the cells tend to strip off the plastic) until the virus control wells are clear. Rinse with distilled H_2O and dry. An alternative staining procedure is given in *Protocol 2* below.

8. To quantitate the results add 1.0 ml of dye elution solution and incubate at room temperature for 2 h. Remove the solution to a tube and rinse out the well with a further 1 ml of dye elution solution and combine. Measure the optical density at 570 nm.

[a] It is a good idea to cover a 1000-fold range of concentration initially using 10-fold steps and use smaller increments in later tests.
[b] We routinely use VSV, but Sindbis or SFV can be used in its place. Mengo virus works well with mouse cell-lines. EMCV can be used with both mouse and human cells. Preparation of virus is described in *Protocol 4*. For some viruses it is advisable to add bovine serum albumin (0.1% w/v) or serum (0.1% v/v) to prevent adsorption to plasticware.
[c] The nature of the cytopathic effect varies from one cell-virus combination to another. In general, cells round up and take on an irregular, vacuolar appearance. Small blebs may appear early in infection and the cells disintegrate to leave granular debris. The time required varies from 24 h for Mengo (or EMCV) to 36–48 h for VSV.
[d] We use a plastic 'squirt' bottle but some cell-lines attach rather weakly to the dishes and a more gentle method is needed.

An alternative staining procedure (6) uses a tetrazolium salt MTT which is a substrate for mitochondrial dehydrogenase activities and thus depends on metabolic activity of live cells. This has the advantage of improved backgrounds since debris sticking to the wells often stains with methyl violet and other dyes. MTT assays are also described in Chapters 17 and 21.

An example of this assay is shown in *Figure 1*. Control wells show strong staining while the virus control wells contain little or no stain. The staining intensity of the experimental wells is decreased as the cytokine, IFNs β and γ in this case, is diluted. In the experiment shown in *Figure 1* the response is essentially all or none. Subsequent titrations using 2-fold serial dilution steps will provide a clearer assessment of the actual potency of the cytokine

units/ml	cc	100	10	1	0	cc

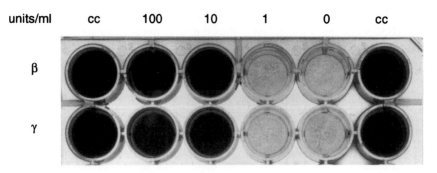

Figure 1. Example of the multi-well assay. Mouse L-929 cells were seeded in a 24-well dish as described in *Protocol 1*. Serial 10-fold dilutions of IFNs β and γ were prepared and added to the cells. After 24 h, the cells were infected with VSV and after a further 24 h stained with MTT as described in *Protocol 2*. cc = cell control.

but this is better performed using the 96-well assay described below (*Protocol 3*). A duplicate set of experimental wells which receive no virus should be included as a control for cytotoxicity of the cytokine. It is useful to define a laboratory end-point as 50% protection for comparative purposes, and international standards are being developed for some cytokines. See ref. 3 for a further description of unit definitions for interferons.

Protocol 2. An alternative staining procedure using MTT

Equipment and reagents

- As *Protocol 1* minus methyl violet stain and dye elution solution
- MTT solution: 6 mg/ml MTT (Thiazolyl Blue, Sigma cat. No. M-5655) in PBS; filter sterilized and stored at 4°C for 1–2 weeks
- 40 μM HCl in isopropanol (mix 2 μl 1 M HCl with 50 ml isopropanol)
- Automated ELISA plate scanner

Method

1. Follow *Protocol 1* to step 6.
2. Add 250 μl of MTT solution to each well. Incubate at 37°C in 5% CO_2 incubator for 3–4 h.
3. Shake out medium into 20% Chlorox and add 0.5 ml of 40 μM HCl in isopropanol.
4. Read the optical density of experimental wells at 570 nm. A background subtraction for wells containing dye alone can be made at OD_{630} if desired.

2.2 Micro-titre plate assay

This assay allows a more effective quantitation of antiviral activities but uses the same principles described above. It requires a little more work to set up.

Protocol 3. Micro-titre plate (96-well) assay

Equipment and reagents

- Normal growth medium
- Cell culture
- Virus (VSV, EMCV, or Mengo; 4×10^4 p.f.u./ml)
- Flat-bottomed 96-well dishes (Falcon 3072)
- Cytokine
- Multi-channel pipetting device
- UV light source
- 5% CO_2 incubator
- Paper towels
- Suction device
- Horizontal shaker
- 20% Chlorox
- PBS (see *Protocol 1*)
- Pasteur pipettes
- Squirt bottle
- Methyl violet stain (see *Protocol 1*)
- Dye elution solution (see *Protocol 1*)
- Automated ELISA plate scanner

Method

1. Seed 3×10^4 cells in 0.1 ml of growth medium in flat-bottomed 96-well dishes (e.g. Falcon 3072) and grow to confluency.

2. In a separate 96-well dish prepare serial dilutions of the cytokine. For a 2-fold dilution series, pipette 0.125 ml of growth medium into all but the top row (horizontal). Dilute the cytokine or test substance in growth medium to the highest concentration to be tested and pipette 0.25 ml to the top row of wells. For each sample prepare four wells in order to provide two sets of duplicate columns. Using a multi-channel pipetting device transfer 0.125 ml from the top row to the next row (i.e. A to B) and mix by pipetting up and down several times. Continue transferring down the rows of the dish. Leave the last two columns of the dish for controls and place 0.125 ml of growth medium in them.

3. If the cytokines were not sterile place the dish uncovered under an ultra-violet light source for 15–20 min and then cover and place in a 5% CO_2 incubator for 30 min to equilibrate the pH.

4. With a sharp flick of the wrist, tap the medium out of the dish containing the cells into a pan lined with paper towels (to prevent splashing). Remove any medium from the edges of the wells with a suction device. Using a multi-channel pipetting device transfer 0.1 ml of the cytokine dilutions directly to the dish containing cells, working from low concentrations to high. Incubate for 16–24 h at 37°C.

5. Add directly to duplicate experimental wells 0.025 ml of virus (VSV, EMCV or Mengo) diluted to 4×10^4 p.f.u./ml in growth medium. Leave one set of duplicate cytokine dilutions and at least two of the medium

---

Protocol 3. *Continued*

controls uninfected (cell control) to test for cytotoxicity. Mix gently by shaking the plate horizontally. Incubate for 24–48 h at 37 °C.

6. Check virus controls for cytopathic effects and when 90% or more of the cells are killed proceed to step 7.

7. Stain the monolayers by shaking the medium carefully into a pan containing 20% Chlorox. Rinse the wells with PBS from either a squirt bottle or a Pasteur pipette (2–3 drops) and shake out again. Add 2 drops (approximately 0.1 ml) of methyl violet stain solution and incubate for 5–10 min on a shaking platform at room temperature. Shake out the stain into a sink with a good sharp wrist movement and rinse several times by plunging into a dish of clean tap-water (ambient temp.). Rinse with distilled-water and dry. An alternative staining procedure using MTT is given in *Protocol 4*.

8. Quantitate by elution into 0.1 ml of dye elution solution as described in *Protocol 1* on an automated ELISA plate scanner.

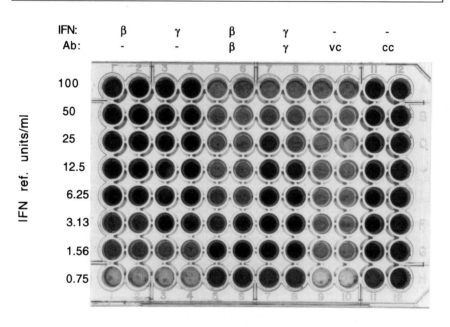

Figure 2. Example of the microtitre plate assay. Mouse L-929 cells were treated as described in *Protocol 3* with 2-fold serial dilutions of mouse IFN-β or mouse IFN-γ. As a control for specificity, neutralizing antibodies for IFNs β and γ were included in a separate set of columns. In this case the antibodies were serially diluted 2-fold into growth medium containing a constant amount (100 units/ml) of interferon β (columns 5 and 6) or γ (columns 7 and 8). After exposure to the cytokines for 24 h followed by infection with VSV the monolayers were stained with MTT (*Protocol 4*).

The protocol described above generally provides excellent quantitation since a large number of closely-spaced dilution intervals can easily be handled. It can also be used to test for specificity by including appropriate antibodies (preferably monoclonal). For this purpose prepare an extra dilution series but replace the normal growth medium diluent with an appropriate fixed concentration of antibody diluted in growth medium. In this way the antiviral agent is diluted at some point into the neutralizing range of the antibody. An example of such an assay including antibody neutralization is shown in *Figure 2*. In this case IFNs β and γ were titrated in 2-fold serial dilutions either in the absence of a neutralizing antibody (columns 1–4) or the presence of specific antibodies (columns 5–8). An end-point was reached where 50% of the cells were protected (corresponding to 1 unit/ml). In columns 5, 6 and 7, 8 each well received 100 units/ml of IFNs but a neutralizing antibody preparation was serially diluted down the columns. In the first several rows no protection was observed owing to the presence of sufficient antibody to neutralize the IFN. In subsequent rows the antibodies were diluted beyond their capacity to neutralize all the IFN. In addition to providing a useful check on the identity of the antiviral substance this provides a means to titre antibody preparations. The example shown here was stained with the MTT procedure (*Protocol 4*).

Protocol 4. Alternative staining procedure using MTT

Equipment and reagents

- As *Protocol 3* minus methyl violet stain and dye elution solution
- MTT solution (see *Protocol 2*)
- HCl in isopropanol (see *Protocol 2*)

Method

1. Follow *Protocol 3* to step 6.
2. Add 25 μl of MTT solution (*Table 1*) and incubate at 37°C for 3 h.
3. Shake out medium and add 0.15 ml of 40 μM HCl in isopropanol.
4. Read optical density at 570 nm on automated ELISA plate scanner.

2.3 Virus yield reduction assay

This procedure can be used for suspension culture cells and is described for monolayer cells elsewhere (3). Cells are infected with virus after exposure to the cytokine and the virus produced is then titred in a second assay using monolayer cells susceptible to the virus. The details vary somewhat according to the type of cells used. Stationary suspension cultures are simple to handle because small volumes can be used. Spinner cultures are more difficult because the minimum volume which can be handled may be 25–50 ml, depending on the type of apparatus available.

Protocol 5. Virus yield reduction or plaque assay

Equipment and reactions

- Normal growth medium
- 6-well dishes (Falcon 3046)
- Centrifuge (bench-top)
- PBS (see *Protocol 1*)
- Cell culture
- Serum
- Virus (VSV, EMCV, or Mengo)
- 6 cm culture dishes (Falcon 3002)
- Sterile tubes (Falcon 2005)

- Vortex mixer
- Pasteur pipette
- Suction device
- Agarose (Seakem ME, FMC Corp., Marine Colloids Division)
- 2 × MEM (powdered, GIBCO or Flow labs)
- Neutral red stain: 0.03% (w/v) Neutral red in PBS (available as a 10 × stock from GIBCO)

Method

1. Prepare a culture of the cells to be tested and divide into several small cultures. A volume of 1 ml of stationary suspension culture (at 5×10^5 cells/ml) is sufficient. For monolayer cultures we generally use 1 ml of inoculum at 3×10^5 cells/ml growing in 6-well dishes (Falcon 3046). Add cytokine to different final concentrations leaving one culture untreated. Incubate at 37°C for 16–24 h.

2. For suspension cultures, concentrate the cells by centrifugation at 1000 g for 5 min and resuspend in PBS containing 0.5% serum at 5×10^6 cells/ml (e.g. about 0.2 ml).

3. Add virus (VSV, EMCV, or Mengo) diluted in PBS to achieve a multiplicity of infection (m.o.i.) of 10 p.f.u./cell. For the suspension cells (1×10^6 cells in 0.2 ml), add 0.1 ml of virus diluted to 1×10^8 p.f.u./ml. For monolayers (3×10^6 cells) add 0.25 ml of virus diluted to 5×10^8 p.f.u./ml. Incubate at 37°C for 1 h.

4. Centrifuge the suspension cells at 1000 g for 5 min and discard the supernatant carefully as virus waste. To remove unadsorbed virus resuspend the cells in 5 ml of growth medium containing serum and centrifuge at 1000 g for 5 min. Wash monolayers twice with growth medium containing serum.

5. Resuspend the suspension cells in growth medium at 1×10^6 cells/ml or add 3 ml of growth medium to monolayers and incubate for 16–24 h.

6. Freeze the cultures at −70°C and thaw rapidly. Repeat this freeze-thaw cycle once to rupture the remaining cells and liberate intracellular or adherent virus. Centrifuge at 5000 g for 10 min at 4°C and collect the supernatant. Store at −70°C for assay.[a]

7. Plaque assays can be performed on any monolayer cell susceptible

to the virus.[b] We routinely use mouse L-929 cells which provide a well-attached, robust monolayer. Seed 6 cm culture dishes (e.g. Falcon 3002) with 1×10^6 cells in 5 ml of growth medium and grow to confluency (1–2 days). For each sample to be titred nine dishes are needed.

8. Prepare serial 10-fold dilutions of the samples in PBS. Pipette 1 ml of PBS to 2 sterile tubes (Falcon 2005) and 0.9 ml to three more. Pipette 0.01 ml of virus sample to the first tube and mix by vortexing (10^2-fold dilution). With a new pipette tip[c] transfer 0.01 ml of this dilution to the next tube (10^4-fold dilution) and mix again by vortexing. Continue the series in 10-fold steps by pipetting 0.1 ml to the next tube (0.9 ml of diluent), and so on.

9. Remove the medium from the culture dishes with a Pasteur pipette attached to a vacuum source. Add 0.25 ml of the virus dilutions in triplicate to the centre of the dishes and allow the inoculum to spread over the whole surface. Rock the dish to spread the inoculum evenly. The dilutions required depend on the efficiency of growth in the cells used in steps 1–6, the concentration of the cells and the degree of antiviral protection. Roughly speaking, 5×10^6 untreated cells will liberate 3×10^8 to 1×10^9 p.f.u. Use three dilutions bracketing the expected titre.

10. Incubate the dishes at 37°C for 1 h and occasionally (e.g. every 15 min) rock them to spread the inoculum over the monolayer.

11. Meanwhile prepare the overlay. Melt a 1% solution of agarose (0.5 g of Seakem ME from FMC Corp., Marine Colloids Division, Rockland, Me. 04841 dissolved previously in 50 ml of distilled H_2O by autoclaving) by microwaving with the cap loosened and equilibrate at 45°C. Equilibrate $2 \times$ MEM (minimal essential growth medium) at 37°C. We prepare this by dissolving the contents of 1 litre sachets of powdered medium (from GIBCO or Flow Laboratories) in 500 ml of H_2O and sterilizing by filtration.

12. Just before use, mix equal volumes of the agarose and the $2 \times$ MEM and add serum to 5% final concentration. Working quickly, remove the virus inoculum and add 7 ml of the agarose mixture from a pipette. A little practice is needed to add the overlay before it solidifies and it is advisable to handle only a few plates at a time until experienced. Allow the agarose to set at room temperature for 5 min and then place the dishes inverted in an incubator.

13. Monitor the plates after 24 h for appearance of plaques.[d] EMCV and Mengo will grow in 24–36 h while VSV takes 48 h. Stain the plates by adding 2 ml of Neutral red dye directly to the top of the

Protocol 5. *Continued*

agarose and incubate for 2 h at 37°C. Pour off the dye and count the plaques.[e]

Virus titre (p.f.u./ml) = No. of plaques $\times$ 4 $\times$ dilution factor

---

[a] Most viruses are sensitive to repeated freeze-thaw cycles so it is a good idea to store aliquots if multiple assays are likely to be performed.

[b] One should be aware of the possibility that the cytokine may induce other antiviral agents such as interferons which will be present in the released virus preparation. Usually the dilutions needed for assay are so great (>10^6-fold) that any such activity will effectively be diluted to undetectable levels.

[c] Success in obtaining reliable quantitation depends on careful manipulation to ensure good mixing and prevent carry-over of virus adhering to the walls of pipettes to lower dilutions.

[d] Hold the plate to the light and look upward at the bottom of the dish. Plaques are seen as clear, circular areas against the hazy background of cells. Spotting them takes practice!

[e] Some workers 'flip' off the agar first. This takes considerable practice. Cool the dishes to 4°C for 30 min. Insert a spatula under the agarose at one edge and, with the dish inverted, flip the agar disc out with a quick but smooth jerk into an autoclave bag.

The procedure described above is the most time-consuming of the assays described here but is effective for non-adherent cells. It also provides the most reliable quantitative data of all the procedures. However, it takes experience to perform the assay well and is expensive of time and materials.

3. Preparation of virus stocks

VSV, Mengo, and EMCV can all be grown in mouse L-929 cells. With VSV it is important to start with a low m.o.i. to avoid the production of defective interfering particles.

Protocol 6. Preparation of stock virus suspensions

Equipment and reagents

- Cell culture
- Virus
- 10 cm dishes (Falcon 3003) or 75 cm² (Falcon 3024)
- PBS (see *Protocol 1*)
- Serum (foetal or new born calf serum, according to cells used)
- Growth medium
- Centrifuge
- Capped centrifuge tubes

Method

1. Seed cells on 10 cm dishes (Falcon 3003) or in 75 cm² flasks (Falcon 3024) and grow to confluency (about 2 $\times$ 10^7 cells).

2. Dilute virus in PBS to 2 $\times$ 10^6 p.f.u./ml. Remove the culture medium and add 1.0 ml of virus dilution per dish or flask. This gives an m.o.i of 0.1 p.f.u./cell. Incubate at 37°C for 1 h.

3. Remove the virus inoculum and add 5 ml of growth medium with 1% serum. Incubate for 20–24 h. At this time the cells should show intense cytopathic effects.

4. Collect the culture supernatants and centrifuge[a] at 10 000 *g* for 10 min at 4°C. Collect the supernatant and store aliquots (e.g. 1.0 ml) at −70°C.

5. Titre the virus stock by plaque assay as described in *Protocol 5*.

---

[a] Use sealed tubes to avoid contamination of the centrifuge. When working with large volumes of infected liquids take especial safety precautions.

References

1. Wang, G. H. W. and Goeddel, D. V. (1986). *Nature*, **323**, 819.
2. Mestan, J., Digel, W., Mittnacht, S., Hillen, H., Blohm, D., Moller, A., Jacobsen, H., and Kirchner, H. (1986). *Nature*, **323**, 816.
3. Lewis, J. A. (1987). In *Lymphokines and interferons: a practical approach* (ed. M. J. Clemens, A. G. Morris, and A. J. H. Gearing), pp. 73–87. IRL Press, Oxford and Washington.
4. Reis, L. F. L., Lee, T. H., and Vilcek, J. (1989). *J. Biol. Chem.*, **264**, 16351.
5. Feduchi, E., Alonso, M. A., and Carrasco, L. (1989). *J. Virol.*, **63**, 1354.
6. Mosmann, T. (1983). *J. Immunol. Met.*, **65**, 55.
7. HHS Publication (NIH)-88-8395. *Biosafety in microbiological and biomedical laboratories*. The Department of Health and Human Services in the USA, National Institutes of Health, Bethesda, MD 20892. USA.

Assays for human B cell responses to cytokines

ROBIN E. CALLARD, KAROLENA T. KOTOWICZ, and
SUSAN H. SMITH

1. Introduction

B lymphocyte responses are determined by complex interactions between antigen, cell-surface ligands, and soluble growth and differentiation factors (cytokines) binding to specific receptors on the cell surface. To date, at least 17 of more than 40 recombinant cytokines are known to stimulate B cell activation, proliferation and/or differentiation (*Table 1*). Many of these cytokines are derived from T cells and are involved in T cell–B cell collaboration, but some are made by B cells, monocytes, and a variety of non-lymphoid cells such as endothelial cells and fibroblasts. Two important properties of cytokines need to be taken into account when considering their action on B cells:

(a) Cytokines usually lack specificity for cells of any one lineage and none are specific for B cells.

(b) Any one cytokine may have several distinct biological activities on B cells, depending on the stage of differentiation and mode of activation.

For example, IL-4 can activate resting B cells, promote cell division by activated B cells and T cells, and stimulate production of IgE and IgG4 in man (IgE and IgG1 in mice) by inducing heavy chain switching (1). Similarly, IL-2 is a growth and differentiation factor for human B cells and T cells (2, 3), whereas TGF-beta inhibits B cell proliferation and IgG secretion (4), but enhances IgA secretion, probably by inducing alpha heavy chain switching (5, 6). This diversity of function is further complicated by the fact that two or more cytokines often act in concert, and the possibility that different B cell subpopulations may not all respond in the same way. Another recently emerging property of cytokines that may have far reaching implications is that several come in pairs with very similar but, not identical properties. These 'doppelganger' cytokines include IL1-α/β, IL-2/IL-15, IL-4/IL-13, and TNF-α/β. The very similar and in some cases indistinguishable activity of

Table 1. Responses of human B cells to cytokines

Cytokine	Type of response				
	Activation	**Proliferation**	**Differentiation**	**TRF**	**Isotype**
IL-1	(+)[a]	(+)	(+)	no	no
IL-2	+	+	+	+	no
IL-3	?	+ (pre-B)	+	no	no
IL-4	+	+	+	no	+ (IgE/Ig)
IL-5[b]	no	no	no	no	?
IL-6	+	+	+	no	no[c]
IL-7	?	+ (pre-B)	?	no	?
IL-10	(+)[d]	+	+	no	+ (IgG/Ig)
IL-13	+	+	+	no	+ (IgE/Ig)
IL-14	?	+	−	?	?
IL-15	?	+	+	+	?
IFN-α	?	+	+/−	no	?
IFN-β	+	+	+	no	?
IFN-γ	+/−	+	+	no	− (IgE)
NGF	?	+	+	?	+ (IgG4)
TGF-β	?	−	−	no	+ (IgA)
TNF-α	?	+	+/−	no	?
TNF-β	?	+	+	no	?

[a] Positive response +; negative response −; no response no; unknown or not reported ?. The (+) for IL-1 indicates some controversy. TRF = T cell replacing factor.
[b] IL-5 has no effect on human B cells but it does on murine B cells.
[c] IL-6 may be required for IgE and IgA responses.
[d] IL-10 increases MHC class II on resting murine B cells, but not on human B cells.

these 'doppelganger' cytokines raises very interesting and important questions about the nature of the receptors involved. Unfortunately, it is beyond the scope of this introduction to discuss the nature of the cytokine receptors and the reader is referred to references (7, 8) for further information.

In this chapter, assays for measuring human B cell activation, proliferation, and differentiation in response to cytokines are described. In appropriate combinations, these assays can be used to distinguish between different B cell growth and differentiation factors, and help determine their action on different populations of normal and malignant B cells.

2. Reagents used for *in vitro* experiments with human B cells

2.1 Culture medium

For most B cell assays use RPM1 1640 supplemented with 25 mM Hepes, 2mM glutamine and 5% fetal calf serum (FCS). Gentamicin at 50 μg/ml should be used to inhibit bacterial growth that can occur with B cells prepared from tonsils. Gentamicin from some sources has been found to be toxic to B cells. In our hands, the most reliable form is cidomycin sterile solution from Roussel Laboratories. For specific antibody responses, use RPMI 1640 with 10% horse serum (EqS). The RPM1 1640 medium should be purchased as either liquid medium with Hepes included (GIBCO cat No, O4I-2400), or powdered medium with Hepes included (GIBCO cat No. 079–3018). In both these formulations, the NaCl concentration is reduced to compensate for the increased osmolarity arising from the addition of Hepes buffer. In some experiments this can be crucial. For example, specific antibody responses are absent or significantly reduced if ordinary RPM1 1640 with added Hepes is used. In some B cell assays, Iscove's modified Dulbecco's medium (IMDM) can give better responses than RPMI 1640.

2.2 Holding medium

For preparing cells, and holding on the bench before culture, use RPM1 1640 supplemented with 25 mM Hepes and 5% fetal calf serum (FCS). Some workers use a balanced salt solution (BSS), but we find that human B cells do not do so well in BSS compared to medium. Gentamicin at 50 µg/ml is routinely used with tonsil cell preparations since these frequently have some bacteria. This medium contains no bicarbonate and is used exclusively for washing and holding cells before culture.

2.3 Lymphocyte separation medium

Ficoll-hypaque or Ficoll-sodium metrozoate at 1.077 kg/l such as Lympho-prep (Flow laboratories cat. No. 1692254) or Ficoll-Paque (Pharmacia cat. No. 17. 0840.03) are used to prepare mononulcear cells from blood or tonsils. Percoll (Pharmacia cat. No. 17.0891.01) is used to separate E-rosetting (E$^+$) cells from non-E-rosetting (E$^-$) cells.

Protocol 1. Preparation of Percoll

Percoll solutions for discontinuous density gradient centrifugation are commonly made to a predetermined percentage by mixing with buffered saline or medium. However, the density of stock Percoll varies slightly, as do the various diluting solutions, and a given percentage of Percoll does not always result in the same density. For this reason, Percoll solutions should be prepared from solutions of known densities (specific gravities). Specific gravities are easily and accurately determined by weighing in specific gravity bottles.

Equipment and reagents

- Percoll
- Sterile 10 × concentrated PBS
- 10 ml specific gravity bottles

- Holding medium (see Section 2.2): RPMI 1640 supplemented with 25 mM Hepes, 5% fetal calf serum, and 50 µg/ml gentamicin (Roussel)

Protocol 1. *Continued*

Method

1. Prepare iso-osmolar stock Percoll suspension by adding 1 vol. of sterile 10 × concentrated PBS to 9 vol. of Percoll. (NB once PBS has been added to Percoll, autoclaving will cause it to polymerize.)

2. Measure the specific gravity (SG) of the stock Percoll and holding medium (Section 2.2) using a 10 ml specific gravity bottle. The SG of the iso-osmolar stock Percoll should be about 1.127, and the diluting medium about 1.008.

$$SG_{Percoll} = \frac{\text{weight of known volume of Percoll}}{\text{weight of an equal volume of water}}$$

3. Make up Percoll to the required specific gravity using the formula:

$$\text{per cent Percoll} = \frac{(\text{SG required} - \text{SG diluting medium})}{(\text{SG Percoll} - \text{SG diluting medium})} \times 100$$

Care must be taken with liquid volumes when making up Percoll. A measuring cylinder is normally adequate, but the same cylinder should be used for both the Percoll and diluting medium to avoid inter-cylinder variation. The diluted Percoll should be checked to ensure that the specific gravity is 1.080 ± 0.002.

2.4 Sheep red blood cells (SRBC)

Sheep's blood in Alsevers solution is a source of SRBC used in E-rosetting. This can be purchased from Tissue Culture Services (cat. No. SB068).

2.5 Phorbol ester

Either phorbol 12-myristate 13-acetate (PMA = TPA) or phorbol 12, 13 dibutyrate (PdB) can be used. A stock solution of phorbol ester is made up to 1 mg/ml in dimethylsulphoxide (DMSO), and stored frozen in the dark (wrapping in aluminium foil will suffice). Working dilutions are made by diluting the stock solution into holding medium without FCS. Because phorbol esters in solution will eventually deteriorate, titrations should be carried out every month or so to establish potency.

2.6 Anti-IgM

Affinity purified rabbit antibody to human IgM coupled to polyacrylamide beads (anti-IgM beads, Bio-Rad cat. No. 170–5120) is used to activate B cells in co-stimulation assays for growth promoting cytokines. The lyophilized beads are reconstituted in H_2O as directed, and stored in 0.01% NaN_3 at 4°C. Before use, the beads should be washed 2–3 times in holding medium, and resuspended to 10 μg/ml in culture medium. Washed beads can be stored for short periods at 4°C in holding medium containing 100 μg/ml of gentamicin. Alternatively, soluble F(ab')$_2$ goat anti-human IgM from Cappel (supplied by Precision Medical Ltd in the UK, cat. No. 55055) or Jackson Laboratories (cat. No. 109-006-129) at concentrations between 25μg/ml and 100μg/ml can be used to activated B cells, but this usually gives smaller responses.

2.7 *Staphylococcus aureus* Cowan 1 strain (SAC)

SAC can be purchased as Pansorbin from Calbiochem-Behring (La Jolla, CA). The bacteria are washed and used in culture at a final concentration of 0.05%.

2.8 Influenza virus

The influenza virus used for antigenic stimulation of antibody production (Section 6) was obtained as a sucrose gradient purified preparation. It was not inactivated, but inactivated virus is perfectly satisfactory. Either A or B strain viruses can be used equally well. As a concentrated suspension (about 20 mg/ml), it can be stored in 0.01% NaN$_3$ at 4°C for several years without loss of antigenic activity. Diluted suspensions are made up to 10 μg/ml in holding medium with 50 μg/ml of gentamicin but without FCS, and can be stored at 4°C for several months. The virus suspension is diluted a further 50 × in the antibody cultures which is usually sufficient to avoid inhibition by the azide present in the concentrated stock suspension. With some preparations, however, it may be necessary to dialyse out the azide. Influenza virus preparations cannot be sterilized by filtration as virus is retained by the membrane.

2.9 ^{3}H-thymidine

^{3}HTdR was purchased from Amersham International (cat. No. TRA 120, specific activity 5 Ci/mmol).

2.10 FITC-goat anti-mIg

Staining of human B cells with mouse monoclonal antibodies such as CD19, CD23, and CD40 may be undertaken with directly conjugated monoclonal antibodies, or require a second layer of FITC conjugated affinity purified F(ab')$_2$ goat antibody to mouse IgG/IgM. To avoid cross-reactions with human Ig, it is best to use a second layer antibody which is both affinity purified and adsorbed with human serum proteins. Several such preparations are available. We use a TAGO antibody cat. No. 6253 (available in the UK from Tissue Culture Services).

2.11 Recombinant cytokines

Purified recombinant human cytokines are available from a variety of different sources advertised in most of the major immunology journals.

2.12 Monoclonal antibodies

Anti-CD40 antibody can be used in co-stimulation assays for B cell proliferation. Purified antibody at 1 μg/ml will give optimal results.

3. Preparation of human B cells

The way in which B cells are prepared from human lymphoid tissues can be critical for assay of B cell growth and differentiation factors. It is important to appreciate that completely pure preparations of B cells can never be obtained from normal tissues, even by cell sorting. The degree of contamination with other cell types may interfere with the assays, and should always be determined. T cell depletion of tonsillar mononuclear cells by rosetting will routinely give preparations of greater than 95% B cells which can be used in assays for proliferation and antibody production without further purification. Similar preparations from venous blood usually yield much lower proportions of B cells (10–20%), but can be used in certain circumstances, for example to test for T cell replacing factor (TRF) activity in specific antibody responses. If required, further purification can be achieved by depletion of adherent cells. Negative selection of other unwanted cells by labelling with monoclonal antibodies and depletion by magnetic beads or complement lysis can also be used to improve B cell purity. Subpopulations of B cells can be obtained by fractionation on density gradients or by sorting after staining with selected monoclonal antibodies. In addition to purified normal B cells, selected B cell-lines can also be used as indicators for human B cell growth and differentiation factors (BCGF and BCDF).

3.1 Mononuclear cell preparation

The first step in the preparation of human B cells is to obtain mononuclear cells free of red blood cells, platelets and polymorphonuclear cells (PMN). These are prepared from peripheral blood or tonsils as follows in *Protocol 2*. Splenic B cells can also be obtained in a similar manner.

Protocol 2. Preparation of mononuclear cells from tonsils (TMC) and venous blood (PBMC)

Equipment and reagents

- Holding medium (see Section 2.2 and *Protocol 1*)
- Syringe fitted with 21 gauge needle
- Preservative-free heparin
- Ficoll-sodium metrozoate: (Section 2.3)
- Benchtop centrifuge
- Sterile plastic Universal tubes, pipettes, pipette tips, Petri dishes
- Microscope and haemocytometer
- Trypan blue solution
- Sterile scalpel
- 70% alcohol

A. *Preparation of tonsillar mononuclear cells (TMC)*

1. Surface sterilize excised tonsil with 70% alcohol for 5 sec, and then rinse in RPM1 1640 holding medium.

2. Put the tonsil into a Petri dish with 20 ml of RPMI 1640 holding medium containing 50 μg/ml of gentamicin and tease out the mononuclear cells by scraping from the connective tissue with a sterile scalpel.

3. Pipette the cell suspension into a plastic Universal tube and allow any clumps of tissue to settle for 1 min.

4. Remove the suspended cells and layer on to 8 ml of Lymphocyte separation medium 1.077 kg/litre (Section 2.3) in a plastic Universal tube. Centrifuge at 1000 *g* for 15 min at room temperature.

5. Collect the mononuclear cells from the interface, taking a portion of the Ficoll-Hypaque layer below, and wash (200 *g* for 10 min).

6. Wash the cells a second time (200 *g*) and resuspend in 10 ml of RPMI 1640 holding medium containing 5% FCS and gentamicin.

7. Remove a small aliquot and count on a haemocytometer. Viability determined by Trypan blue exclusion should always be in excess of 90%.

B. *Preparation of peripheral blood mononuclear cells (PBMC)*

1. Take blood from volunteer donors into a syringe wetted with preservative-free heparin.

2. Dilute blood with an equal volume of RPMI 1640 holding medium containing 10 IU/ml of preservative-free heparin but without FCS, and layer on to lymphocyte separation medium (Section 2.3) (2 vol. of diluted blood to 1 vol. of separation medium) and centrifuge at 1000 *g* for 15 min at room temp. The centrifuge brake should be off to prevent disturbance of the interface.

3. Collect the mononuclear cells from the interface, taking a portion of the Ficoll-Hypaque layer below, and wash (≤200 *g* for 15 min) with at least an equal volume of holding medium containing 10 IU/ml of preservative-free heparin, but without FCS. The slow wash in the presence of heparin will leave most of the platelets in the supernatant and prevent clumping of the mononuclear cell pellet.

4. Wash the cells a second time (200 *g*) in RPMI 1640 holding medium with 5% FCS, and resuspend to 5×10^6/ml.

5. Remove a small aliquot and count on a haemocytometer. Viability determined by Trypan blue exclusion should be in excess of 95%.

3.2 Depletion of T cells by E-rosetting

Human T cells and some natural killer (NK) cells express cell surface CD2 antigen which binds to specific receptors on sheep erythrocytes (E) resulting in the formation of E rosettes. Separation of E-rosette-forming cells (E^+) from non-rosette-forming (E^-) cells by density gradient centrifugation is the simplest and most effective method of depleting T cells from mononuclear cell preparations.

Protocol 3. Depletion of T cells by E-rosetting with AET-treated sheep red blood cells (SRBC)

Equipment and reagents

- Gey's Haemolytic Balanced Salt Solution:
  Solution A. NH$_4$Cl 35.0 g, KCl 1.85 g, Na$_2$HPO$_4$.12H$_2$O 1.5 g, KH$_2$PO$_4$ 0.119 g, Glucose 5.0 g, Phenol red 0.005 g, gelatine (Difco) 25.0 g, made up to 1 litre in double distilled-water;
  Solution B. MgCl$_2$.6H$_2$O 4.2 g, MgSO$_4$.7H$_2$O 1.4 g, CaCl$_2$ 3.4 g, made up to 1 litre in double distilled-water;
  Solution C. NaHCO$_3$ 22.5 g, made up to 1 litre in double-distilled water. All three solutions are sterilized by autoclaving and stored at 4°C. Make up Gey's solution freshly when required, by mixing 7 vol. of sterile double distilled-water with 2 vol. of Solution A (warmed at 37°C for 5 min before use to melt the gelatine), 0.5 vol. of Solution B and 0.5 vol. of Solution C.
- AET: S-2 aminoethylisothiouronium bromide hydrobromide (AET) 40.2 mg/ml in water, pH 9.0 adjusted with 1 M NaOH.
- Sheep's blood in Alsevers (Section 2.4)
- Sterile 0.14 M NaCl
- Holding medium (Section 2.2 and *Protocol 2*) with and without FCS
- Percoll (Section 2.3 and *Protocol 1*); (Pharmacia cat. no. 17.0891.01)
- Sterile pipettes, pipette tips, plastic Universal tubes
- Refrigerated bench centrifuge

A. *Preparation of AET-treated SRBC*

E rosettes are best formed using sheep red cells treated with S-2-aminoethylisothiouronium bromide hydrobromide (AET) to stabilize rosettes as described by Kaplan and Clark (9).

1. Wash sheep's blood (stored in Alsevers solution for up to 3 weeks) three times in sterile 0.14 M NaCl. Remove all the supernatant and any buffy coat on each wash.

2. After the last wash, remove all residual supernatant and incubate 1 vol. of packed SRBC pellet with 4 vol. of freshly prepared AET solution at 37°C for 15 min.

3. Wash five times in sterile 0.14 M saline and resuspend in RPMI 1640 holding medium (without FCS) to give 10% AET-SRBC. These may be stored at 4°C for up to three weeks.

B. *E-Rosette formation and separation*

1. Mix 10 ml of mononuclear cells at 5×10^6/ml with 2.5 ml of 10% AET-SRBC and 1 ml of FCS in a plastic Universal tube. Centrifuge at 200 *g* for 15 min with the brake off. Ensure that a good cell pellet is obtained with no SRBC in the supernatant. If necessary, centrifuge for a longer time but not at higher speed.

2. Incubate for 60 min on ice.

3. Resuspend the rosettes by gentle rotation of the centrifuge tube. Do not resuspend by pipetting as this will disrupt some rosettes.

4. Layer on to 7–8 ml of Percoll (SG 1.080) (Section 2.3) in a plastic Universal tube and centrifuge at 1000 *g* for 20 min at room temp. with the brake off.

5. Remove the E$^-$ fraction from the interface along with about 75% of the Percoll. Be careful not to disturb the red cell pellet.

6. Dilute with at least an equal volume of holding medium and wash twice.

7. To recover the E$^+$ cells, remove all of the Percoll above the SRBC and resuspend the pellet with 5 ml of Gey's haemolytic balanced salt solution for 1–2 min. Immediately after red cell lysis, dilute with holding medium, and wash twice.

8. Resuspend all fractions in 2–5 ml of holding medium, and count the cells. It is most important to check each preparation for percentages of B cells, T cells, and monocytes (Section 3.7). Preparations with unacceptable numbers of contaminating T cells (more than 2%) should be discarded.

More effective T cell depletion can be obtained if Percoll rather than Ficoll-Hypaque is used to separate E-rosette-forming (E$^+$) cells from non-rosette-forming (E$^-$) cells (10). The E$^-$ fraction of TMC obtained by this method should contain less than 1% of CD3$^+$ (T) cells. Two cycles of E-rosetting may be required to reduce the proportion of T cells to this level when PBMC are used. T cell contamination must be reduced to this level in E$^-$ preparations used in most B cell assays to avoid indirect action through residual T cells. Poor T cell depletion may be obtained with some batches of SRBC even after AET treatment. In this case it is best to discard the offending batch of SRBC and start afresh. Analyses of typical E$^-$ cell preparations from TMC and PBMC cells are given in *Table 2*.

3.3 Depletion of monocytes by adherence

Blood E$^-$ cells contain 60–80% monocytes compared with only 1–2% in tonsillar E$^-$ cell preparations. These can be depleted by adherence on tissue

Table 2. Phenotype of fractionated E⁻ lymphocytes

Cell type	% Fluorescence postive cells				
	CD20	CD3	CD14	CD57	CD16
Blood E⁻ (2 × rosetted)	21	<1	75	<1	6
(light fraction)	13	<1	78	<1	10
(heavy fraction)	92	<1	1	<1	3
Tonsil E⁻ (1 × rosetted)	95	<1	2	2	1
(light fraction)	98	<1	1	n.d.	n.d.
(heavy fraction)	98	<1	<1	n.d.	n.d.

Antibodies used were: B1 (CD20); UCHT1 (CD3); UCHM1 (CD14); HNK-1 (CD57); Leu 11 (CD16). n.d. = not done.

culture plastic Petri dishes. If recovery of the monocytes is required, plates coated by micro-exudate (extracellular matrix) from the BHK cell-line should be used. Contamination by monocytes in the non-adherent fraction is usually reduced to between 2–10% using this method. Better depletion may be obtained by passing cells through columns of Sephadex G-10, or by cell sorting or cytotoxicity using anti-monocyte monoclonal antibodies, but this is not usually necessary. In some assays, for example the antigen specific antibody response to influenza virus (Section 6), small percentages of monocytes (about 0.5%) are essential, presumably for antigen presentation.

Protocol 4. Depletion of adherent cells on micro-exudate plates

Equipment and reagents
- Holding medium and culture medium (Sections 2.1 and 2.2)
- Holding medium containing 3 mM EDTA
- PBS containing 10 mM EDTA
- BHK cell line
- Sterile plastic 90 mm Petri dishes, pipettes, pipette tips
- Bactericidal UV light source

A. *Preparation of BHK micro-exudate coated Petri dishes*

1. Prepare micro-exudate dishes by growing baby hamster kidney (BHK) cells to confluence on 90 mm tissue-culture plastic Petri dishes.

2. Remove the BHK cell monolayer with 10 mM EDTA in PBS.

3. Wash dishes vigorously several times with PBS, and sterilize with a bactericidal UV light source for 5 min. The treated dishes may be sealed with tape and stored at 4°C.

B. *Depletion of adherent cells*

1. Incubate 2 × 10⁷ E⁻ cells in 10 ml of holding medium containing 10% FCS on the micro-exudate dishes for 45 min at 37°C.

---

2. Resuspend the non-adherent cells by gently rocking the dishes, and remove by pipette. Repeat this procedure at least twice with 10 ml of warm medium carefully added to the dish.

3. To remove the adherent cells, incubate for 15 min with medium containing 3 mM EDTA at 37°C, and pipette vigorously.

3.4 Preparation of heavy and light B cells

In many B cell activation and proliferation experiments, E^- cells are fractionated on discontinuous Percoll gradients to give light (<1.074 kg/litre) and heavy (>1.074 kg/litre) populations. The small heavy B cells are generally considered to be resting (G_0) cells, but there is some evidence to suggest that some of these cells are already activated (11). Because of this, some investigators prefer even higher density B cells to exclude those in G_0^* (12).

Protocol 5. Density gradient fractionation of human B lymphocytes

Equipment and reagents

- Holding medium (Section 2.2)
- Percoll (Section 2.3)
- Trypan blue solution
- Sterile pipettes, pipette tips, Universal tubes

- Refrigerated bench centrifuge
- Microscope and haemocytometer
- Conical centrifuge tubes

Method

1. Layer up to 3×10^7 cells in 2–3 ml of holding medium on to 3 ml of Percoll (SG 1.074) in a 10 ml conical centrifuge tube, and centrifuge at 1000 g for 20 min. If using blood E^- cells, monocytes should be depleted first.

2. Remove the low-density B cells from the interface, and discard most of the Percoll supernatant leaving about 0.2 ml. Take care not to remove all the supernatant or disturb the pellet as the cells in the pellet tend to be very loose. Resuspend the high-density B cells in the remaining Percoll.

3. Wash each fraction twice, and resuspend the cells in 1–2 ml of holding medium containing 5% FCS.

4. Count the cells using Trypan blue.

3.5 Separation of B cell subpopulations with monoclonal antibodies

Various methods have been described for preparing B cell subpopulations defined by monoclonal antibodies. These include complement-mediated cytotoxicity, panning on antibody-coated plastic plates, rosetting with antibody-coated Ox red blood cells, magnetic beads and fluorescence activated cell sorting. A major disadvantage with most of these methods is the difficulty in recovering labelled cells (positive selection) in sufficient numbers and condition suitable for cell culture. The magnetic separation technique (MACS) uses very small (100 nm) paramagnetic beads which make positive selection as easy as negative selection. Moreover, the beads are biodegradable so they do not interfere with the cells in culture, and their small size does not noticeably change the light-scattering properties of the cells on FACS analysis. The method for MACS separation of B cell subpopulations is as follows.

Protocol 6. MACS separation of B cell subpopulations identified by monoclonal antibodies

The MACS magnetic cell sorter obtained from Miltenyi Biotec GmbH, Gladbach 1, West Germany (British agents are Becton Dickinson) comes with magnet, columns and biotinylated paramagnetic beads. Cell labelling is achieved with a sandwich of biotinylated monoclonal antibody followed by fluoresceinated streptavidin and then biotinylated beads. The streptavidin is FITC (or PE) conjugated to allow labelling to be checked by FACS analysis.

Equipment and reagents

- Holding medium (Section 2.2)
- PBS containing 0.01% sodium azide and 5 mM EDTA at pH 7.4 (PBS-EDTA)
- PBS containing 1% BSA and 5 mM EDTA at pH 7.4 (PBS-BSA)
- 70% and 95% ethanol
- Mild detergent
- Double distilled or Milli-Q water
- Biotinylated monoclonal antibodies, FITC or PE streptavidin

- 10 ml syringe, 25 gauge short-needles and 21 gauge needles
- Three-way stopcock to fit column and syringe
- MACS apparatus with biotinylated paramagnetic beads and columns of appropriate size
- Sterile pipettes and Universal tubes

A. *Column preparation*

1. Attach a three-way stopcock to the MACS column. Fill column with 70% ethanol slowly from the bottom using a syringe fitted to the stopcock, tapping frequently to dislodge bubbles. Replace the ethanol by running in PBS-BSA from the top. Leave at 4°C until required. This filling method avoids bubbles in the column.

2. Connect a 25 gauge short needle to the three-way stopcock attached to the column. Connect a 10 ml syringe of PBS-BSA to stopcock to flush through the needle if it becomes blocked.

3. Cool column by passing through 3 column vol. of ice-cold PBS-BSA before use.

4. To regenerate the column after use, rinse by suction with 500 ml of mild detergent followed by 1 litre of double-distilled water then 500 ml of 95% ethanol. Dry the column completely. The columns can be sterilized by autoclaving at 120°C, but this does shorten column life. Columns can normally be used about 10 times.

B. *Cell staining and separation*

1. Wash cells in ice-cold PBS-EDTA. Keep an aliquot for FACS analysis.

2. Add 100 μl of biotinylated monoclonal antibody to the washed pellet of 10^6 cells, mix, and incubate for 10 min at 4°C.

3. Wash cells in PBS-EDTA then add 10 μl of FITC-streptavidin and incubate for 10 min at 4°C. The aliquot put aside in step 1 as a negative control should be treated in the same way.

4. Wash cells once in PBS-EDTA and resuspend in 200 μl of PBS-EDTA (if using more than 4×10^7 cells, add an extra 50 μl per 10^7 cells). Add 1 μl of concentrated biotinylated MACS beads to each 100 μl of cell suspension, mix, and incubate for 5 min at 4°C.

5. Add 400 μl of PBS-BSA to each 100 μl of cell suspension and apply immediately to the MACS column fitted with a short 25 gauge needle. Note that the column flow rate is determined by the needle gauge. Keep an aliquot of unseparated cells at this stage for FACS analysis.

6. As soon as the cells have been applied to the column, place column in the magnet. Open the stopcock and add 5 column vol. of PBS-BSA to the top of the column. Allow the unbound cells to pass through the column and collect into a container on ice. Do not let the column run dry.

7. Replace the 25 gauge needle with a 21 gauge needle (to increase the flow rate) and run through 5–10 column vol. of PBS-BSA. Discard this wash.

8. To remove bound cells, remove column from the magnet, attach a 10 ml syringe to each end of the column (one filled with PBS-BSA) and vigorously squirt PBS-BSA back and forth through the column to dislodge the bound cells. Wash this bound fraction into a container on ice.

Note: the use of sodium azide as described and keeping the cells at 4°C is important to prevent capping and shedding of the antibody. The azide is removed by washing the cells and does not affect subsequent viability or function.

Separation purity should always be checked by FACS analysis of cells incubated with FITC-streptavidin only (background), stained unseparated cells, stained unbound cells, and stained bound cells. Better purity of unbound cells can sometimes be obtained by passing through the column twice.

3.6 Immunofluorescence analysis of B cell preparations

It is essential to monitor all human B cell preparations used to investigate responses to cytokines for purity since unacceptable contamination can sometimes occur and interfere with the assay. This is best done by indirect immunofluorescence using well-defined monoclonal antibodies to B cells, T cells, monocytes, and NK cells. For small numbers of cells, staining is carried out in round-bottomed microtitre wells, or round-bottomed flexible PVC microtitration plates which can be cut to the required number of wells and supported on a rigid microtitre tray. Larger numbers of cells can be stained in 12 × 75 mm Falcon tubes.

Protocol 7. Immunofluorescent staining of B cell preparations

Equipment and reagents

- Holding medium (Section 2.2) containing 2% FCS and 0.01% NaN$_3$
- Monoclonal and second-layer antibodies (see below)
- Plastic round-bottomed microtitre trays, Falcon tubes (cat. No. 2052), pipettes, and pipette tips
- Refrigerated benchtop centrifuge fitted with microtitre plate holder
- Whirlimixer
- Flow cytometer such as the FACScan or fluorescence microscope
- Paper tissue

A. *Monoclonal antibody staining of cells in microtitre wells*

1. Dispense between 1 and 3 × 10^5 cells per well of a non-sterile round-bottomed microtitre tray. Include wells for each monoclonal antibody as well as negative and positive controls.

2. Centrifuge the cells into a pellet at 200 *g* for 2 min using a microtitre plate attachment available for most centrifuges.

3. Remove supernatants by inverting the plate with a *single* flicking motion over a sink. Do not repeat this action as the cells will become resuspended and lost. Carefully wipe any droplets of medium from the surface of the plate using a paper tissue, then resuspend the cells by holding the plate firmly on to a vortex Whirlimixer.

4. Add 50 μl of monoclonal antibody diluted in holding medium containing 2% FCS and 0.01% NaN$_3$. When checking for B cell purity, include an anti-CD19 for B cells, an anti-CD3 for T cells, and a monocyte specific antibody such as UCHM1 (CD14). For the negative control, use medium alone or a monoclonal antibody known not to react with human leukocytes. For the positive control, use an anti-common leukocyte (CD45). Mix each well individually when adding the antibody. Do not use the Whirlimixer as antibody will spill over the sides into adjacent wells.

5. Incubate on ice for 30 min.

6. Wash the cells three times by adding 200 μl of medium to each well, centrifuging at 200 *g* for 2 min, and removing the supernatant by inverting the plate with a *single* flicking motion over a sink. Resuspend the cells by holding the plate firmly on a vortex Whirlimixer. After each wash, carefully remove any droplets remaining on the surface of the plate with a paper tissue. Extra care must be taken at this stage to prevent any spill-over of antibody from one well to the next.

7. If directly conjugated antibodies are used, go to step 10. Otherwise, to the resuspended cells add 50 μl of pretitrated FITC conjugated F(ab')$_2$ goat anti-mouse Ig absorbed against human serum proteins (Section 2.10) in medium containing 2% FCS, 0.01% NaN$_3$, and 2% normal goat serum (NGS) to inhibit Fc receptor binding.

8. Incubate on ice for 30 min.

9. Wash three times as in step 6.

10. Resuspend the cells and make up to 0.5 ml in Falcon tubes with holding medium containing 2% FCS and 0.01% NaN$_3$. The stained cells can be kept on ice for 2–3 h before analysis on a flow cytometer such as the FACScan. If it is necessary to delay the FACS analysis, the cells can be fixed as described in *Protocol 8*, and kept for several days without loss of fluorescence.

B. *Monoclonal antibody labelling of larger numbers of cells*

For larger numbers of cells, staining can be done in Falcon tubes.

1. Incubate 0.5 × 10^6 cells with 100 μl of pre-titrated monoclonal antibody in Falcon tubes for 30 min on ice.

2. Wash twice in cold medium containing 2% FCS and 0.01% NaN$_3$.

3. For directly conjugated antibodies go to step 4. Otherwise incubate the cells with 100 μl of pre-titrated FITC conjugated goat anti-mIg for a further 30 min on ice.

4. Wash twice, then resuspend the cells in 0.5 ml of holding medium and analyse samples on a FACScan, or fluorescence microscope.

Note: in some cases, phenotypic analysis of B cells which have been cultured in the presence of another monoclonal antibody is required. To avoid detection of this antibody one can use either directly conjugated monoclonal antibodies for staining, or a second layer specific for the IgG subclass of the staining (e.g. IgG2a) but not the other (e.g. IgG1) monoclonal antibody.

Protocol 8. Fixing FITC-labelled cells

Equipment and reagents
- Paraformaldehyde
- 1 M sodium hydroxide solution
- PBS
- Holding medium (Section 2.2)
- Waterbath

- Plastic pipettes, pipette tips, Falcon tubes (cat. No. 2052)
- Benchtop centrifuge
- FACScan

Method

1. Make up 0.1 g of paraformaldehyde with 2 drops of 1 M NaOH and 0.2 ml of H_2O.
2. Heat to 80°C in a waterbath. Do not boil.
3. Cool, and add 9.8 ml of PBS.
4. Wash labelled cells in PBS and then resuspend pellet in 200 μl of fixative at 4°C for 20 min, then wash twice in holding medium.
5. Wash the cells and resuspend to 0.5 ml in holding medium containing 5% FCS before reading on the FACScan.

3.7 Human B cell lines for assaying B cell growth and differentiation factors

In some circumstances, continuous B cell lines may offer certain advantages over purified B cells for measuring growth and differentiation in response to cytokines.

(a) Continuous B cell-lines are free of the non-B cell contamination which can cause problems even with FACS selected preparations.

(b) B cell-lines are more homogeneous than normal B cell preparations which generally consist of ill-defined B cell subpopulations and/or B cells at different stages of activation.

These advantages are offset to some extent by the abnormal physiological status of B cell-lines and the uncertain relevance of their responses to normal B cell growth and differentiation. For this reason, B cell-lines are best used in conjunction with other assays.

By judicious selection of indicator B cell-lines, it is possible to distinguish between different factors. For example, the mouse B9 hybridoma cell line proliferates in response to very low concentrations of IL-6, but not to any other known cytokine (13), see Chapter 21. In addition, responses to particular cytokines can be confirmed by using specific blocking antibodies. Of the numerous B cell-lines which respond to cytokines, we have selected four on the basis of their responses in different assays. The best characterized of these is CESS, a lymphoblastoid line used for detecting B cell differentiation factors such as IL-6 (14). In addition, we have identified three other lines (HFB1, L4 and BALM 4) which proliferate and/or differentiate (secrete immunoglobulin) in response to different cytokines. There are many other lines which can be used according to the investigator's needs and preferences.

3.7.1 Maintenance of B cell lines used in assays for B cell growth and differentiation.

The B cell-lines used in proliferation and differentiation assays (Sections 5 and 6) are grown in medium RPMI 1640 supplemented with 25 mM Hepes, 2 mM glutamine, and 10% FCS in 25 cm^2 or 75 cm^2 flasks at 37°C in an atmosphere of 5% CO_2 in air. Small numbers of cells can be grown in 24-well Costar plates. Antibiotics are not normally necessary, but gentamicin at 50 μg/ml can be used without interfering with the assays. The lines should be split regularly to keep the cells in log-phase growth.

One important problem which must be closely monitored is that of mycoplasma contamination which can interfere with responses to B cell growth factors. Interpretation of [^{3}H]TDR incorporation results may be impossible if the responding cell-lines are contaminated. Under these circumstances, cell proliferation can be determined with the MTT test (15). The EBV transformed line, CESS, is often contaminated with mycoplasma, but this does not seem to inhibit its ability to respond to IL-6. Mycoplasma-free lines should be checked routinely for infection. If possible, contaminated and mycoplasma-free lines should be dealt with completely separately to reduce the possibility of cross-contamination. If separate facilities are not available, mycoplasma-free cultures should be handled only after fumigation of the laminar flow cabinet with formaldehyde, preferably overnight. This is most easily done by leaving an open dish of formaldehyde in the closed cabinet with the fan off. It is also important to use separate stocks of medium and other reagents, and to keep the incubator clean and free of mycoplasma infected cultures. There are fewer problems with contamination, especially fungi and mycoplasmas, if the incubators are kept dry.

4. B cell activation

Activation of B cells by ligand-binding to receptors on the cell membrane such as anti-Ig or IL-4 involves a series of biochemical events. These start with receptor-mediated signal transduction and stimulation of second-messenger cascades, followed by activation and/or synthesis of transcription factors responsible for gene regulation, leading to synthesis of RNA and expression of activation antigens. As a result, the cell leaves G_0 and enters G_1, but in the absence of a second signal does not normally synthesize DNA or divide. The early events of B cell activation, such as the hydrolysis of phosphatidyl inositol bisphosphate (PIP$_2$) and protein tyrosine phosphorylation in response to anti-Ig (16), and the elevation in cAMP following stimulation with IL-4 (17) occur within minutes and are dealt with in Chapter 6. Later events, such as c-myc expression, synthesis of RNA, size increase, and expression of cell-surface activation antigens involve more complex cell functions including activation of the nucleus, and occur within hours or days.

4.1 Measurement of RNA synthesis

Activation of B cells to leave G_0 and enter G_1 is accompanied by RNA synthesis. Most of this is ribosomal RNA not mRNA and is part of the cells preparation for protein synthesis. *De novo* RNA synthesis can be measured easily by incorporation of [³H]uridine.

Protocol 9. Measurement of RNA synthesis

Equipment and reagents

- Culture medium (Section 2.1)
- [³H]uridine Amersham International cat. No. TRK 178, specific activity 25–30 Ci/m mole
- Sterile flat-bottomed microtitre trays for tissue culture
- Sterile pipettes, pipette tips, Universal tubes
- CO_2 Incubator
- Cell harvester with glass-fibre filters
- Liquid scintillation counter, scintillant, and capped counting tubes

Method

1. Prepare small resting B cells (*Protocol 5*) and culture at 10^6 cells/ml in 200 μl of culture medium in flat-botttomed microtitre wells with and without the cytokine to be investigated. Control cultures in medium alone and with a B cell activator such as PMA (10 ng/ml) should also be included.

2. Incubate for 18 h at 37°C in an atmosphere of 5% CO_2 then add 1 μCi of [³H]uridine to each well and incubate for a further 6 h at 37°C.

3. Harvest the labelled cells on to glass-fibre discs with an automated cell harvester, and count incorporated [³H]uridine on a liquid scintillation counter. Results should be corrected for quenching and efficiency and expressed as disintegrations per minute (d.p.m.).

4.2 Measurement of B cell activation by expression of surface activation antigens such as CD23 and surface IgM

Activation of B cells with cytokines such as IL-4 or IL-13 results in increased expression of activation antigens including CD23, surface IgM, CD40, and in some cases MHC class II. These can easily be measured by monoclonal antibody labelling and flow cytometry. The most commonly used B cells for activation experiments are small heavy tonsillar E⁻ cells isolated on discontinuous Percoll gradients (*Protocol 5*). Other sources of B cells such as spleen can be used, but peripheral blood B cells are not very satisfactory because of the relatively small numbers of B cells obtained, and the difficulties in removing the high proportion of monocytes, NK and null cells (*Table 2*). The technique for activating B cells and measuring the expression of activation antigens is described in *Protocol 10*.

Protocol 10. Expression of human B-cell activation antigens

Equipment and reagents

- Culture medium (Section 2.1)
- Lymphocyte separation medium (e.g. Ficoll-Hypaque) at 1.077 kg/litre (Section 2.3)
- PMA
- Monoclonal antibodies to activation antigens (e.g. CD23 IgM)
- Sterile 24-well Costar plates, 12 × 75 mm Falcon tubes, Universal tubes, pipettes and pipette tips, Pasteur pipettes
- CO_2 incubator
- Bench centrifuge
- Flow cytometer (FACScan)

Method

1. Prepare T cell depleted E⁻ cells as described in Section 3, and resuspend to 1×10^6 cells/ml with culture medium containing 5% FCS, 25 mM Hepes and 50 µg/ml of gentamicin.

2. If small numbers of activated cells are required, use 2 ml cultures at 10^6 cells/ml in 24-well Costar plates. For activation with cytokine such as IL-4, add 50 µl of an appropriate dilution in culture medium. A final concentration of IL-4 of 100 units/ml is optimal. Use PMA at a final concentration of 0.5–10 ng/ml for positive control cultures. Normally, 10 ng/ml of PMA will both activate and induce cell division whereas 0.5 ng/ml will activate without significant DNA synthesis. To prevent drying of the cultures, put sterile water into the outer wells, and place the Costar plate inside an unsealed plastic box. For larger numbers of cells, set up the cultures at 10^6 cells/ml in tissue culture flasks.

3. Incubate at 37°C for 72 h in a dry incubator with an atmosphere of 5% CO_2 in air.

4. At the completion of the culture period, resuspend the cells and centrifuge at 1000 g for 20 min over Ficoll-Hypaque (1.077 kg/litre) to remove dead cells and debris. For small numbers of cells, this can be done in 12 × 75 mm Falcon tubes containing 1.5 ml of Ficoll-Hypaque, otherwise use plastic Universal containers with at least 7 ml of Ficoll-Hypaque.

5. Carefully remove the cells from the interface with a Pasteur pipette, then wash twice and resuspend to 2×10^6 cells/ml in holding medium.

---

6. To assess the expression of activation antigens, stain the cells with the appropriate monoclonal antibody as in *Protocol 7*, and read on the FACScan.

Note: Appropriate negative controls for B cells activated with cytokine or PMA are B cells cultured under the same conditions but without the activation signal. About 20% more cells are required in control cultures as they do not survive as well as activated cells. Freshly prepared E$^-$ cells are not suitable controls.

An example of B cell activation with IL-4 is given in *Figure 1*. After stimulation with IL-4 for 72 h the cells were stained for surface CD23 and IgM expression. The FACScan histograms clearly show an increase in expression of CD23 and surface IgM by B cells activated with IL-4. Much lower concentrations of IL-4 are required for optimal surface IgM expression compared with CD23 expression (18).

The enhanced expression of surface activation antigens detected by FACS analysis can be expressed in three different ways:

(a) The fluorescent channel subdividing the histogram into two equal proportions (median) can be compared. This is usually known as the median fluorescence intensity (MFI).

(b) The fluorescent channel showing the peak fluorescence (mode) can be compared.

(c) The proportion of activated cells labelled can be given by setting the marker at the base of the peak obtained with control cells.

Of these, MFI is the preferred measurement. Changes in cell size on activation can be determined by comparing forward angle scatter histograms in the same way.

5. Assays for B cell proliferation in response to cytokines

Cytokines which promote the growth of B cells can be assayed in different ways. The two most common methods are co-stimulation of normal (tonsillar) B lymphocytes, and enhancement of cell division of indicator B cell-lines.

5.1 Cytokine stimulation of B cell proliferation

This assay depends upon the co-stimulation of small heavy (resting) B cells with cytokine and a second activation signal such as anti-IgM, anti-CD40, or PMA. Under the right conditions, co-stimulation with both signals will result in B cell proliferation whereas each signal alone will have little or no effect. Several different methods for activating resting B cells to respond to growth promoting cytokines have been described. Most of these use either soluble or particular preparations of anti-immunoglobulin, *Staphylococcus aureus*

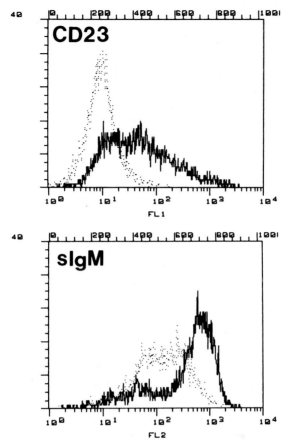

Figure 1. FACS histogram of small heavy tonsillar B cells cultured with medium alone or IL-4 for 3 days and then stained for CD23 and sIgM expression.

Cowan 1 preparations (SAC), anti-CD40 monoclonal antibodies, or phorbol esters (PMA). We have found the most reliable to be anti-IgM coupled to polyacrylamide beads and anti-CD40 monoclonal antibodies. These are used in the co-stimulation assay for B cell growth factors as follows.

Protocol 11. Proliferation of resting B cells in response to co-stimulation with cytokine and anti-IgM or anti-CD40

Equipment and reagents

- Culture medium (Section 2.1)
- Anti-IgM beads or CD40 monoclonal antibody (Sections 2.6 and 2.12)
- [^{3}H]TdR (Section 2.9)
- Sterile 96-well flat-bottomed microtitre plates for tissue culture
- Hamilton Stepper syringe
- CO_2 Incubator
- Cell harvester with glass-fibre filters
- Liquid scintillation counter, scintillant, and capped counting tubes
- Saline

163

Protocol 11. *Continued*

Method: co-stimulation of tonsillar B cells

1. Prepare small resting tonsillar E⁻ as described in Section 3.2, and resuspend to 10^6 cells/ml in culture medium supplemented with 5% FCS and 50 µg/ml of gentamicin.

2. Dispense 100 µl (10^5) of cell suspension into sterile flat-bottomed microtitre wells. To triplicate wells, add 50 µl of each cytokine to be tested plus 50 µl anti-IgM beads (10 µg/ml final concentration) or anti-CD40 monclonal antibody (1 µg/ml). For each experiment, control cultures with medium alone, and anti-IgM beads (or anti-CD40 monoclonal antibody) alone, should be included. If using a dry incubator, fill the outer wells with sterile water and place the plate in an unsealed plastic box to minimize evaporation.

3. Incubate for 72 h at 37°C in an atmosphere of 5% CO_2 in air.

4. Add 1 µCi of [³H]TdR in 10 µl of saline to each culture well using an 0.5 ml Hamilton Stepper syringe, and incubate at 37°C for 6–8 h.

5. Harvest the labelled cells on to glass-fibre discs with an automated cell harvester, and count incorporated [³H]TdR on a liquid scintillation counter. Results should be corrected for quenching and efficiency and expressed as disintegrations per minute (d.p.m.).

A typical result from an IL-4 co-stimulation experiment is given in *Table 3*. In this case, tonsillar B cells were stimulated with anti-IgM beads and IL-4 (100 units/ml). A 10-fold stimulation index with counts up to 30 000 is typical for this sort of experiment.

It is important to note that the choice of activation signal in these experiments is not just a matter of convenience. Different B cell activators may have quite different effects and prime B cells to respond in different ways to the various B cell growth promoting cytokines. For example, PMA, but not anti-IgM, is a powerful inducer of the CD23 antigen. It may, therefore, be important to try different activation signals when screening for unknown B cell growth factors.

Table 3. Co-stimulation of B cells with anti-IgM and IL-4

Anti-IgM beads	IL-4	Response (³HTdR d.p.m.)
–	–	1286 ± 187
+	–	2612 ± 297
–	+	1879 ± 241
+	+	87 345 ± 1223

5.2 Cytokine stimulation of proliferation by B cell lines

When grown at low cell-densities, some B cell-lines depend upon the addition of exogenous growth factors for continual proliferation. Using selected B cell-lines such as HFB1, L4, and BALM 4, this property can be exploited to assay for growth promoting cytokines.

Protocol 12. Cytokine stimulation of growth by B cell lines

Equipment and reagents

- Culture medium (Section 2.1)
- [^{3}H]TdR (Section 2.9)
- Sterile 96-well flat-bottomed microtitre plates for tissue culture
- Hamilton Stepper syringe
- CO_2 incubator
- Cell harvester with glass-fibre filters
- Liquid scintillation counter, scintillant, and capped counting tubes

Method: measurement of proliferation by B cell lines

1. Harvest cells from a vigorous log-phase growth culture of L4, BALM 4 or HFB1. The cells should be subcultured 24–48 h beforehand. Cultures which contain many dead cells, or are growing slowly, will not perform well in this assay. Wash the cells once and resuspend to 10^6 cells/ml in culture medium.

2. Add 100 µl of cells to give final cell concentrations in six replicate microtitre wells ranging from 10^3 to 10^5 cells/ml. Add 100 µl of cytokine to six replicate cultures at each cell concentration. Add medium only to one set of six wells as a negative control.

3. If using a dry incubator, fill the outer wells with sterile water, and put the plate in an unsealed plastic box to minimize evaporation. Incubate for 48 or 72 h at 37°C in an atmosphere of 5% CO_2 in air.

4. Add 1 µCi/well of [^{3}H]TdR in 10 µl of saline using a 0.5 ml Hamilton Stepper syringe and incubate for a further 6–8 h.

5. Harvest the cells on to glass-fibre discs with an automatic cell harvester, and count incorporated [^{3}H]TdR on a liquid scintillation counter.

Note: in these experiments the starting cell concentration is critical. If the cell concentration is too high, cells proliferate well without any exogenous factor. On the other hand, if the concentration is too low, the cells will die even in the presence of added cytokine. We routinely use at least three cell concentrations of each cell-line. With the normal day-to-day variation, any of the three concentrations may give an optimal response. A typical set of results obtained with low molecular weight BCGF (BCGF$_{low}$) is shown in *Table 4*.

Table 4. Response of selected B cell lines to BCGF$_{low}$

Cell line	Cells/well ($\times 10^{-3}$)	[^{3}H]TdR incorporation (d.p.m.)	
		Medium	BCGF
BALM 4	20	38 681 ± 11 217	93 226 ± 12 119
	10	6700 ± 1474	35 614 ± 4630
HFB1	15	40 275 ± 5236	95 734 ± 13 403
	5	3191 ± 925	37 446 ± 1123
L4	12	7441 ± 1359	115 329 ± 3059
	4	1206 ± 296	38 451 ± 570

6. Assays for cytokines which stimulate human B cell differentiation (Ig secretion)

Cytokines which stimulate antibody secretion by human B cells (B cell differentiation factors) can be measured in co-stimulation assays with SAC or anti-Ig activated B cells (19) and on indicator B cell-lines (14). Other assays have also been developed for cytokines which stimulate the production of immunoglobulin isotypes and those which can replace T cells in specific antibody responses called T cell replacing factors (TRF). It is worth bearing in mind that B cell differentiation could in principle also refer to other B cell responses such as memory cell production or antigen presentation. Assays to measure cytokine control of these other functions have not yet been developed.

6.1 Co-stimulation assays for cytokines which stimulate immunoglobulin secretion (B cell differentiation factors)

These are set up in exactly the same way as described for proliferation (Section 5.1) except that SAC and anti-CD40 monoclonal antibodies or CD40 ligand (CD40L) are more commonly used than anti-Ig to activate the B cells, and the cultures normally run for 5–7 days rather than 3–4 days. At the end of the culture period, supernatants are removed and assayed for Ig as described in Section 8. A typical result using SAC activated B cells and IL-6 or IL-2 as B cell differentiation factors is given in *Table 5*. Note that timing of addition of the cytokine may be important in these assays. For example, IL-4 does not stimulate Ig secretion in co-stimulation assays with SAC when the IL-4 and SAC are added together at the initiation of the cultures, but it is now known to enhance IgM and IgG secretion by SAC preactivated B cell blasts (20).

6.2 Assays for B cell differentiation factors using B cell lines

Immunoglobulin secretion by B cell-lines L4, BALM 4 and CESS (*Table 6*) may be increased in the presence of B cell differentiation factors (21). With

166

Table 5. SAC co-stimulation assay for B cell differentiation factors

			Response	
SAC	IL-2	IL-6	Proliferation ([^{3}H]TdR d.p.m.)	IgG (ng/ml)
−	−	−	800 ± 280	18 ± 6
+	−	−	2000 ± 780	52 ± 11
+	−	+	1850 ± 540	1876 ± 230
+	+	−	66 400 ± 6260	2360 ± 420

Note: that IL-2 stimulates both [^{3}H]TdR uptake (proliferation) and Ig secretion whereas IL-6 stimulates only Ig secretion. IL-6 is therefore a true B cell differentiation factor whereas in this experiment the increased Ig secretion obtained with IL-2 may have been due to an increase in the number of B cells and not differentiation. This distinction should be taken into account when assaying for B cell differentiation factors.

Table 6. Assay for IL-6 on B cell lines

		IgG production (ng/ml)	
Cell-line	Cells/well ($\times$ 10^{-3})	Medium	IL-6
CESS	3	28 ± 3	526 ± 49
L4	12	11 ± 6	185 ± 35
	4	<5	14 ± 5
BALM 4	20	4 ± 1	32 ± 8
	10	<5	11 ± 4

CESS it was originally reported that the minor population of surface IgG positive cells were more responsive to BCDF than unseparated CESS cells (14), but we have not found it necessary to isolate this subpopulation. The assay is carried out in exactly the same way as described for cytokines which stimulate proliferation (*Protocol 12*) except that the cultures are extended to 5–7 days. Supernatants are then harvested and assayed for immunoglobulin content as described in Section 8. CESS, L4 and BALM 4 are all suitable cell lines for measuring immunoglobulin secretion in response to cytokines. Lower concentrations of immunoglobulins are secreted by L4 and BALM 4, and it may be necessary to culture for a longer period (6–7 days) and at higher cell concentration.

An example of the response of each cell-line to IL-6 is shown in *Table 6*. IgG secretion by CESS cells is typically increased 5–50 times by IL-6. When using L4 or BALM 4, it is advisable to use two different cell concentrations as the response is cell dose dependent (*Table 6*).

6.3 Assays for cytokine regulation of Ig isotype responses

A number of different assays have been used to investigate cytokine regulation of Ig class and IgG subclass responses in humans. Culture of B cells with T cells and IL-4 has been used to show that IgE production depends on IL-4, activated T cells and other cytokines (IL-6) (22–24). More recently, antibody or CD40 ligand (CD40L) binding to CD40 on B cells has been shown to be an essential signal for heavy chain switching by human B cells (25, 26). Co-stimulation of B cells with anti-CD40 antibody (or CD40L) and cytokine has since emerged as a suitable assay for Ig class and IgG subclass regulation. IgE secretion can also be obtained by polyclonal activation of B cells with EBV and IL-4 in the absence of CD40 activation (27). Other polyclonal B cell activators such as PMA have also been used to investigate cytokine regulation of Ig isotype regulation (28, 29). The effect of cytokines on isotype and IgG subclass production can be determined by specific EIA described in Section 8.

Protocol 13. Assays for cytokine regulation of immunoglobulin isotype production

Equipment and reagents

- IMDM culture medium (Section 2.1)
- EBV-containing supernatant from the marmoset B95–8 cell-line: Grow the B95–8 line in RPMI culture medium until the cultures are bright yellow, then centrifuge to remove the cells and filter the supernatant through a 0.45 micron filter. The EBV-

containing supernatant can be aliquoted and stored at $-70°C$ until required.
- CD40 antibody or CD40L
- CO_2 incubator
- Sterile flat-bottomed 96- and 48-well microtitre plates, pipettes, pipette tips

A. *Co-stimulation with CD40 antibody and cytokine*

1. Prepare T cell depleted E$^-$ cells from PBMC as described in *Protocol 3* and make up to 10^6 cells/ml in IMDM culture medium (Section 2.1) with 5% FCS.

2. Dispense 100 µl (10^5 cells) into each well of a 96-well flat-bottomed microtitre plate.

3. Add 50 µl of purified anti-CD40 antibody to give a final concentration of 1 µg/ml and 50 µl of cytokine to be tested. Include control cultures with CD40 antibody alone. If available, CD40L can be used instead of CD40 antibody.

4. Culture at 37°C in 5% CO_2 for 8–9 days then collect supernatant for Ig class and IgG subclass assay as described in Section 8.

B. *Ig isotype responses by B cells stimulated with EBV and cytokine*

1. Prepare T cell depleted E⁻ cells from TMC or PBMC in RPMI with 5% FCS as described in *Protocol 3*.

2. Centrifuge at 200 g for 7 min and resuspend pellet to 12.5×10^6 cells/ml in EBV-containing B95–8 supernatant and incubate at 37°C in 5% CO_2 for 1–2 h.

3. Wash the cells twice to remove unbound EBV and resuspend to 2×10^6 cells/ml in IMDM culture medium with 5% FCS (Section 2.1).

4. Dispense 0.5 ml (10^6 cells) into each well of 48-well Costar plate and add 0.5 ml of cytokine to be tested.

5. Culture at 37°C in 5% CO_2 for 12–14 days.

6. Harvest the supernatant for Ig class or Ig subclass assay as described in Section 8.

6.4 T cell replacing factors

In addition to stimulating B cell growth and differentiation, some cytokines can also replace T cells in specific antibody responses *in vitro*. In man, these T cell replacing factors (TRF) can be readily assayed on thoroughly T cell depleted E⁻ cell preparations obtained from blood, tonsil, or spleen (30, 31). In this assay, E⁻ cells are cultured with antigen (influenza virus in the example described below) in the presence and absence of TRF, and antibody production determined by specific EIA. Antibody responses obtained in the presence of TRF are usually comparable with those obtained in the presence of T (E⁺) cells.

Protocol 14. Assay for T cell replacing factors (TRF) in specific antibody responses

Equipment and reagents

- RPMI 1640 culture medium supplemented with 10% horse serum (Section 2.1)
- Holding medium (Section 2.2)
- Purified influenza virus antigen (Section 2.8)

- Sterile 12 × 75 mm Falcon culture tubes (cat. No. 2054), pipettes, pipette tips
- CO_2 incubator
- Bench centrifuge

Method

1. Prepare T cell depleted E⁻ (B) cells from TMC or PBMC as described in *Protocol 3* and resuspend to $1–1.5 \times 10^6$/ml in RPMI 1640 culture medium supplemented with 10% horse serum and 50 μg/ml of gentamicin (Section 2.1), and dispense 0.5 ml ($0.5–0.75 \times 10^6$ cells) into 12 × 75 mm culture tubes. Note that the horse serum is stable for many months at −20°C, but loses activity after a few days at 4°C.

Protocol 14. *Continued*

2. Dilute purified influenza virus (Section 2.8) to 10-times the final con-centration (usually 0.2 μg/ml) in the same medium, and add 100 μl to each culture tube.

3. Add 0.5 ml of medium to three tubes for a negative control, and 10^6 E$^+$ (Th) cells in 0.5 ml of medium to three tubes for a positive control.

4. Dispense 0.2–0.5 ml of factor to be assayed into the remaining tubes, and make up the total volume of each culture to 1 ml with culture medium.

5. Loosely replace the caps and incubate for 7 days at 37°C in a dry incubator with an atmosphere of 5% CO_2 in air.

6. At the end of the 7 day culture period, wash the cells once and re-suspend in 0.5 ml of culture medium containing 5% FCS. Cover the tubes with aluminium foil rather than recap them, and incubate for a further 12–18 h at 37°C. This step minimizes background in the EIA assay by removing free virus and horse serum present in the culture medium.

7. Collect supernatants for assay of specific antibody (Section 7.3). IgG subclass antibody can also be measured (Section 7.4). Supernatants for EIA may be stored at −20°C before assay.

A typical result from a TRF induced antibody response to influenza virus is given in *Table 7*.

The antibody cultures are usually carried out in triplicate in 12 × 75 mm capped Falcon tubes, but a micromethod is also available using 5-times fewer cells in round-bottomed microtitre plates (32). In this case, the responses obtained are more variable, and more replicates (>6) should be used. The micromethod is essentially the same except that 10^5 E$^-$ cells in 200 μl are cultured in round-bottomed microtitre wells. Washing of the cells in micro-

Table 7. Specific antibody responses obtained with TRF

Responding cells	Antigen	TRF	Antibody to X31 (ng/ml)	
			IgM	IgG
[a]PBM E$^-$	−	−	<1	<1
PBM E$^-$	+	−	<1	2
PBM E$^-$	−	+	<1	<1
PBM E$^-$	+	+	2	220 ± 80
PBM E$^-$ + E$^+$	+	−	2	241 ± 41

[a] Depleted of T cells by two cycles of E-rosetting with AET-SRBC (Section 2.2).

titre wells (see step 6 above) is best carried out by sucking off the medium from each well using a 21 gauge needle attached to a suction pump. If the needle is slid down the side of the microtitre well as far as the top of the curvature, the cell pellet will not be disturbed. The cells can then be washed by resuspending in medium and repelleted by centrifugation at 200 g for 5 min on a centrifuge with a microtitre tray adapter.

7. Enzyme immunoassays for immunoglobulin secreted by human B cells

All the assays we use for detection of human Ig and specific antibody are based on solid phase enzyme immunoassays (EIA).

7.1 Measurement of Ig (IgM, IgG, IgA, and IgE) secretion

The effect of B cell differentiation factors is easily monitored by measuring Ig production with an ELISA assay. The example that follows is for IgG, but the other isotypes or total Ig can be measured in same way by substituting the appropriate antibodies for capture in step 1 and detection in step 8 as shown in *Table 8*. The assays for IgE and IgG subclasses have an extra step 10.

Table 8. Conditions for Ig class and IgG subclass assays

Assay	First layer (step 1)	Second layer (step ?)	Third layer (step ?) IgE only
IgM	Goat anti-human IgM (Sigma I-0759) 1 μg/ml	HPO-goat anti-human IgM (Sigma A-6907) 1 μg/ml	—
IgG	Goat anti-human IgG (Sigma I-3382) 1 μg/ml	HPO-goat anti-human IgG (Sigma A-6029) 1 μg/ml	—
IgA	Goat anti-human IgA (Sigma I-0884) 1 μg/ml	HPO-goat anti-human IgA (Sigma A-7032) 1 μg/ml	
IgE	Goat anti-human IgE (Sigma I-0632) 2 μg/ml	Rabbit anti-human IgE (MIAB Ab 105) 0.1 μg/ml	HPO-goat anti-rabbit IgG (Sigma A-0407) 1 μg/ml
IgG1	Rabbit anti-human G (H+L) (Jackson 309-005-082) 1 μg/ml	Biotin-mab anti-IgG1 (Zymed 05-3340) 0.5 μg/ml	HPO (or AP) streptavidin (Amersham RPN-1231 or RPN-1234) 1/1000
IgG2	Rabbit anti-human G (H+L) (Jackson 309-005-082) 1 μg/ml	Biotin mab anti-IgG2 (Zymed 05-3540) 1 μg/ml	HPO (or AP) streptavidin (Amersham RPN-1231 or RPN-1234) 1/1000
IgG3	Rabbit anti-human G (H+L) (Jackson 309-005-082) 1 μg/ml	Biotin mab anti-IgG3 (Zymed 05-3640) 2 μg/ml	HPO (or AP) streptavidin (Amersham RPN-1231 or RPN-1234) 1/1000
IgG4	Sheep anti-human IgG4 (Binding Site AB009) 2 μg/ml	Biotin mab anti-IgG4 (Zymed 05-3840) 2 μg/ml	HPO (or AP) streptavidin (Amersham RPN-1231 or RPN-1234) 1/1000

Protocol 15. Measurement of IgG by ELISA

Equipment and reagents

- 4% normal goat serum
- PBS/BSA/Tween: PBS containing 1% BSA and 0.05% Tween 20
- 3 M NaOH
- PBS/Tween: PBS containing 0.05% Tween 20
- Bicarbonate coating buffer pH 9.6: 1.59 g Na_2CO_3, 2.93 g $NaHCO_3$, 0.2 g NaN_3 made up to 1 litre in double distilled-water and adjusted to pH 9.6
- 2 M H_2SO_4
- *p*-nitro phenyl phosphate (1 mg/ml) (Sigma 104–105) in alkaline phosphatase buffer: 1.59 g Na_2CO_3, 2.93 g $NaHCO_3$, 9.52 mg $MgCl_2$ made up to 1 litre in double distilled-water and adjusted to pH 9.6
- Affinity-purified goat anti-human IgG (Sigma cat. No. I-3382)
- Horseradish peroxidase (HPO) or alkaline

- phosphatase (AP)-conjugated affinity-purified goat anti-human IgG (Sigma cat. No. A-6029)
- Rabbit-anti-human IgE (MIAB cat. No. AB 105)
- ELISA plate reader
- HPO-goat anti-rabbit IgG (Sigma cat. No. A 0407)
- Phosphate/citrate buffer, pH 5.0
- *O*-phenyl-diamine (OPD) (Sigma P1526) in phosphate citrate buffer: 1 M Na_2HPO_4 (28.4 g/litre) and Citric acid (21.0 g/litre) made separately. Mix equal volumes just prior to use and add 0.5 mg/ml OPD and 0.015% v/v H_2O_2
- Non-sterile flat-bottomed 96-well ELISA plates (Immunolon II and Dynatech M1298)
- Incubator

Method

1. Dispense 75 μl of affinity purified goat anti-human IgG at 1 μg/ml in bicarbonate coating buffer and incubate at room temp. overnight. For other Ig classes use antibodies as shown in *Table 8*. Note that individual batches of antisera should be pre-titrated to determine the optimal signal to noise ratio with high, medium and low concentrations of IgG.

2. Wash three times with PBS/Tween.

3. To block remaining protein binding sites, add 100 μl PBS/BSA/Tween to each well and leave at room temp. for 30–90 min. 4% normal goat serum (NGS) diluted in PBS can also be used in this step.

4. Wash three times with PBS/Tween.

5. Add 75 μl of test supernatant and standards in duplicate. An 11 point standard curve using doubling dilutions of either pooled normal human serum, or partially purified IgG at 1000 ng/ml in PBS/BSA/Tween is set up in duplicate on each plate along with a buffer only zero standard.

6. Incubate at room temp. for 1–2 h. When assaying for IgE, incubate overnight at room temp.

7. Wash three times with PBS/Tween.

8. To each well, add 75 μl of horseradish peroxidase (HPO) or alkaline phosphatase AP-conjugated affinity purified goat anti-human IgG diluted in PBS/BSA/Tween. For IgE, use rabbit anti-human IgE. Individual batches of antisera should be titrated to determine the optimal dilution.

9. Incubate at room temp. for 1–2 h (4 h for IgE) then wash three times with PBS/Tween and once with distilled water.

10. For IgE, add 75 μl of HPO-goat anti-rabbit IgG, incubate for 1–2 h at room temp., then wash twice with PBS/Tween and once with distilled water.

11. To each well, add 100 μl of OPD (HPO substrate) in phosphate/citrate buffer and incubate for 15–60 min at RT for colour development. Stop with 40 μl of H_2SO_4. If AP-conjugated antibodies are used, add 100 μl of *p*-nitrophenyl phosphate in alkaline phosphatase buffer and allow the colour to develop for about 1 h at 37°C. Stop with 40 μl of 3 M NaOH.

12. Read absorbance on an automatic ELISA plate reader at 492 nm for OPD or 405 nm for AP. The results should be expressed as ng/ml of Ig calculated from the standard curve. Most automatic plate readers now come with computer software which will plot a standard curve obtained by one of several choices of curve fitting methods and will calculate the sample concentrations. Alternatively, a standard curve obtained by plotting the logit OD against $\log_2$ of antibody concentration where:

$$\text{logit}_{OD} = \ln \frac{P}{1 - P}$$

where $P = OD/OD_{max}$ and OD_{max} is the optical density obtained when all the NPP is hydrolysed, usually about 11.0.

Note that antisera can be stored undiluted in small aliquots at −70°C, and frozen and thawed a maximum of three times. Immunoglobulin standards should be diluted in assay buffer to 10 μg/ml and stored frozen in small aliquots. By substituting class-specific antisera at stages (1) and (8) above, it is also possible to measure levels of IgM, IgA and IgE in culture supernatants. The different antisera and conditions for IgM, IgG, IgA and IgE are shown in *Table 8*.

7.2 Measurement of IgG subclasses by ELISA

Measurement of IgG subclasses in tissue culture supernatants by ELISA has proved quite difficult. A number of methods have been reported but we have found that some of these are subject to high backgrounds and in some cases lack of specificity. In our experience some of these difficulties have arisen from the complex sandwich methods employed. To minimize this problem, a method was developed by Karolena Kotowicz in this laboratory which employs IgG subclass specific polyclonal antibody as a capture antibody and biotinylated mouse monoclonal antibodies to the human IgG subclasses (Zymed, distributed by Cambridge Biosciences in UK), followed by HPO or

AP-conjugated streptavidin for detection. The different antibodies used for capture and detection are shown in *Table 8*. The individual steps of this assay are essentially the same as in Section 7.1 above except for the use of these reagents, and longer incubation times. In addition, PBS/Tween with 1% normal mouse serum (NMS) is used in blocking step 3, and PBS/Tween with 0.2% NMS is used as a diluent in steps 5, 8 and also 10 when streptavidin is used instead of second antibody. Note that BSA is never used as this causes serious cross-reactivity problems with mouse antibodies. Optimal incubation times and concentrations of reagents should be determined to suit individual circumstances.

7.3 ELISA for specific antibody production

Specific antibody production *in vitro* can be readily measured by solid phase enzyme immunoassays (ELISA). In each assay, a standard curve is constructed to enable the results to be expressed as ng/ml. The standard is obtained by measuring the IgG concentration (Section 7.1) in pooled supernatants from influenza stimulated PBM which contains only specific antibody. Secondary standards from normal human serum can also be used after calibration. Specific antibody containing supernatants are assayed as follows.

Protocol 16. ELISA assay for specific antibody production

Equipment and reagents

- Influenza virus
- PBS
- PBS/BSA: PBS containing 1% BSA (Fraction V from Sigma)
- Alkaline phosphatase conjugated affinity-purified goat anti-human IgG (Sigma cat. No. A-3150)
- p-nitro phenyl phosphate (1 mg/ml) (Sigma 104–105) in alkaline phosphatase buffer: 1.59 g Na_2CO_3, 2.93 g $NaHCO_3$, 9.52 mg $MgCl_2$ made up to 1 litre in double distilled-water and adjusted to pH 9.6
- Non-sterile flat-bottomed 96-well ELISA plates
- Alkaline phosphatase conjugated streptavidin
- p-nitrophenyl phosphate substrate (NPP) in bicarbonate buffer, pH 9.6, containing 10^{-4} M $MgCl_2$
- Incubator
- Multiskan automatic EIA plate reader

A. *ELISA assay for specific IgG*

1. Prepare a suspension of influenza virus (same strain as used for stimulation *in vitro*) at 20–100 µg/ml in PBS containing 0.02% azide, and dispense 75 µl into each well of a flat-bottomed non-sterile 96-well microtitre tray.

2. Incubate for 1 h at 37°C.

3. Recover the virus (which may be used at least 20 times) and store at 4°C.

4. Wash plates twice with PBS and once with PBS 1% BSA.

5. Add 100 μl of 1% BSA in PBS to each well and incubate for 1 h at 37°C to block the remaining non-specific binding sites.

6. Wash once with PBS containing 1% BSA.

7. Add 75 μl of standard (serially diluted from 1:1 to 1:128) to eight consecutive wells for the standard curve, and 75 μl of medium or PBS containing 1% BSA to three wells for the negative control. Add 75 μl of test supernatant to each of the remaining wells.

8. Incubate for 1 h at 37°C.

9. Wash the plate twice with PBS and once with PBS containing 1% BSA.

10. Add 75 μl of alkaline phosphatase conjugated affinity purified goat anti-human IgG, diluted in PBS containing 1% BSA. This particular antibody can normally be used at 1:1000, and stored at 4°C. To detect specific IgM, IgA or IgE, use the equivalent detection antibody of the desired specificity.

11. Incubate at 37°C for 1 h.

12. Wash twice in PBS and twice in distilled water.

13. To each well, add 100 μl of *p*-nitrophenyl phosphate substrate (NPP) in bicarbonate buffer pH 9.6 containing 10^{-4} M $MgCl_2$ and allow the colour to develop (approx. 1 h at 37°C).

14. Read absorbance on a Multiskan automatic EIA plate reader at 405 nm. Results are expressed as ng/ml obtained from the standard curve as described in Section 7.1 above.

B. *ELISA assay for specific IgG subclasses*

1. Coat non-sterile flat-bottomed microtitre wells with 75 μl of purified influenza virus, block, then add standards and test supernatants as described in *Protocol 16A*, steps 1–9.

2. Add 75 μl of biotinylated monoclonal antibody to human IgG subclass (described in Section 7.2 and *Table 8*) diluted appropriately in PBS containing 1% BSA. Optimal dilutions should be determined for each batch of monoclonal antibody.

3. Incubate at 37°C for 1 h, then wash the plate twice in PBS and once in PBS containing 1% BSA.

4. Add 75 μl of alkaline phosphatase conjugated streptavidin and incubate at 37°C for 1 h, then wash twice in PBS and twice in distilled water.

5. Add 100 μl of *p*-nitrophenyl phosphate substrate for alkaline phosphatase (NPP) in bicarbonate buffer pH 9.6 and incubate for 1–2 h at 37°C, or overnight at 4°C for colour development.

6. Read absorbance on a Multiskan plate reader at 405 nm.

Protocol 16. *Continued*

Note: In the absence of calibrated standard preparations of specific IgG subclass antibodies to influenza virus, it is necessary to express the results simply as OD_{405}, or as units per ml calculated from a standard curve using serum or purified human Ig. Under these conditions, it is not possible to make quantitative comparisons between IgG subclasses, but by using units/ml, it is possible to make valid intra- and inter-experimental comparisons for any one IgG subclass.

References

1. Banchereau, J. (1991). In *The cytokine handbook* (ed. A. W. Thomson), pp. 119–48. Academic Press, London.
2. Jelinek, D. F., Splawski, J. B., and Lipsky, P. E. (1986). *Eur. J. Immunol.*, **16**, 925.
3. Teranishi, T., Hirano, T., Lin, B.-H., and Onoue, K. (1984). *J. Immunol.*, **133**, 3062.
4. Kehrl, J. H., Roberts, A. B., Wakefield, L. M., Jakowlew, S., Sporn, M. B., and Fauci, A. S. (1986). *J. Immunol.*, **137**, 3855.
5. Coffman, R. L., Lebman, D. A., and Shrader, B. (1989). *J. Exp. Med.*, **170**, 1039.
6. Islam, K. B., Nilsson, L., Sideras, P., Hammarstrom, L., and Smith, C. I. E. (1991). *Int. Immunol.*, **3**, 1099.
7. Gearing, D. P. and Ziegler, S. F. (1993). *Current Opin. Haematol.*, in press.
8. Callard, R. E. and Gearing, A. J. H. (1994). *The cytokine facts book*. Academic Press, London.
9. Kaplan, M. E. and Clark, C. (1974). *J. Immunol. Methods* **5**, 131.
10. Callard, R. E. and Smith, C. M. (1981). *Eur. J. Immunol.*, **11**, 206.
11. Proust, J. J., Chrest, F. J., Buchholz, M. A., and Nordin, A. A. (1985). *J. Immunol.*, **135**, 3056.
12. Walker, L., Guy, G., Brown, G., Rowe, M., Milner, A. E., and Gordon, J. (1986). *Immunology*, **58**, 583.
13. Aarden, L. A., de Groot, E. R., Schaap, O. L., and Lansdorp, P. M. (1987). *Eur. J. Immunol.*, **17**, 1411.
14. Muraguchi, A., Kishimoto, T., Miki, Y., Kuritani, T., Kaieda, T., Yoshizaki, K., and Yamamura, Y. (1981). *J. Immunol.*, **127**, 412.
15. Mosmann, T. (1983). *J. Immunol. Methods*, **65**, 55.
16. Cambier, J. C., Monroe, J. G., Coggeshall, K. M., and Ransom, J. T. (1985). *Immunology Today*, **6**, 218.
17. Finney, M., Guy, G. R., Michell, R. H., Dugas, B., Rigley, K. P., Gordon, J., and Callard, R. E. (1990). *Eur. J. Immunol.*, **20**, 151.
18. Rigley, K. P., Thurstan, S. M., and Callard, R. E. (1991). *Int. Immunol.*, **3**, 197.
19. Saiki, O. and Ralph, P. (1981). *J. Immunol.*, **127**, 1044.
20. Defrance, T., Vanbervliet, B., Pene, J., and Banchereau, J. (1988). *J. Immunol.*, **141**, 2000.
21. Shields, J. G., Smith, S. H., Levinsky, R. J., Defrance, T., de Vries. J. E., Banchereau, J., and Callard, R. E. (1987). *Eur. J. Immunol.*, **17**, 535.

22. Pene, J., Rousset, F., Briere, F., Chretien, I., Paliard, X., Banchereau, J., Spits, H., and de Vries, J. E. (1988). *J. Immunol.*, **141**, 1218.
23. Vercelli, D., Jabara, H. H., Arai, K.-I., and Geha, R. S. (1989). *J. Exp. Med.*, **169**, 1295.
24. Vercelli, D., Jabara, H. H., Arai, K., Yokota, T., and Geha, R. S. (1989). *Eur. J. Immunol.*, **19**, 1419.
25. Shapira, S. K., Vercelli, D., Jabara, H. H., Fu, S. M., and Geha, R. S. (1992). *J. Exp. Med.*, **175**, 289.
26. Zhang, K., Clark, E. A., and Saxon, A. (1991). *J. Immunol.*, **146**, 1836.
27. Thyphronitis, G., Tsokos, G. C., June, C. H., Levine, A. D., and Finkelman, F. D. (1989). *Proc. Natl. Acad. Sci. USA*, **86**, 5580.
28. Lundgren, M., Persson, U., Larsson, P., Magnusson, C., Edvard Smith, C. I., Hammarstrom, L., and Severinson, E. (1989). *Eur. J. Immunol.*, **19**, 1311.
29. Flores-Romo, L., Millsum, M. J., Gillis, S., Stubbs, P., Sykes, C., and Gordon, J. (1990). *Immunology*, **69**, 342.
30. Callard, R. E. (1979). *Nature*, **282**, 734.
31. Callard, R. E., Booth, R. J., Brown, M. H., and McCaughan, G. W. (1985). *Eur. J. Immunol.*, **15**, 52.
32. Zanders, E. D., Smith, C. M., and Callard, R. E. (1981). *J. Immunol. Methods*, **47**, 333.

11

T cells and cytokines

S. B. A. COHEN, J. CLAYTON, M. LONDEI, and M. FELDMANN

1. Introduction

The functions of T cells and cytokines are intimately interlinked. It was not until T cell growth factors were defined and subsequently purified, and the cDNA encoding it cloned as 'interleukin 2' (1), that abundant non-malignant T cells could be cultured. The isolation of this and other T cell growth factors, now included in the broader category of cytokines, has permitted the development of long-term T cell-lines and clones. T cells can be isolated from normal healthy individuals or patients with disease, according to their antigen specificity and/or phenotype, and then perpetuated in long-term culture by cytokine stimulation. IL-2 is the major T cell cytokine used but certain subsets of T cells will proliferate in response to other cytokines, some independent of IL-2. For example IL-3 will stimulate most $\alpha\beta$ T cell receptor expressing $CD4^-$ $CD8^-$ T cells to proliferate, in man and mouse (2, 3).

T cells are important immunoregulatory cells involved in the activation of B cells, macrophages and other T cells, and in suppression and cytotoxicity of certain target cells. Many of these functions are mediated (or in the case of cytotoxicity enhanced) by cytokines. More than 20 different cytokines are known to be produced by T cells upon activation, the range depending on the phenotype of the cell. These include $CD4^+$ T cells (originally thought to be the major T cell producers of cytokines), $CD8^+$ T cells (4), and cells expressing neither CD4 nor CD8 T cell markers (called double negative cells) (Katsikis *et al.*, submitted). The cytokine profiles of these cells are discussed below. Although cytokines were first defined as secreted proteins, an increasing number have also been detected in a cell surface form, including IL-1α, TNF-α, TNF-β (lymphotoxin) complexed to LTβ, TGF-β, c-kit ligand, TGF-α, and M-CSF. The relative functions of the secreted surface and cell forms are not completely understood.

2. Isolation of T cells for analysis of cytokine production

Peripheral blood mononuclear cells (PBMC) consist of T cells (approximately 50–70%), B cells (10%), monocytes (10–25%) and natural killer (NK) cells

(10%). For a routine mitogen-induced or specific antigen-induced proliferation assay of fresh blood derived T cells, PBMC can be used without further purification of the T cell fraction (*Protocol 1*). This provides autologous monocytes as antigen presenting cells for antigen specific proliferation assays. The effect of various cytokines or anti-cytokine antibodies can then be tested on the immune response (see below).

To determine the ability of T cells to produce cytokines within the PBMC population, the T cells should receive a specific stimulus such as anti-CD3, as cells other than T cells within this population can also produce cytokines. Alternatively, the T cells can be separated from this population, for example by rosetting the T cells with sheep red blood cells (SRBC) (*Protocol 2*) or removal of monocytes by adherence to plastic followed by depletion of B cells with anti-B cell antibodies (*Protocol 3*). The T cells can then be stimulated in isolation, or with the addition of fixed or irradiated autologous monocytes as antigen presenting cells. In addition to isolating T cells from PBMC, T cells can be isolated from other body fluids such as synovial fluid (*Protocol 4*) or from solid tissue (*Protocol 5*). Methods for isolating T cells from all these sources are described below.

2.1 T cell separation using a one-step Ficoll gradient

Protocol 1. Separation of mononuclear cells from peripheral blood (at room temp.)

Equipment and reagents

- 15 and 50 ml conical tubes (Falcon) or 25 ml Universal tubes (Sterilin)
- Heparin (Leo Laboratories)
- Hank's balanced salt solution (HBSS) (Falcon) + 2% heat-inactivated fetal calf serum (FCS) (Biological Industries)
- Lymphoprep sodium metrozoate-Ficoll solution (Nycomed)

- Benchtop centrifuge
- Complete RPMI medium: RPMI 1640 (Gibco), supplemented with 10% AB^+ normal human serum (e.g. from National Blood Transfusion Centre), 2 mM L-glutamine, 100 U/ml penicillin/streptomycin (Gibco)
- Haemocytometer

Method

1. Fill conical or Universal tubes (depending on blood volume) to one-third full.

2. Add heparin to blood at 5 U/ml. Mix heparinized blood with an equal volume of HBSS and layer gently over Lymphoprep by pipetting.

3. Centrifuge at 200 *g* for 20 min.

4. Aspirate off the white buffy coat containing the PBMC at the serum/Lymphoprep interface and pipette into another tube, to no more than half volume. Discard the remaining Lymphoprep with the red cell and granulocyte pellet beneath.

5. Fill aspirated PBMC suspension to the top of the tube with HBSS + 2% FCS.

6. Centrifuge at 400 *g* for 15 min, to pellet the PBMC.

7. Discard the supernatant, then wash by resuspending PBMC in HBSS + 2% FCS. Centrifuge at 200 *g* for 10 min. Repeat.

8. Finally, resuspend the cells in 5–10 ml complete RPMI; count, using a haemocytometer, then dilute further to required concentration (e.g. 10^6/ml).

Expected yield: we normally expect to obtain $1-2 \times 10^6$ PBMC/ml of blood, although this can vary, particularly in some disease states. These cells can then be separated further into subsets as detailed below.

Protocol 2. SRBC-rosette formation

Equipment and reagents

(In addition to *Protocol 1*):
- S-2-aminoethylisothiouronium bromide hydrobromide (AET; Sigma)
- Sheep red blood cells (SRBC) in Alsever's solution (Gibco)
- 0.2 μm filter (Millipore)
- Sterile distilled water

Method

All steps are at room temp. unless otherwise stated.

1. Centrifuge SRBC at 400 *g*, discard the supernatant and resuspend in HBSS+5% serum. Repeat three times.

2. Make up AET at 40 ng/ml, i.e. 2 μg in 50 ml distilled water. Adjust the pH to 9.0 and filter to sterilize. This reagent must be made fresh for each use.

3. Add approximately 50 ml AET solution to the SRBC pellet and leave for 30 min.

4. Centrifuge at 400 *g*.

5. Discard the supernatant and resuspend in HBSS + 5% serum. Wash 3 times.

6. Resuspend in 20 ml of complete RPMI.

7. Adjust the concentration of Lymphoprep-separated PBMC to 10^7/ml in RPMI + 10% human serum.

8. Add a 1/10 to 1/20 vol. of the AET-SRBC suspension and mix gently but thoroughly. Leave for 1 h.

Protocol 2. *Continued*

  9. Centrifuge at 400 *g* for 10 min.

10. Leave the pellet at 4°C overnight.

11. Decant the supernatant, swirl the pellet by hand and resuspend in 15 ml of medium.

12. Overlay on to 30 ml of Lymphoprep and spin at 800 *g* for 20 min.

13. Harvest the E$^+$ cell fraction (T cells), which is the pellet, by lysing the SRBC in ammonium chloride lysis buffer.

2.2 Negative B cell depletion using immunomagnetic beads

We recommend use of the protocol provided by Dynal with their reagents. Essentially this involves direct coupling of the magnetic beads with monoclonal antibodies which distinguish lymphocyte subsets. This is followed by separation in a magnetic field to remove the targeted cells. In our experience, this method of cell separation works better for negative rather than positive selection.

Protocol 3. Preparation from synovial fluid (SF)

Equipment and reactions

- 25 ml heparinized Evans bottles
- Heparin
- Centrifuge

- HBSS + 2% FCS (see *Protocol 1*)
- Ficoll gradient (see *Protocol 1*)

Method

1. Collect SF in 25 ml heparinized Evans bottles. Add extra heparin to SF to 5 U/ml to prevent clotting.

2. Centrifuge hard at 400 *g* for 15 min to pellet the white cell infiltrate.

3. Aspirate off the supernatant (which can either be stored at −20°C or discarded).

4. Wash cells by resuspending the pellet in the same tube in 20 ml of HBSS + 2% FCS.

5. Centrifuge at 200 *g* for 10 min. Repeat.

6. Separate mononuclear cells from other cells on a one-step Ficoll gradient as in *Protocol 1*.

Protocol 4. Preparation from solid tissue, e.g. synovial membrane, by enzymatic digestion

Equipment and reagents

(In addition to *Protocol 1*):
- Enzymes: 5 mg/ml collagenase (Boehringer) + 0.15 mg/ml DNase (Sigma) i.e. for a 20 ml solution, weigh 100 mg of collagenase and 3 mg of DNase
- Dulbecco's MEM + 1% Gentamicin (Gibco) + 1% FCS
- RPMI + 1% Gentamicin + 5% FCS

- Sieve beaker (glass beaker with nylon mesh taped over)
- Conical flask
- 20 ml and 5 ml syringes (Becton Dickinson)
- Scissors/forceps
- 9 cm Petri dish (Nunc)
- 0.2 μM filter (Millipore)
- 50 ml conical tube

Method

1. Wash the tissue with minimal essential medium (MEM), if contaminated with blood, then chop up into approximately 0.5 mm pieces, in 10 ml MEM in a Petri dish.

2. Put the tissue into the flask and filter enzyme mixture over it.

3. Incubate in a 37°C waterbath, shaking intermittently, for 1–2 h depending on consistency (keep checking).

4. Remove flask when tissue is broken down and shake vigorously by hand to break up remaining clumps for approx. 5 min. Add approximately 30 ml MEM to the flask to prevent further enzyme activity.

5. Pipette the suspension in small volumes on to the sterile sieve beaker, and wash through with MEM. It is usually necessary to use the plunger of a 5 ml syringe to push the cells through the mesh.

6. Pipette the suspension into a 50 ml conical tube, dilute further with MEM and centrifuge at 200 g for 10 min at 4°C.

7. Discard the supernatant and resuspend the cells in MEM. Centrifuge and wash twice.

8. Resuspend the cells in MEM, and count.

9. If there is sufficient cell yield (see below), mononuclear cells can now be separated on Ficoll as in *Protocol 1*.

Expected yield: this can vary enormously depending on the size, cellularity and quality (e.g. freshness) of the tissue sample.

---

3. Analysis of T cell cytokine production

The wide spectrum of T cell cytokine production, which include IL-1α, IL-2, IL-3, IL-4, IL-5, IL-6, IL-7, IL-8, IL-9, IL-10, IL-13, GM-CSF, TGF-α, TGF-β1, TGF-β2, TNFα, TNFβ, and IFNγ, sometimes makes their analysis difficult compared to those of other cell types such as macrophages. This problem is accentuated by the lesser quantity of cytokines sometimes produced by T cells.

One type of assay useful for screening purposes is that of cytokine mRNA. For normal T cells, a good correlation has been detected between mRNA levels and protein produced (5). The exceptions (mRNA, but little protein) are due to immunoregulatory effects (e.g. effects of TGF-β2 or of drugs).

The advantage of mRNA assays is their relative simplicity: multiple cytokine mRNAs can be assayed using the same Northern blots if the appropriate probe is available. The blots can be reprobed many times, e.g. up to ~ 10 times, and stored for use in the future with cytokines which may not yet be defined. A further advantage is that the mRNA assays are molecule-specific, in marked contrast to most of the currently used bioassays which demonstrate the presence of protein.

mRNA assays also have certain disadvantages. Northerns require a larger amount of mRNA, from 10^6–10^7 cells (usually 10^7), depending on mRNA abundance, while slot-blots require fewer cells, about 1–2×10^6. Northerns however provide more precise information (6), for example, the length of the mRNA can be established by Northern but not by slot- or dot-blot analysis, to verify the identity, and cytokines with different message sizes can be detected simultaneously. In addition, slot- or dot-blots can be marred by non-specific hybridization. Most importantly, the presence of mRNA does not unambiguously demonstrate the release of functional protein available for signalling. Detailed protocols for analysis of cytokine mRNA are given in Chapters 3, 4, and 5.

The polymerase chain reaction (PCR) has several advantages over the study of mRNA by Northern blot analysis. Chiefly, very little mRNA is needed, and therefore very few cells, sometimes only one. In addition, the appropriate probes, specific oligonucleotide primers, are easily and relatively cheaply obtainable, and are now commercially available in kits which cover most of the cytokines produced by T cells (e.g. from Stratagene). A unique advantage to this technique is the possibility of storing cDNA obtained from different cell types for use in the cloning of as yet undefined cytokines, particularly when novel activities are described in supernatants or biological fluids. But the extreme sensitivity of the PCR technique also creates disadvantages. A small contamination of the test sample will lead to false positive results. Protocols for RT-PCR measurement of cytokine mRNA are given in Chapters 4 and 5.

Additional techniques which provide useful information are the study of mRNA by *in situ* hybridization, intracellular protein accumulation by immunofluorescence, using flow cytometry or immunohistochemistry. Using double staining these techniques can identify which cells are transcribing, producing and/or secreting the relevant cytokine. These techniques can be used to analyse cytokine/cellular interactions in tissue, in addition flow cytometry is useful for quantitating cytokine production by specific cells (7). Detailed protocols for this are found in Chapters 4 and 5.

Immune binding assays have become increasingly popular, due to the availability and use of recombinant cytokines to generate specific polyclonal antisera and monoclonal antibodies. With the appropriate calibration, these can be sensitive to 10 pg/ml or less, well within biologically relevant ranges. These assays (radioimmunoassays, ELISA, or more recently luminescence or fluorescence assays) have many advantages. These can be used to detect cytokines in biological fluids in the presence of potentially toxic constituents. If appropriate standards are used, they are quick and sensitive, and are molecule-specific, e.g. IL-1α and IL-1β can (and need to) be assayed separately. Detailed protocols are given in Chapter 22.

These binding assays are all quantitative, and allow the analysis of cytokine production by a specified number of stimulated cells. A more qualitative method of determining cytokine production is the ELISPOT, which will detect cytokine production at the single cell level (8). Using this method 'spots' of colour represent cytokine production by one cell, therefore the proportion of cells producing a particular cytokine in response to either an antigenic or mitogenic stimulus can be determined. As these spots can be enumerated, ELISPOT is quantitative in terms of the number of high producing cells, but not in terms of total cytokine production. A detailed protocol is given in Chapter 20.

In general, however, binding assays are restricted in their use. Although they quantitatively detect levels of cytokine protein produced, their major limitation is that they provide no indication of the **bioactivity of the cytokine**. An important difference can readily be detected by bioassays and binding assays, e.g. with TNFα, which can be inhibited by TNF-binding molecules present in normal and pathological body fluids derived from the extracellular domain of the TNF receptors. For other cytokines there are also discrepancies between levels of immunoreactive protein and bioassay levels which again may be due to the production of soluble receptors now known for IL-2, IL-4, IL-6, IL-7 (9), IFN-γ (10), and TNF (11). It seems likely that as more receptors are defined and cloned, more cytokine binding proteins and inhibitors will be defined. In addition, in rheumatoid arthritis, for example, rheumatoid factors (anti-IgG) may interfere with binding assays, especially those involving rabbit antibodies. Binding assays are described in Chapter 22.

For the above reasons, the most important cytokine assays remain *bioassays*, which evaluate the levels of functional and active protein, but not cytokines denatured or neutralized by inhibitors. These assays are also not without their limitations. The first is molecular specificity. The well-known murine thymocyte co-stimulation assay, initially thought to be an IL-1 assay, can detect murine IL-1α and β, IL-2, IL-4, IL-6, IL-7, and TNF-α and almost as many human cytokines. However, bioassays can be rendered molecule-specific by the use of neutralizing antibodies. These can establish cytokine identity but if multiple cytokines are present, accurate quantitation is very difficult. In addition it is important to maintain a wide range of sensitive cells for different assays: cell lines may drift in sensitivity, and may need to be recloned and selected which can be time-consuming and expensive.

The methods described for cytokine detection and analysis all have advantages and disadvantages, and thus, when possible, use of several assays will provide more useful and reliable information. For instance, there are conditions in which small amounts of a biologically active cytokine are detected, but a comparatively large amount of mRNA expressed (12).

4. Cytokine production by T cells of different phenotypes

T cells are a varied set of cells comprised of numerous phenotypes, according to their stage in maturation, activation state or lineage. For example normal PBMC consist of a majority of CD45RO-positive T cells expressing either CD4 or CD8 markers and a minority of double-negative (CD4$^-$ CD8$^-$) and double-positive (CD4$^+$ CD8$^+$) cells. These different subsets may respond to, or secrete different cytokines. Monoclonal antibodies specific for subset markers can be used for their identification and isolation.

4.1 Cytokine production by CD4$^+$ cells

By analysing the cytokine production profile (at both the mRNA and protein levels) of a number of well-established murine CD4$^+$ T 'helper' cell clones, Mosmann and colleagues described two groups of clones (13). One was named 'Th$_1$' which secrete IL-2, IFN-γ and TNF-β, but not IL-4, IL-5, or IL-6 while the other, labelled 'Th$_2$' cells, had a reciprocal cytokine secretion profile. These different subsets display functional differences which correlate with their cytokine profile; Th$_1$-type cells mediate delayed type hypersensitivity reactions and help B cells produce IgG2a, whereas Th$_2$-type cells help B cells particularly in the production of IgG1 and IgE antibodies. Precursors to the Th$_1$/Th$_2$-type cells have subsequently been described according to their cytokine profile. So called 'Th$_0$' cells secrete all the above cytokines, and are likely to be a consequence of short-term activation of 'Th$_P$' (precursor) cells, which only secrete IL-2 on activation [see (14) for review]. Although these functionally distinct T cell subsets were originally described in the mouse, human T cell clones have also been shown to have a similar, but not identically restricted cytokine profile. Th$_1$- and Th$_2$-type cells in the human are usually defined according to the ratio of IFN-γ to IL-4 production, as other cytokines which are restricted to a Th$_1$/Th$_2$ subset in the mouse show no similar restriction in the human. IL-10 for example, is produced exclusively by Th$_2$-

type cells in the mouse (15), but is produced by both Th_1 and Th_2 subsets in the human (16, 17). This has interesting implications: Th_1 cells, usually thought of as pro-inflammatory, become anti-inflammatory if the ratio of IL-10 to IFNγ (Katsikis *et al.*, submitted) is high.

4.2 Cytokine production by CD8$^+$ T cells

Despite the earlier belief that CD4$^+$ cells were the chief cytokine producers, it is now well established that CD8$^+$ cells can also produce abundant cytokines [4]. Indeed, CD8$^+$ cells can also be split according to their cytokine profile into 'Th$_1$/Th$_2$' in both mouse and man. Whether the levels of cytokines produced are equal to those of CD4$^+$ cells has yet to be established. As with CD4$^+$ cells the production of cytokines by CD8$^+$ T cells will also determine their function. For example, IFN-γ production by Th$_1$-type CD8$^+$ cells is likely to synergize with their cytotoxic function, since IFN-γ will upregulate class II [18] rendering the target cells more susceptible to lysis.

4.3 Cytokine production by CD4$^-$ CD8$^-$ $\alpha\beta$ T cells

We have recently shown that the majority of human CD4$^-$ CD8$^-$ $\alpha\beta$ T cells have a Th$_2$ phenotype in that they can produce high levels of IL-4 and low levels of IFN-γ. In addition these cells can make IL-10 (Katsikis *et al.*, submitted), when given the approximate stimulation.

4.4 Cytokine production by $\gamma\delta$ T cells

Like $\alpha\beta$ T cells, $\gamma\delta$ T cell receptor-expressing cells can be stimulated to secrete a variety of cytokines which include IL-2, IFN-γ, IL-5, TNFα, and GM-CSF. However, it has yet to be established whether they fall into distinct categories like the Th$_1$/Th$_2$/Th$_0$ subsets of the $\alpha\beta$ T cell lineage (19).

4.5 Optimization of cytokine production in response to various stimuli

T cells have exquisite specificity in terms of recognition of antigen. However it is commonly believed that in terms of effector function, T cells are not specific in their response, liberating whatever cytokines they are capable of producing. Conversely we (unpublished observations) and others (20) have noted differential production of cytokines depending on the nature of the stimulus, suggesting that different cytokines use at least partly separate signalling pathways.

Further considerations in optimizing T cell stimulation conditions are the length of time the cells are stimulated and the concentration of cells used. Different cytokines peak in their mRNA and protein production at different times, after which the cytokine may be stable in solution or removed, for example by being consumed by proliferating cells or binding to soluble receptor. Therefore the optimal time for harvesting cells after stimulation is dependent on the cytokine being analysed. In addition, although a higher concentration of cells will produce a high concentration of cytokines there will also be a higher rate of cytokine consumption. In general we use 1×10^6 T cells/ml and harvest 24 h after stimulation, since this time point is sufficient to detect most cytokines.

5. Cytokine production of antigen specific T cells

T cells can vary in cytokine production according to their antigen specificity (e.g. see *Protocol 5*). For example, mycobacterial antigen specific clones are predominantly high IFN-γ, low IL-4 producers (Th$_1$ type) (21) whereas T cells reactive to allergens such as house dust mite are predominantly high IL-4 low IFN-γ producers (Th$_2$ type) (20).

6. Effect of cytokines on T cells

Cytokines have profound effects on T cells. The parameter most clearly documented is T cell proliferation, but that is not the only effect. T cell activation, detected through cytokine production and cytotoxicity, is also influenced either in a positive way, for example by IL-2 and IL-7, or a negative way by TGF-β and IL-10.

6.1 Regulation of T cell growth

IL-2 was purified, defined and cloned on the basis that it induced T cell growth very effectively. Indeed, for several years it was the accepted dogma that IL-2 was the only T cell growth factor. However, it has become increasingly clear that other cytokines can regulate growth depending on the type of cell. For example the CD4$^-$ CD8$^-$ $\alpha\beta^+$ population can grow independently of IL-2 (2, 3). Within the murine CD4$^+$ T cell subsets a cross-regulatory system has been identified; Th$_1$-type cell growth is enhanced by Th$_1$-type cytokines such as IFN-γ and inhibited by Th$_2$-type cytokines such as IL-4 (22, 23). Conversely, Th$_2$-type cell growth is enhanced by IL-4 and inhibited by IFN-γ. Similarly, in humans cytokines produced by both subsets can also regulate the T cell response. For example, human IL-10, produced late in T cell activation, can directly inhibit Th$_1$- and Th$_2$-type cell proliferation (24) and IL-3, produced by both Th$_1$/Th$_2$-type cells, acts as a growth factor for CD4$^-$ CD8$^-$ $\alpha\beta^+$ T cells, but not CD4$^+$ or CD8$^+$ cells (2).

Protocol 5. Establishment of tetanus antigen-specific T cell lines

Equipment and reagents

(In addition to *Protocol 1*):
- Tetanus toxoid vaccine (TT; Wellcome), dialysed for 24 h against PBS to remove the thiomersal BP toxic preservative, filtered to re-sterilize and stored at 4°C
- [³H]-thymidine (Amersham), stock solution made to 1 mCi/ml
- Recombinant human interleukin-2 (rIL-2) T cell growth factor

- Humidified CO_2 incubator
- ^{137}Cs source
- Benchtop centrifuge
- 24-well tissue culture plate (Falcon)
- 96-well (round-bottomed) microtitre plate (Nunc)
- 20 Terasaki tissue-culture plates (Nunc)
- 6-well plates (Falcon)
- 15 ml conical tube

A. *Primary T cell-line*

1. Adjust concentration of PBMC, prepared as in *Protocol 1*, to 10^6/ml in complete RPMI (for example, a total of 10^7 cells in 10 ml). These PBMC can be either freshly prepared or thawed from liquid nitrogen storage.

2. Pipette 2 ml of the cell suspension into each of five wells of the 24-well plate (i.e. 2×10^6 cells/well).

3. Add 1 TT to four of the five wells, with a different dilution to each; for example, 1/50, 1/100, 1/200, 1/400 of the stock solution. Leave the fifth well as a control, i.e. no tetanus. This is because individuals vary in their sensitivity to TT.

4. Incubate the cells at 37°C in a humidified CO_2 incubator for 6 days.

5. Pipette the cells of each well gently to re-suspend and remove a 100 μl aliquot from each, including the fifth 'control' well, for measurement of T cell proliferation.

6. Place each aliquot into one well of a 96-well microtitre plate.

7. Pulse each well with 10 μl of (3H) thymidine (10 μl Ci) and incubate at 37°C for 8 h.

8. Harvest and measure incorporation of radioactivity by scintillation counting.

9. Select the well containing cells with the highest proliferation rate to establish a long-term line.

10. Add recombinant IL-2 to the selected culture to a final concentration of 10 ng/ml. This will promote the growth and expansion of T cells previously activated by tetanus antigen.

11. Return cells to the incubator for a further 7 days.

The cells are now ready for secondary stimulation with antigen and feeder cells.

Protocol 5. *Continued*

B. *Preparation of autologous feeder cells for antigen presentation*

1. Use as autologous feeder cells prepared PBMC of the same donor as the T cell-line. Irradiate the PBMC cell suspension (at 5×10^6 cells/ml in complete RPMI) with a sublethal dose of 4000 rads from a ^{137}Cs source. This will prevent feeder cell growth and division which would contaminate the T cell-line. The ratio of feeder cells to T cells used is 2–5: 1 (i.e. 2–5 $\times$ 10^6 feeder cells to 10^6 T cells).

2. Adjust feeder cell suspension to the required concentration (e.g. 2 $\times$ 10^6 cells/ml).

3. Pipette 1 ml vol. into each of four wells of a 24-well plate (i.e. 2 $\times$ 10^6 cells/well).

4. Add the appropriate concentration of tetanus antigen, determined on day 6.

C. *Restimulation of the primary T cell-line*

1. Suspend the 2 wells of T cell-line by gentle pipetting, transfer to a 15 ml conical tube and centrifuge at 400 *g*.

2. Re-suspend the cells in 4 ml fresh complete RPMI and divide this into the four wells of the 24-well plate containing autologous feeder cells and antigen (tetanus) to a total volume of 2 ml. No IL-2 is added at this stage.

3. Incubate for 7 days.

4. Add rIL-2 to 10 ng/ml.

D. *Long-term maintenance of the T cell-line*

The cells can now be maintained by alternate weekly cycles. In the first week the cells receive antigen plus feeder cells prepared above in (B) together with rIL-2. In the second week the cells receive IL-2 alone.

Protocol 6. Establishment of T cell clones by limiting dilution

N.B. Incubator for cloning needs to be optimally humidified to prevent desiccation

Equipment and reactions

(In addition to *Protocol 1*):
- rIL-2 (10 ng/ml)
- 15 ml conical tube
- Benchtop centrifuge
- Universal tubes

- Tetanus antigen
- Terasaki plates
- Phase-contrast microscope

Method

1. Re-suspend the primary T cell culture selected for the highest proliferation.

2. Count the cells and pipette the suspension into a 15 ml conical tube.

3. Centrifuge at 400 g for 10 min.

4. Discard the supernatant and resuspend the cells in fresh complete RPMI, to a concentration of 3×10^4 cells/ml.

5. Dilute an aliquot of these cells by serial dilution to obtain cells at a working concentration of 300 and 30 cells/ml (i.e. 100 μl in 10 ml, then a further 1/10 dilution).

6. Prepare autologous feeder cells as above and adjust the concentration to 10^6 cells/ml.

7. Add T cells and feeder cells together in a ratio of 1:1 in two Universal tubes (i.e. 10 ml feeder cells +10 ml of T cells per tube). Feeder cells are now at 5×10^5 cells/ml; and T cells at 1.5×10^2 and 1.5×10^1 cells/ml. Invert several times to thoroughly mix the cells.

8. Add tetanus antigen to each tube to working dilution.

9. Pipette 20 μl of the suspension into each well of the Terasaki plates. There should now be 10 plates for each final concentration of either 3 or 0.3 T cells/well, with 10^4 feeder cells/well. Incubate the cells as before for 7–12 days.

10. Screen by phase contrast microscopy from day 7 for growing colonies and note the positions of the positive wells.

11. Resuspend the cells in positive wells and transfer to wells of a 96-well plate for expansion.

12. Add to each well 30 μl complete RPMI, and a 50 μl mixture of feeder cells at 2×10^6 cells/ml (i.e. 10^5 cells/well), tetanus antigen, and rIL-2 (10 ng/ml), final volume 100 μl.

13. Incubate for 7 days.

14. Add 50 μl rIL-2 (of 3 × working dilution) to each well. Incubate for 7 days.

15. Add 50 μl mixture of feeder cells at 2×10^6/ml, tetanus antigen and rIL-2 (at 4 × working concentration). Incubate for 7 days.

16. Check wells at each stage for T cell growth. Further expansion into 24-well plates followed by 6-well plates can be done with appropriate numbers of cells and reagents at any alternate feeding stage if cells have reached maximum growth. For 24-well plates, transfer in a final volume of 2 ml, approx. $2–5 \times 10^5$ T cells and 10^6 feeder cells/well, antigen, and rIL-2 to 10 ng/ml.

Protocol 6. *Continued*

17. When T cell clones have reached sufficient numbers of cells they can be examined for mitogen and antigen reactivity as well as their cytokine physiology.

Comments

The starting frequency of tetanus-specific T cells in the PBMC population will greatly influence the number of positive colonies in the concentration range recommended for placing into Terasaki plates. This can only be determined by trial and error with blood from individual donors. Recently primed donors (less than 1 year) are usually highly responsive.

6.2 Variations on cloning technique

(a) An alternative cloning method after selection of an antigen-primed culture is to expand clones non-specifically using the anti-T cell monoclonal anti-body OKT3 hybridoma available from American Tissue Culture Collection (ATCC). This is pre-absorbed on to the irradiated feeder cells which, because the T cells are separated clonally, can be from an irrelevant donor mismatched for HLA. This can be prepared, for example, by a 30 min incubation of 10^8 feeder cells with 5 µl of a 100 µg/ml stock solution at 4°C, followed by one wash in complete RPMI.

(b) The pan-T cell mitogen phytohaemagglutinin (PHA) is also used instead of OKT3 by some workers for non-specific expansion of primed T cells and it appears that both methods work equally well. This non-specific expansion gives greater flexibility to the choice of feeder cells, which is vital when donor supplies are short. Antigen-specific reactivity can then be tested using autologous feeder cells when convenient. This procedure is useful if the antigen is either unknown or rate limiting, as in the case of autoantigens.

(c) With large numbers of feeder cells available, it is possible to clone directly into 96-well plates, with the same T cell numbers as in *Protocol 6*, but with feeders at 10^5 cells/well. For 10 plates, for example, this would require 10^8 feeder cells for each feeding cycle.

(d) It may also be preferable to expand non-specifically with PHA from the start if antigen reactivity is not a priority and T cell subsets have been isolated for clonal expansion by, for example, FACS. If there are insufficient cells for FACS sorting from the starting PBMC, these can be expanded first for 7 days with PHA + IL-2.

6.3 Isolation of autoreactive T cells

T cells reactive to self antigens such as thyroglobulin and thyroid peroxidase (TPO) (autoantigens specific to the thyroid gland) may be at least in part responsible for initiation and or perpetuation of autoimmune disease. Isolation and characterization of these cells is therefore of importance. However, these T cells are difficult to isolate for various reasons. At the time of disease presentation, these T cells may be in too low a frequency to detect, or may be suppressed by factors in the surrounding environment. For example, TGF-β (25), and IL-10 [18] are both produced in abundance by synovial membrane cells of rheumatoid arthritis patients and are implicated in T cell suppression. These factors may therefore inhibit detection of autoantigenic T cells. Novel methods may therefore be required for the detection of autoantigen-specific cells. We have recently described a protocol which enables the detection of TPO-specific clones, using autologous Epstein–Barr virus (EBV) transformed B cells, transfected with TPO antigen as antigen presenting cells (26). Using this method enabled the detection of TPO responsive clones which were previously undetectable using autologous EBV transformed B cells as antigen presenting cells and recombinant TPO. This method may be useful in the detection and isolation of other autoreactive T cells which have previously proved difficult to isolate.

6.4 Regulation of activation

The cytotoxic activity of T cells is a complex phenomenon. Elucidation of its mechanism has preoccupied many groups, and has led to the definition of three mechanisms which are not mutually exclusive. Cytotoxic activity can involve release of pore forming molecules, 'perforins', which introduces itself into the membrane and assembles to form pores, granule enzymes, 'granzymes', which are serine proteases, or cytotoxic mediators such as TNF-α and TNF-β. Cytostatic cytokines such as the IFNs may also play a part.

Cytotoxic activity is not constitutive, and has been reported to be inducible by triggering of the T cell receptor. However, IL-2, IL-4, IL-7, and IFN-γ have also all been shown to be able to induce cytotoxic activity in suitable cells.

6.5 Measurement of T cell growth in response to cytokines and antigen stimulation

T cell growth can be measured in a variety of ways depending on the assay involved. T cell proliferation assays are often performed in flat- or round-bottomed 96-well microtitre plates in which different stimulating cytokines or antigens can be placed, in varying concentrations. These 96-well plates are especially useful since they allow multiple replicates and serial dilutions to be tested.

6.6 T cell proliferation—direct [^{3}H]thymidine incorporation

Measurement of [^{3}H]thymidine incorporation is commonly taken as an indicator of T cell activation, growth, and cell division. This can be following cytokine, mitogen, and antigen stimulation. The tetanus response in *Protocol 5* is an example of the latter. It is usually wise to optimize concentration of the stimulus and the time of the peak of the response.

6.7 IL-2 release bioassay

Antigen-induced T cell activation can also be determined by measurement of IL-2 release. The culture supernatants of each well can be removed and added to a second 96-well plate containing the mouse cell-line HT2, which proliferates in the presence of human IL-2. This proliferation can then be measured by [³H]thymidine incorporation. The HT2 cells should be prepared as follows:

Protocol 7. IL-2 stimulation of CTLL-2 cells (ATCC)

Equipment and reactions

- RPMI + 10% FCS, 2 mM sodium pyruvate (Sigma), 50 U/ml IL-2, 2 mM ʟ-glutamine, 100 U/ml penicillin/streptomycin
- CTLL-2 cells
- [³H]thymidine
- Benchtop centrifuge
- 96-well plates
- T cell culture supernatant
- Humidified CO_2 incubator
- Liquid scintillation counting system

Method

An alternative cell line is the HT2 cell-line (ATCC)

1. Grow CTLL-2 cells in RPMI + 10% FCS, 2 mM sodium pyruvate (Sigma), 50 U/ml IL-2, 2 mM ʟ-glutamine, and 100 U/ml penicillin/streptomycin.

2. Grow cells to a concentration of 1×10^4 cells/well and feed at 3 to 4 day intervals.

3. Two to three days after the last feeding centrifuge at 400 g for 10 min.

4. Resuspend in fresh RPMI/FCS. Repeat.

5. Place 100 μl aliquots of CTLL-2 cells at 4×10^4 cells/ml into wells of a 96-well plate (to give 4×10^3 cells/well).

6. Add 100 μl of T cell culture supernatant (neat, or in serial dilution) and IL-2.

7. Incubate at 37°C in a humidified CO_2 incubator for 24 h.

8. Pulse each well with 10 μl containing 10 μCi of [³H]thymidine for 48 h.

9. Harvest and count radioactivity by liquid scintillation counting.

10. Compare the values obtained with a standard curve of CTLL-2 cells assayed with rIL-2 at concentrations ranging from 0.001–200 μg/ml. Note: this cell line also responds to murine IL-4, but not human IL-4.

---

7. Conclusions

T cells are a major source of cytokine production, and their function as 'leader of the immunological orchestra', as envisaged by Dick Gershon, is chiefly mediated by cytokines. The way in which this is achieved is as complex as the properties of the T cells themselves. As yet we have only a very rudimentary knowledge of how cytokines influence T cells and how T cells selectively release certain cytokines in response to diverse stimuli. Achieving an understanding of these activities will considerably augment our understanding of how the immune system is regulated.

References

1. Smith, K. A. (1984). *Ann. Rev. Immunol.*, **2**, 319.
2. Londei, M., Verhoef, A., de Berardinis, P., Kissonerghis, M., Grubeck-Loebenstein, B., and Feldmann, M. (1989). *Proc. Natl. Acad. Sci. USA*, **86**, 8502.
3. Kubuta, H., Okazaki, H., Onuma, M., Kano, S., Hattori, M., and Minato, N. (1992). *J. Immunol.*, **149**, 1143.
4. Salgame, P., Abrams, J. S., Clayberger, C., Goldstein, H., Convit, J., Modlin, R. M., and Bloom, B. R. (1991). *Science*, **254**, 279.
5. Cherwinski, H. M., Schumacher, J. H., Brown, K. D., and Mosmann, T. R. (1987). *J. Exp. Med.*, **166**, 1229.
6. Buchan, G., Barrett, K., Turner, M., Chantry, D., Maini, R. N., and Feldmann, M. (1988). *Clin. Exp. Immunol.*, **73**, 449.
7. de Caestecker, M. P., Telfer, B. A., Hutchinson, I. V., and Ballardie, F. W. (1992). *J. Immunol. Methods*, **154**, 11.
8. Kabilan, L., Andersson, G., Ekre, H.-P., Olsson, T., and Troye-Blomberg, M. (1990). *Eur. J. Immunol.*, **20**, 1085.
9. Kaczmarski, R. S. and Mufti, G. J. (1991). *Blood*, **5**, 193.
10. Novick, D., Engelmann, H., Wallach, D., and Rubinstein, M. (1989). *J. Exp. Med.*, **170**, 1409.
11. Seckinger, P., Zhang, J. H., Hauptmann, B., and Dayer, J. M. (1990). *Proc. Natl. Acad. Sci. USA*, **87**, 5188.
12. Feldmann, M., Brennan, F. M., and Maini, R. N. (1993). In *Clinical applications of cytokines: role in pathogenesis, diagnosis and therapy* (ed. A. Gearing, J. Rossio, and J. Oppenheim), p. 109. Oxford University Press.
13. Mosmann, T. R., Cherwinski, H., Bond, M. W., Giedlin, M. A., and Coffman, R. L. (1986). *J. Immunol.*, **136**, 2348.
14. Street, N. E. and Mossman, T. R. (1991). *FASEB*, **5**, 171.
15. Mosmann, T. R. and Moore, K. W. (1991). *Immunol. Today*, **12**, 49.
16. Yssel, H., de Waal Malefyt, R., Roncarolo, M. G., Abrams, J. S., Lahesmaa, R., Spits, H., and de Vries, J. E. (1992). *J. Immunol.*, **149**, 2378.
17. Del Prete, G., De Carli, M., Almerigogna, F., Giudizi, M. G., Biagiotti, R., and Romagnani, S. (1993). *J. Immunol.*, **150**, 353.
18. Trinchieri, G. and Perussia, B. (1985). *Immunol. Today*, **6**, 131.
19. Haas, W., Pereira, P., and Tonegawa, S. (1993). *Ann. Rev. Immunol.*, **11**, 637.

20. Yssel, H., Johnson, K. E., Schneider, P. V., Wideman, J., Terr, A., Kastelein, R., and De Vries, J. E. (1992). *J. Immunol.*, **148**, 738.
21. Haanen, J. B. A. G., deWaal Malefijt, R., Res, P. C. M., Kraakman, E. M., Ottenhoff, T. H. M., de Vries, R. R. P., and Spits, H. (1991). *J. Exp. Med.*, **174**, 583.
22. Swain, S. L., Weinberg, A. D., English, M., and Huston, G. (1990). *J. Immunol.*, **145**, 3796.
23. Le Gros, G., Ben-Sasson, S. Z., Seder, R., Finkelman, F. D., and Paul, W. E. (1990). *J. Exp. Med.*, **172**, 921.
24. Taga, K. and Tosato, G. (1992). *J. Immunol.*, **148**, 1143.
25. Brennan, F. M., Londei, M., Jackson, A. M., Hercend, T., Brenner, M. B., Maini, R. N., and Feldmann, M. (1988). *J. Autoimmun.*, **1**, 319.
26. Mullins, R., Chernajovsky, Y., Dayan, C., Londei, M., and Feldmann, M. (1994). *J. Immunol.*, **52**, 5572.

The generation and quantitation of cell-mediated cytotoxicity

ELIZABETH GRIMM and WILLIAM LOUDON

1. Introduction

Cell-mediated cytoxicity (CMC) is defined as the process by which a cell of the immune system directly kills another cell. CMC is composed of a specific series of events occurring between an effector cell (lymphocyte or monocyte) and a susceptible target cell which ultimately results in the delivery of a 'lethal hit' by the effector to the target cell. An obligate event in CMC, which distinguishes it from other modes of immunologically mediated target cell killing (e.g. complement-mediated cytolysis) is intimate cell membrane contract between the effector and the target, referred to as conjugate formation. This brief chapter will be restricted solely to the description of lymphocyte-mediated CMC systems. Many excellent reviews cover the complicated processes which encompass effector cell-mediated recognition, conjugate formation, triggering, and delivery of the lethal hit ultimately resulting in the death of the specific target (1, 2).

Lymphocyte populations capable of expressing CMC can be divided into two major categories based upon the target structures recognized by the effector population: major histocompatibility complex (MHC)-restricted and MHC-unrestricted killer cells.

Classical cytotoxic T lymphocytes (CTL) mediate antigen-specific, MHC-restricted cytolytic activity, presumably through the interaction of the T cell receptor with a specific processed target antigen presented by the appropriate MHC structure. The generation of these MHC-restricted effector lymphocytes requires the co-ordinated assistance of accessory cell populations in order to educate and mature the T killer. In contrast, MHC-unrestricted cytotoxic effectors exhibit the inherent capacity to recognize an as yet undetermined characteristic(s) of certain 'abnormal' cell populations. As the name implies, the recognition process is independent of MHC, demands no prior antigen presentation, and requires no known interaction with accessory cells. Natural killer (NK) and lymphokine-activated killer (LAK) cells represent the major effector types expressing lymphocyte-mediated,

MHC-unrestricted cytotoxicity. Considerable controversy remains surrounding attempts to unambiguously distinguish NK and LAK, based upon phenotypic or morphologic criteria. We and others have proposed that a definition based solely upon activation requirements, and the resultant expression of differential susceptible target spectra, offers the most practical distinctions (3). Natural killer cells are endowed with the inherent capacity to spontaneously kill a defined target spectrum including some haematological neoplasms (*Figure 1a*) (NK may play an additional role in inhibiting the vascular

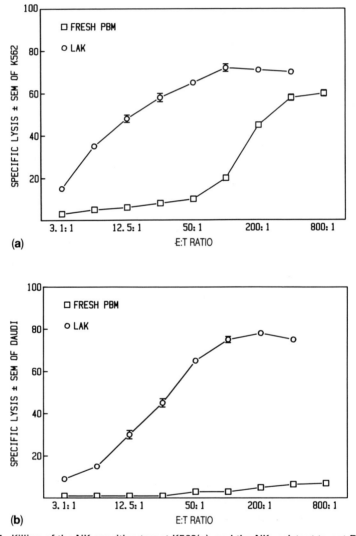

Figure 1. Killing of the NK-sensitive target K562(a), and the NK-resistant target Daudi (b) by LAK (PBL+IL-2, 4 days of activation) and by fresh PBMC from the same donor.

spread of metastatic emboli), virally infected tissues, and may play a role in controlling haematopoiesis (4). Although NK cells are not known to require any activation prior to expression of their lytic activity, cytokine augmentation of NK potency is well-documented. In contrast, LAK manifest an absolute dependency upon activation prior to the induction of their unique capacity to recognize both NK-sensitive (*Figure 1a*), as well as a broad spectrum of NK-resistant targets (*Figure 1b*), including both fresh and cultured tumours growing in suspension or adherently, embryonic tissues, virally infected tissues, tissue culture- and chemically-modified cells. Maintenance of this lytic potential as well as expansion of the lytic population is likewise dependent upon a continuing source of activation stimuli. We therefore hypothesize that LAK may represent one part of an inducible limb of the immune system, in which, when the local environment provides the appropriate quality and quantity of activation signals, prior effector functions may be overridden and a new, broad spectrum target recognition system may be evoked. As such, even antigen-specific CTL and NK may be driven to acquire lytic activity against an entirely new, MHC-unrestricted target spectrum.

2. Techniques for the isolation of lymphocyte populations

2.1 Isolation of PBMC, PBL, or LGL

Peripheral blood mononuclear leukocytes (PBMC), derived from venupuncture or leukophoresis in the presence of anti-coagulants (e.g. heparin, EDTA, etc.) can be isolated by sedimentation on gradients such as Histopaque 1077 for human (or rat) cells or Lympholyte-M for mouse blood (Sigma). (See also *Protocol 2*, Chapter 10 and *Protocol 1*, Chapter 11.)

Protocol 1. Isolation of PBMC, PBL, or LGL

Equipment and reagents

- Anti-coagulated blood
- Ca^{2+}/Mg^{2+} deficient Hank's balanced salt solution (HBSS)
- Sedimentation gradient (Histopaque 1077 for human or rat cells, Lympholyte-M for mouse blood; Sigma)
- Cell centrifuge

Method

1. Dilute the anticoagulated blood (1:2 blood to Ca^{2+}/Mg^{2+} deficient Hank's balanced salts solution (HBSS) for venupuncture-derived and 1:4 for leukophoresis-derived blood) and then layer over the gradient material as per manufacturer's instructions, and centrifuge.

2. Aspirate the 'buffy coat' cellular interface and wash twice in HBSS.

Protocol 1. *Continued*

3. To eliminate contaminating platelets, the cell suspension may be subjected to slow-speed centrifuge runs until contaminating platelets are no longer evident microscopically (150 g for 10 min; aspirate supernatants to avoid losing the weakly packed pellet).

We find that platelet contamination can significantly inhibit lymphocyte activation, presumably through platelet-derived immunosuppressive agents such as the transforming growth factors. PBMC can be directly cultured or further enriched for effector cell precursors. Peripheral blood lymphocytes (PBL) are prepared from PBMC by plastic and nylon wool depletion of monocytes and B cells, respectively (5).

Protocol 2. Preparation of peripheral blood lymphocytes

Equipment and reagents

- Complete medium (complete defined serum-free medium and/or serum containing complete medium)
- Tissue-culture grade plastic flasks (e.g. 175 cm^2)
- Cell centrifuge
- Sterile, pre-washed nylon wool

Method

1. Dilute PBM to approximately 10^7 cells/ml in complete medium (either serum-supplemented growth medium or complete, defined serum-free medium) and incubate on tissue culture grade plastic (e.g. 20 ml into a 175 cm^2 flask) for 1 h at 37°C.

2. Aspirate non-adherent cells and gently wash adherent cells with warm medium to collect remaining non-adherent cells.

3. Pellet the non-adherent cells and resuspend to 5×10^7 cells/ml.

4. Incubate for 1 h at 37°C on sterile, pre-washed nylon wool which has been pre-incubated with serum-containing complete medium (0.6 g of nylon wool per 10^8 cells).

5. Following incubation, aspirate the non-adherent cells, and gently wash the nylon wool with warm medium to recover residual non-adherent cells. Plastic adherence followed by nylon-wool depletion significantly enriches for lymphocytes expressing T cell and NK phenotypic markers.

At this point, PBL can be further segregated into populations with differing morphologies based upon cell densities following the procedure of Timonen *et al.* (6).

Protocol 3. Cell density purification

Equipment and reagents
- Percoll (Pharmacia) adjusted to 290 osmol/kg with 10 × PBS
- 50 ml conical polystyrene centrifuge tubes
- Cell centrifuge
- RPMI medium containing 0.75% BSA
- Trypan blue

Method

1. Layer PBL on to multi-step, discontinuous Percoll gradients. The choice of the densities and the number of steps composing the gradient will be determined by the cell populations desired, the purities desired, and by the total number of cells needing separation. As a general starting point, adjust Percoll (Pharmacia) to 290 osmol/kg with 10 × phosphate-buffered saline (PBS).

2. Prepare a 4-step gradient in a 50 ml conical, polystyrene centrifuge tube by diluting the prepared Percoll with RPMI medium containing 0.75% BSA as follows. As an aid to monitoring the quality of layering as well as helping to identify the gradient interfaces, add a drop of Trypan blue to alternating steps.

Fraction	% Percoll	Medium (ml)	Percoll (ml)
1	41.1	5.83	4.18
2	45.8	5.42	4.58
3	50.0	3.0	3.0
4	66.6	2.0	4.0

   Up to 5×10^8 cells can be loaded per gradient

3. Centrifuge gradients at 500 g for 30 min, using slow acceleration and no breaking for deceleration.

4. Harvest individual fractions which collect at the interfaces, wash three times and resuspend for cell count. Low-density cells (large granular lymphocytes) collect at the 41.1/45.8% interface, mixed granular and T cells at the high-density interface. NK cell activity can be greatly enriched by isolating the large granular lymphocyte (LGL) fraction. In contrast, LAK cells can be generated, albeit not at the same relative efficiency, from lymphoid cells of significantly disparate densities, indicating the profound heterogeneity of LAK precursors (7).

2.2 Isolation of specific lymphoid populations based upon phenotypic markers or light-scatter characteristics

2.2.1 Fluorescence-activated cell sorting

An alternative approach to the isolation of either precursor or mature effector populations utilizes the technology of fluorescence-activated cell sorting

(FACS). Cell sorting can provide a highly enriched population (> 95% purity) of moderate numbers of cells ($\leq 10^7$) within a few hours of sorting. The quality and quantity of the yield is dependent upon many factors including the condition of cells in the starting population, the relative frequency of the cell of interest within the starting population, the working condition of the sorter, and the skill of the operator. Many commercially available antibodies are available for analysis of phenotypic markers expressed on lymphocyte subpopulations. Since the first description of T cells, NK, and LAK, phenotypic markers have been sought which could unambiguously describe the population of interest. It is important to remember that the vast majority of phenotypic markers describing lymphocyte populations recognize epitopes which are not uniquely relevant to the effector function. As a result, phenotypic markers have consistently proven to be shared by multiple cell types, and caution must be exercised in interpreting results based on phenotypic analysis (8, 9). Another concern of cell separation by phenotype involves the ability of many antibodies to directly modulate the activity of cell populations expressing the corresponding epitope (e.g. anti-CD2, anti-CD3, anti-CD16, etc.) (10). Because of the lack of functional specificity and the concerns of artefactual stimulation, we reported a method for cell sorting of a unique interleukin-2 (IL-2) responsive population based upon characteristic morphologic alterations associated with IL-2 activation (11). These unique light-scatter and forward-scatter characteristics permit isolation of all LAK activity as well as the cycling population involved in the maintenance and further expansion of LAK effector activity. This approach proved useful for human and mouse samples, and allows for the further phenotypic analyses of a population highly enriched for LAK cells.

2.2.2 Separation using antibody-coated magnetic beads

An alternative approach to lymphocyte separation based upon phenotypic markers utilizes magnetic beads coated with specific antibodies. Magnetic separation methodology represents a relatively simple and efficient means for generating moderate numbers of enriched cells (10^6–10^8), and the initial investment is considerably less than acquiring a FACS facility with a trained operator. After allowing the bead-anchored antibodies to attach to the appropriate cell-surface associated phenotypic epitope (or secondary antibody labelled beads binding to primary antibody labelled cells), the bead:cell conjugates are sedimented with a strong magnet, allowing the non-conjugated cells to be removed by pouring, and gentle washing. Both bound and unbound populations can be repeatedly absorbed by the antibody-bead preparation, rapidly generating purities greater than 80–90%. Prepared beads are commercially available, as well as starter beads which can then be coated with the antibody(ies) of interest. Since magnetic separation is based upon phenotypic markers, it suffers from the same potential pitfalls described above. Magnetic separation also presents unique problems, such as a possible difficulty in removing the antibody-anchored beads from positively selected cells. Overall, this technique is best suited for enrichment by negative selection. We have successfully used magnetic separation to remove B cells and monocytes (a rapid alternative to plastic and nylon wool depletion) from PBM suspensions, and for negative depletion of Leu11a presenting cells. We have also used magnetic beads to present anti-OKT3 for co-stimulation with IL-2 (see below) in macrophage-depleted lymphocyte populations. Rosenberg has described the use of antibody-conjugated magnetic beads to directly enrich for lymphocytes (12). Phenotypic analyses should always be performed in order to verify the resultant purities, as well as to control for unexpected alteration in the cell populations.

2.3 Isolation of tumour infiltrating lymphocytes

With the hope of learning more about the dynamic interaction of the cellular immune response to cancer, methods for obtaining lymphocytes which have migrated into tumour tissues have been devised. Preliminary reports suggested that these tumour-infiltrating lymphocytes (TIL) might exhibit unique qualities which could prove beneficial for clinical applications (13). We routinely isolate TIL from a variety of tissues. For tough, or firm tissues, a combination of mechanical and enzymatic disaggregation is used (14). In contrast, for very soft, fragile tissues (e.g. CNS neoplasms), mechanical disaggregation alone often provides superior results.

Protocol 4. Isolation of tumour infiltrating lymphocytes

Equipment and reagents

- Sterile complete medium
- Disaggregation medium: RPMI 1640 supplemented with 300 mg/ml L-glutamine, 100 mg/ml penicillin, 100 mg/ml streptomycin, 50 mg/ml gentamicin sulphate, 0.25 mg/ml amphotericin
- Sterile dissection scissors, forceps, and scalpels
- Enzymatic cocktail: 0.002% DNase (Sigma, type I), 0.1% collagenase (Sigma, type IV), 0.01% hyaluronidase (Sigma, type V)
- Sterile sealed containers
- Magnetic stirrer or shaker
- Serum-containing complete medium
- 25, 10, 5 ml pipettes
- Syringes fitted with 18- and 20-gauge needles
- Trypan blue
- Sterile Nytex gauze
- Histopaque gradient (see *Protocol 1*)

Method

1. Transport tissue samples obtained during surgery or at autopsy, completely submerged in sterile, complete medium.

2. Manipulate tissue samples in a disaggregation medium (serum is not included in sample which will be enzymatically disaggregated).

3. Disaggregate solid tissue to single-cell suspensions as soon as possible to achieve maximal viable recoveries. Use sterile dissection scissors, forceps, and scalpels to reduce the tissue into small cubes (≤1 mm).

Protocol 4. *Continued*

4. Incubate the cubed tissue in an enzymatic cocktail in a sealed sterile container. Use either a magnetic stirrer or shaker to gently swirl the suspension. Treat tissue until the majority of solid cubes have been digested (0.5–1 h at 37°C or overnight at room temp.) and then wash in serum-containing complete medium to remove residual enzyme.

5. For soft tissues, cut the sample likewise into the smallest possible pieces. Aspirate the tissue pieces through progressively smaller pipettes (25, 10, 5 ml pipettes). Allow the larger pieces in this crude suspension to settle at unit gravity for 5 min. Pass the remaining suspension through needle bores of decreasing diameter (i.e. 18- and then 20-gauge; smaller bore needles tend to significantly reduce viability). Draw tissue suspensions into the syringe, attach the needle, and express the suspension. Repeat the entire process three times with each needle size. Check for cell viability.

6. This suspension can now be directly cultured under appropriate activation conditions to grow out the TIL. Alternatively, the sample can be processed further to enrich for TIL and/or for tumour cells.

7. Remove residual tissue clumps by passing the suspension through sterile Nytex gauze.

8. The resultant single-cell suspension can be passed over a Histopaque gradient in order to remove dead cells and debris.

9. Lymphocytes can often be enriched from the tumour population by preparing a discontinuous density gradient using diluted steps of Histopaque. Carefully layer 10 ml of cells ($\leqslant 5 \times 10^7$) over three 10 ml steps of (25, 75, and 100%) Histopaque, and centrifuge at 400 g for 30 min. Tumour cells usually collect at the 25–75% interface and lymphocytes collect at the 75–100% interface.

Even after enrichment procedures, the relative frequency of TIL may still be below the level of immediate detection, and may require significant expansion over an extended culture period before they become the predominant cell type.

3. Generation of lymphokine-activated killers

3.1 The generation of IL-2 activated killers

Lymphokine-activated killers (LAK) have been successfully cultured from virtually all lymphocyte compartments including peripheral blood, thoracic duct, skin, thymus, cerebrospinal fluid, and malignant tissues. LAK are readily generated by short-term tissue culture in the presence of exogenously added IL-2. Culture lymphocytes at a concentration of 5×10^5–2×10^6 cells/

ml, with a total volume of 10 ml in 25 cm^2, 25 ml in 75 cm^2, or 50 ml in 175 cm^2 flasks. Incubate upright at 37°C, 5% CO_2 in 98% humidity. We have tested numerous media, sera, etc., for the optimization of LAK generation. Mouse, rat, and human LAK are readily generated in RPMI 1640 medium supplemented with glutamine (300 mcg/ml), Hepes (10 mM), penicillin and streptomycin (100 mcg/ml of each), and 5–10% serum. Fetal calf (FCS) and newborn calf serum (NCS) are generally equivalent, however, each new lot should be individually tested for its ability to generate LAK. After identifying a good lot, it should be reserved in order to use the same source throughout the course of a given series of experiments. Although FCS or NCS work well, we prefer to use human AB for the generation of human LAK. Serum lots should likewise be tested to identify good lots and exclude inhibitory ones. All serum samples should be heat inactivated (thawed serum heated to 55°C for 30 min) and then aliquoted into single-use samples (e.g. 50 ml) and frozen at −20°C until needed.

Motivated by potential clinical applications for the adoptive transfer of activated lymphocytes, several manufacturers have introduced serum-free, defined media for use in the activation of lymphocytes. Serum-free medium offers several advantages including reproducibility and the assurances of no risk of infectious agents. However, of all the products we have tested to date, only the Gibco product, AIM-V, has proved to be a suitable replacement for serum-supplemented medium. We consider AIM-V the medium of choice, and use it exclusively for all clinical work. It is important to note that the ingredients in AIM-V as well as in other serum-free formulations are generally not disclosed and may contain components which may have significant effects upon the system of study (e.g. indomethicin, insulin growth factors, steroid hormones, etc.). We suggest that data generated exclusively in serum-free mediums be reproduced in serum-supplemented complete medium.

We and others have published that IL-2 concentrations from 22 pM–22 nM induce LAK activity (10). Lower concentrations (<200 pM) may satisfactorily activate lymphocytes in serum-supplemented medium. We have not found any consistent differences between purified human and human recombinant forms of IL-2 from several different producers, and routinely use recombinant products. As a general guide-line, lymphocytes are initially cultured in 2–10 nM IL-2. IL-2 induction of LAK activity can be detected within 24 h of culture, but we routinely culture for 4–6 days for the generation of a mature effector population. Cultures usually required feeding by day 6 with replacement of IL-2 for sustained growth and cytolytic activity. We have maintained potent lytic populations for greater than three months, although many cultures lose much of their lytic activity after several weeks (> 28 days).

3.2 Anti-CD3/IL-2 activated 'T cell' killers

Ochoa *et al.* first reported the use of the T cell activating antibody anti-OKT3 (anti-CD3) to augment IL-2 induction of effectors expressing LAK activity.

Activation with anti-CD3 presumably stimulates a transmembrane signalling structure associated with the T cell antigen receptor. We and others have reported that when used in conjunction with IL-2, anti-CD3 generates a significantly increased yield ($10-1000 \times$) of cytolytic lymphocytes relative to IL-2 stimulation alone (15). By multiplying the lytic potency (expressed as lytic units) of the population by the increased total cellular yield, the anti-CD3/IL-2 activation scheme generated significantly increased lytic potential relative to IL-2 stimulation alone. We first used this activation method to generate large numbers of LAK (10^9-10^{10} total yield) from very small volumes (<10 ml) of peripheral blood for use in adoptive immunotherapy protocols in which the patients could not afford to donate large numbers of fresh lymphocytes (15). We later applied the same activation scheme for the generation of TIL in which small numbers of lymphocytes were initially obtained or from tumours that proved highly immunosuppressive (i.e. glioblastoma) to normal IL-2-mediated outgrowth of LAK (16). This activation method involves incubation of PBM plus monocytes or some other means of antibody presentation. The cells should be adjusted to 10^6/ml and incubated with anti-CD3 (e.g. Ortho-Clone OKT3, approved for clinical use) at 10 ng/ml for 48 h. The cells are then thoroughly washed and then resuspended in IL-2 supplemented, complete medium. Rapid cell growth requires cultures to be split often, frequently every other day. One pitfall of this activation mechanism is that cultures often begin to lose lytic activity after three weeks in culture. However, cell expansion in excess of 10^6-fold will often be achieved within this time, providing sufficient cell numbers for virtually any application.

3.3 Alternative cytokine activation strategies

Since this initial description of LAK, in which the obligate role for exogenously added IL-2 was determined, the involvement of several endogenously produced cytokines have been described (17). The cytokine tumour necrosis factor (TNF) has proven to play a central role in virtually all CMC systems (18). The addition of anti-TNF neutralizing antibody during IL-2 activation eliminates $>90\%$ of LAK-mediated tumour lysis. Given the problematic toxicity associated with the high-dose IL-2 administration protocols used in the first LAK clinical trials, alternative methods which might require less IL-2 were sought. We have reported that TNF-α, TNF-β, or interleukin-1 (IL-1) can all be used in conjunction with low-dose IL-2 to generate LAK of equivalent, or even augmented, lytic potency relative to high-dose IL-2 stimulation alone (17, 19). There is preliminary evidence suggesting that TNF prestimulation prior to IL-2 activation, generates increased lytic potential at the population level, resulting from both increased effector cell frequency and effector cell efficiency (i.e. decreased recycling times). Preliminary data from our laboratory, confirming the report by Widmer, suggests that the exogenous addition of IL-7 is sufficient to stimulate LAK activity from both PBL and PBM cultures. Exogenously added IL-7 appears to be capable of inducing the endogenous production of IL-2 (Stephen Yang, personal communication). Yang also finds that neutralizing antibody against IL-2 inhibits LAK activity by IL-7 activated lymphocytes. Undoubtedly, many other stimulatory and inhibitory cytokine products serve to modulate LAK induction and potency. With a better understanding of the dynamic interplay occurring at the cellular level between these products, new approaches for the *in vitro* generation and maintenance of these effector populations should be defined.

4. Cryopreservation of lymphocytes and tumours

Precursor, as well as mature effector populations and tumour cells, may be easily stored frozen for later use. We find it useful to expend cultures of both stock LAK and common tumour targets (i.e. Daudi, Raji), to be frozen in single-use aliquots (e.g. LAK at 10^7/tube, tumour targets at 10^6/tube). Samples can then be thawed and used when needed. Whenever possible, cultured cells should be harvested in early log phase with viability greater than 90%.

Protocol 5. Cryopreservation

Equipment and reagents

- Complete medium
- Trypan blue
- Cell centrifuge
- Freezing medium: 90% heat-inactivated serum, 10% dimethylsulfoxide (DMSO)
- Freezer vials
- Aluminium boxes
- 100% ethanol
- Liquid nitrogen freezer

Method

1. Wash cells thoroughly in complete medium, and perform a viable cell count.

2. Pellet cells and resuspend in freezing medium to give a final concentration of 10^6–10^8 cells/ml.

3. Once the cells have been suspended in the freezing medium, they must be aliquoted into freezer vials and transferred to the freezer as rapidly as possible ($\leq$ 1 min) to maintain maximal viability.

4. Place the freezer vials into aluminium boxes containing sufficient 100% ethanol to submerge the cell suspensions.

5. Transfer the freezer boxes into a −80°C freezer overnight, and then into the vapour phase of a liquid-nitrogen freezer.

6. To use, rapidly thaw frozen samples by constant shaking in a 37°C water bath, and then dilute 1/10 in warm, complete medium.

7. Wash cells extensively to remove DMSO and use (e.g. plated as effectors or ^{51}Cr-labelled as targets). Excellent viability can be maintained for at least two years if cells are carefully frozen and thawed.

---

5. Cytotoxicity assays

In 1960, A. Govaerts transplanted a kidney from a donor dog X to a recipient dog Y. When the transplanted kidney was rejected, thoracic duct lymphocytes were obtained from the recipient dog X and added to a culture of kidney (epithelial) cells from dog Y. 48 h after the addition of the primed lymphocytes, lesions were noted in the confluent monolayer indicating lymphocyte-mediated killing of the allogeneic cells (20). This represented one of the first reports of direct lymphocyte-mediated toxicity. This *in vitro* detection of CMC was adapted by several other investigators, including K. Theodor Brunner, Jean-Charles Cerrotini, and Borris D. Brondz into what have ultimately become the current methods for detecting killing.

5.1 ^{51}Cr-release cytotoxicity assay

Brunner and Cerrotini introduced two major changes to Govaerts assay: they used tumour cells as targets for their sensitized effectors, and they adapted the ^{51}Cr-labelling technique to the study of target cell death (21). With minor modifications, this ^{51}Cr-release assay has remained the standard technique for measuring *in vitro* CMC. The assay represents a relatively rapid, simple to perform, and reliable method for the measure of *in vitro* CMC. The human Daudi (HLA class I deficient), and Raji Burkitts lymphoma cell-lines are commonly used NK-resistant LAK targets, while the K562 erythroleukaemia line and the U937 lines are excellent NK targets. The use of fresh uncultured tumour cells (prepared as described above) as LAK targets may represent the closest approximation of *in vivo* interactions between LAK and tumour, and should be included in definitive studies, especially if autologous targets are available for clinical studies. We recommend that targets be stored in frozen aliquots rather than being maintained in long-term culture, on order to avoid the possibility of tissue-culture induced changes in susceptibility to killer cells. Mycoplasma testing should also be routinely performed.

Protocol 6. ^{51}Cr-release assay

Equipment and reagents

- Complete medium
- Appropriate pelleted target cells
- Effector cells
- Sodium ^{51}Cr in saline (1 mCi/ml at 650 mCi/mg)
- 5% CO_2 humidified incubator
- 96-well, round-bottomed plates
- 0.1 M HCl
- Centrifuge fitted with multi-well plate rotor
- Skatron filter-rack harvesting system (Skatron Inc.)
- Gamma counter windowed for ^{51}Cr

Method

1. Wash appropriate targets in complete medium and label the pelleted cells with 400 μCi sodium 51-chromate in saline per 10^7 targets. Label

---

the targets for 1–2 h in a 37°C, 5% CO_2, humidified incubator, with occasional shaking to inhibit pelleting.

2. Wash targets 3–4 times in complete medium and then adjust to 5×10^4 cells/ml medium.

3. Wash effector cells in complete medium, adjust to their final plating concentration, and aliquot into replicate wells.

4. Add effector cells at multiple ratios with a fixed number of targets, to obtain the characteristic sigmoidal killing curve. For example, a 0.6 ml suspension containing 3.84×10^6 total effector cells can be added in 0.1 ml aliquots to triplicate wells (96-well, round-bottom plates), and the remaining 0.3 ml repeatedly diluted with 0.3 ml medium, followed by adding 0.1 ml aliquots of targets at 5×10^4/ml in order to achieve effector to target ratios of 128:1, 64:1, 32:1, 16:1, 8:1, 4:1, 2:1, 1:1.

5. Determine spontaneous and maximal release by adding 0.1 ml aliquots of medium or 0.1 M HCl, respectively, on to replicate sets (n=3–6) of 0.1 ml of targets alone.

6. Centrifuge the plates at 150 g for 5 min, and incubate in a humidified, 37°C, 5% CO_2 incubator for 4 h.

7. At the end of the incubation period, centrifuge the plates again and harvest aliquots of the supernatants from each well. The Skatron filter-rack harvesting system represents a major improvement in both time and accuracy over manual harvesting.

8. Count samples in a gamma-counter, appropriately windowed for chromium-51. The entire sigmoidal killing curve can be described if sufficient E:T ratios, starting at a suitable high effector number, are assayed. We recommend that at least four and preferably more E:T ratios be assayed, whenever possible. Thorough characterization of this curve will greatly facilitate data interpretation.

5.1.1 Data analysis for the ^{51}Cr-release assay

[51]Cr-release assays generate data expressed as c.p.m. By convention, this is transformed into per cent specific lysis (%SL) using the following equation:

$$\%SL = [(E) - (S)]/[(M) - (S)] \times 100$$

(E) refers to the mean of the experimental wells at a given E:T ratio, (S) is the mean of the spontaneous release wells in which targets were incubated with medium only, and (M) is the mean of the maximal release wells generated by targets incubated in 0.1 M HCl. We generally plate six spontaneous and six maximal wells, with experimental wells plated in triplicate.

Unfortunately, no standard method for data presentation has emerged. The simplest, and perhaps best way is to provide the %SL ± SEM for each

E:T ratio, either in a tabular format or graphically with %SL on the Y axis vs. log of *E:T* on the *X* (*Figure 1a* and *b*). It is extremely important to present sufficient data to accurately describe the killing curve generated by a given effector/target combination. We attempt to plate at least eight *E:T* ratios starting at a sufficient high effector cell-number to describe both the maximal killing plateau (abscissa) as well as the rest of the curve. Unfortunately, effector cell-number often represents the limiting factor in an assay, significantly limiting the maximal achievable *E:T* ratio. In some cases, by decreasing the number of targets per well (e.g. from 5000 to 2000–2500), the assay can be modified to achieve the desired ratios. Presenting experiments in terms of %SL becomes problematic when large amounts of data must be presented and/or when several curves must be directly compared within one experiment. Lytic units have been used to avoid these problems by representing the potency of an effector population in terms of a single numerical value. Lytic units (LU) are generally defined as the inverse of the number of effector cells required to achieve a specific %SL against a given target (e.g. 30% SL for LAK and 15% for NK). Although the lytic unit, when applied correctly, may represent a useful means for data reduction, several potential problems can be associated with its use.

The first problem involves the types of data which can and cannot be presented in terms of lytic units. Lytic units should only be used as a means for comparing relative lytic efficiency of effector populations assayed in parallel, which are exhibiting similar potency against a given target. LU should not be used as a method of assigning absolute potencies, and should not be used to assess susceptibilities of differing relative targets to a given effector population. A second class of problem involves the mathematical model used to translate %SL data into lytic units. Since ^{51}Cr-release data describes a unique sigmoidal curve for each effector–target pair, attempting to describe the line with a linear regression fit can at best grossly approximate the curve. Curve fitting with a linear equation is further hampered if the point chosen for extrapolation (e.g. 15 or 30% SL) does not fall on the linear rise portion of the curve, or if too few points exist to accurately describe the linear rise portion. The sigmoidal curve is a better fit using non-linear models.

Pross *et al.* have published extensively on exponential fit models for describing NK killing data, and will generously provide investigators with ^{51}Cr-release cytotoxicity software for personal computer use (22) (Dr Hugh Pross, Dept. of Radiation Oncology and Microscopy and Immunology, Queen's University, Kingston, Ontario, Canada; please include blank discs and stamped, addressed return envelope). However, significant controversy surrounds this approach as well, with several reports presenting extremely complicated arguments supporting specific non-linear models. A recent reappraisal of linear regression, exponential fit, and Von Krogh models for data analysis by Pollock *et al.* (23) encourages the reader to test the given models on data sets provided in order to better understand the associated problems. We highly recommend that investigators and their statistical support groups thoroughly review this and related literature, in order to better plan and process the ^{51}Cr-release cytotoxicity data for their unique applications.

5.2 Single-cell CMC assay

In contrast to the ^{51}Cr-release assay which only describes the net lytic outcome of effector–target interactions, this simple approach provides additional information by directly measuring the frequency of effectors which can bind (binding frequency) as well as measuring specific killing (killing frequency) of a given target (24).

Protocol 7. Single-cell CMC assay

Equipment and reagents

- Effector and target cell populations
- Complete medium
- Cell centrifuge
- Pasteur pipette
- 0.5% agarose
- Tissue culture receptacles
- 5% CO_2 humidified incubator
- 0.1% Trypan blue
- 0.2% formaldehyde in saline
- Inverted microscope
- Fluorescein diacetate (FDA, Sigma)

Method

1. Mix equal numbers of effector and target cell populations in complete medium and place into a water bath at 30°C for 10 min.

2. Centrifuge the sample at 250 g at room temp. for 5 min. Prepare control tubes of target cells alone in parallel.

3. Discard supernatants and resuspend pellets in a minimum volume by six gentle aspirations with a Pasteur pipette (aspiration technique represents a crucial step for experimental accuracy and reproducibility).

4. Remove a small portion of this suspension and dilute for direct counting in order to measure the frequency of effector–target conjugates (the number of lymphocytes bound to target cells per every 100 total lymphocytes).

5. Mix the remaining suspension with aliquots of 0.5% agarose (maintained liquified just above its melting point) and pour as a thin layer (⩽2 mm, thicker beds will require long focal length lenses) into tissue-culture receptacles.

6. Overlay with complete medium to avoid dehydration, and place the cultures directly into 37°C, 5% CO_2, humidified incubators for 1–4 h.

7. Assay samples for killing frequency by aspirating the medium overlay, adding a small volume of 0.1% Trypan blue for 5 min at room temp., aspirating residual stain and fixing with 0.2% formaldehyde in saline.

Protocol 7. *Continued*

8. The conjugates are counted, corrected for the spontaneous lysis of target cells plated alone, can be directly visualized under normal light on an inverted microscope. If effectors and targets are difficult to distinguish, effectors can be prestained with agents such as the fluorescent dye fluorescein di-acetate (FDA, 1 μg/ml for 10 min at 37°C in the dark, followed by three washes; available from Sigma) which will allow easy discrimination without adversely affecting binding or killing frequencies.

5.3 Alternative approaches for measuring CMC, including non-radioactive assays

Many variables affect the ability of a given effector to specifically recognize, bind, and ultimately lyse a target cell. New and better ways of describing these complicated interactions are needed to expand our understanding of CMC. Alternative approaches already being studied include flow-cytometric assays for binding/killing frequency determinations in which both the kinetics as well as the phenotypic profiles of effectors and targets can be simultaneously described (25).

Methods to perform cytotoxicity assays without the use of radioactively-labelled target cells have been developed (26, 27), and are identical to the chromium release assay in their performance, except that the targets are not pre-labelled, and that the measurement of cytotoxicity is via a post-labelling. One common approach is to measure lactate dehydrogenase (LDH) which is released upon cell lysis. The LDH is determined via the conversion of a tetrazolium salt into a red formazan product, and the amount of colour formed is proportional to the number of lysed cells. This method is available in kit form from Promega, Madison, WI. As we continue to reduce radionuclide usage, such non-radioactive procedures are expected to gain in popularity. Our preliminary experience with the LDH method has been very favourable, and we intend to convert completely to such a method within the next few years.

References

1. Berke, G. (1989). In *Fundamental Immunology* (2nd edn), (ed. W. Paul), p. 735. Raven Press, New York.
2. Henkart, P. A. (1985). *Ann. Rev. Immunol.*, **3**, 31.
3. Wolf, J. A. and Grimm, E. A. (1988). *Annals of the Institute Pasteur Immunology*, **139**, 433.
4. Lotzova, E. and Ades, E. W. (1989). *Natural Immunity and Cell Growth Regulation*, **8**, 1.
5. Julius, M. H., Simpson, E., and Herzenberg, L. A. (1973). *Eur. J. Immunology*, **3**, 645.

6. Timonen, T., Ortaldo, J. R., and Herberman, R. B. (1981). *J. Exp. Medicine*, **153**, 569.
7. Grimm, E. A. and Rosenberg, S. A. (1984). In *The Lymphokines* (ed. E. Pick), Vol. 9, p. 279. Academic Press, New York.
8. Ortaldo, J. R., Mason, A., and Overton, R. (1986). *J. Exp. Medicine*, **164**, 1193.
9. Damle, N. K., Doyle, L. V., and Bradley, E. C. (1986). *J. of Immunol.*, **134**, 2814.
10. Ochoa, A. C., Gromo, G., Alter, B. J., Sondel, P. M., and Bach, F. H. (1987). *J. of Immunol.*, **138**, 2728.
11. Loudon, W. G., Abraham, S. R., Owen-Schaub, L. B., Hemingway, L. I., Hemstreet, G. P., and DeBault, L. E. (1988). *Cancer Research*, **48**, 2184.
12. Topalian, S. L., Solomon, D., and Rosenberg, S. A. (1989). *J. Immunol.*, **142**, 3714.
13. Rosenberg, S. A., Spiess, P., and Lareniere, R. (1986). *Science*, **233**, 1318.
14. Rong, G. H., Grimm, E. A., and Sindelar, W. F. (1985). *Journal of Surgery and Oncology*, **28**, 131.
15. George, R. E., Loudon, W. G., Moser, R. P., Brunner, J. M., Steck, P. A., and Grimm, E. A. (1988). *J. Neurosurgery*, **69**, 403.
16. Grimm, E. A., Brunner, J. M., Carinhas, J., Koppen, J. A., Loudon, W. G., Owen-Schaub, L. B., Steck, P. A., and Moser, R. P. (1991). *Cancer Immunology and Immunotherapy*, **32**, 193.
17. Owen-Schaub, L. B., Gutterman, J. U., and Grimm, E. A. (1988). *Cancer Research*, **48**, 788.
18. Shalaby, M. R., Espevik, T., Rice, G. C., Ammann, A. J., Figari, I. S., Ranges, G. E., and Palladino, M. A. (1988). *J. of Immunol.*, **141**, 499.
19. Crump, III W. L., Owen-Schaub, L. B., and Grimm, E. A. (1989). *Cancer Research*, **49**, 149.
20. Govaerts, A. (1960). *J. of Immunol.*, **85**, 516.
21. Brunner, K. T., Mauel, J., Cerottini, J. C., and Chapuis, B. (1968). *Immunology*, **14**, 181.
22. Pross, H. F., Baines, M. G., Rubin, P., Shragge, P., and Patterson, M. S. (1981). *J. of Clin. Immunol.*, **1**, 51.
23. Pollock, R. E., Zimmerman, S. O., Fuchshuber, P., and Lotzova, E. (1990). *J. of Clin. and Lab. Analysis*, **4**, 274.
24. Grimm, E. A. and Bonavida, B. (1979). *J. of Immunol.*, **123**, 2861.
25. Wolf, J. A., Carinhas, J., and Grimm E. A. (1989). *Proc. Amer. Assoc. Cancer Res.*, **30**, 238.
26. Kolber, M. A. (1988). *J. of Immunol. Methods*, **108**, 255.
27. Decker, T. and Lohmann-Matthes, M.-H. (1988). *J. of Immunol. Methods*, **15**, 61.

13

Assays for chemotaxis

JO VAN DAMME and RENÉ CONINGS

1. Introduction

In view of the recent developments of cytokine research, interest in chemotaxis assays has received a real revival. Indeed, within a short time a number of novel chemotactic cytokines have been identified. These low molecular weight proteins are different from other cytokines such as interleukin-1 (IL-1) and tumour necrosis factor (TNF), previously reported to be chemotactic. They belong to a novel supergene family and include proteins exerting their chemotactic activity specifically on granulocytes or monocytes (1–4). Some members of this family of inflammatory proteins were already biochemically characterized, but their chemotactic effect has only recently been demonstrated. The lack of sensitive and specific bioassays for measuring chemotaxis, as well as the complexity of these tests, can in part explain the late discovery of this family of chemotactic cytokines. However, the assays that were used to detect their activity during purification were basically identical to those already used for a long time, such as the Boyden chamber assay (5) and the agarose method (6) described in this chapter. Since many substances can exert chemotactic activity, their identification rather had to wait for techniques to produce sufficient amounts and to purify them to homogeneity.

2. Chemotaxis under agarose

This method for chemotaxis originally described by Nelson *et al.* (6, 7), uses the migration distance of cells under agarose as a parameter to measure the chemotactic effect of cytokines. Since the assay can be performed in tissue culture dishes, the technique allows handling of a large number of samples (e.g. column fractions from purification runs) and a fast microscopic score of their chemotactic potency. When the preparation of the test cells involves time consuming isolation and purification (e.g. granulocytes from peripheral blood) the reader is recommended to prepare the agarose plates the day before the actual chemotaxis assay is performed.

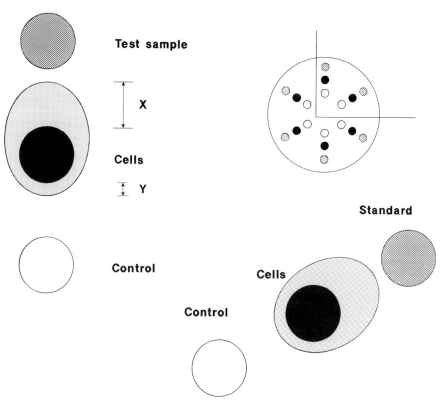

Figure 1. Chemotaxis under agarose. Configuration of wells: six series of three wells, each series containing a central well for cells and two surrounding wells for the test sample (or standard) and control medium, respectively. Cells migrate towards the chemotactic gradient of the test sample (distance *X*) and the control medium (random migration distance *Y*). *X–Y* represents the effective migration towards the test sample.

Protocol 1. Preparation of agarose plates

Equipment and reagents

- *Solution A*: 10 ml pre-warmed (50°C) medium, consisting of 2 ml fetal calf serum (FCS), 2 ml 10 × concentrated Eagle's minimum essential medium (EMEM) with Earle's salts, L-glutamine and sodium bicarbonate, and 6 ml pyrogen-free distilled water
- *Solution B*: 0.18 g agarose (Indubiose, IBF) boiled in 10 ml distilled water until completely dissolved and cooled to 50°C
- Solution AB: equal volumes of A and B at 50°C
- 6 cm (i.d.) plastic tissue culture dishes (Nunc)
- Stainless steel punch with inside bevel and template
- Vacuum system
- CO_2 incubator

Method

1. Pour 6 ml solution AB into plastic tissue culture dishes and allow to cool (30 min) before transfer to refrigerator (4°C) until further processing.

> **2.** Cut six series (per dish) of three wells (3 mm i.d., 3 mm inter-space) in the gel, using a stainless steel punch with inside bevel and template to align wells as shown in *Figure 1*.
>
> **3.** Remove agarose cores with a pipette using vacuum and incubate plates at 37°C in a CO_2 incubator until samples and cells are prepared.

The final steps in preparing the agarose plates, i.e. the punching of wells is best postponed until immediately before the chemotaxis assay, since leakage of fluid from the gel into the wells should be prevented. In cutting wells, care should be taken to avoid damage to the surface of the dish, because scratches in the plastic can inhibit cell migration. Do not lift the gel or damage the wells during suction of the punched cores.

The agarose assay for chemotaxis is applicable to different leukocyte types. In our laboratory this technique has been successfully used in isolating two novel chemotactic cytokines, specific for neutrophils and monocytes, respectively (8, 9). Thus, in order to discriminate between these two activities, the purity of the test cells determines the specificity of the test system. Human peripheral blood neutrophils can be purified according to *Protocol 2*.

Protocol 2. Isolation of granulocytes from human peripheral blood

Equipment and reagents

- Fresh human blood
- Heparinized tubes
- PBS
- Hydoxyethyl starch (Plasmasteril, Fresenius AG)
- 50 ml measuring cylinder
- Beckman L_{5-50} centrifuge fitted with fixed angled rotor R30
- Double-distilled water
- 3.6% NaCl
- Hank's balanced salt solution (HBSS)
- Human serum albumin (HSA)
- Percoll gradient ($d = 1.054$, Pharmacia)
- Trypan blue

Method

1. Collect 10 ml of fresh human blood in a heparinized tube and dilute 1:2 in phosphate-buffered saline (PBS).

2. Mix 20 ml cell suspension with 10 ml hydroxyethyl starch (Plasmasteril, Fresenius AG, FRG) and place in a cylinder for 30 min at 37°C to allow sedimentation of erythrocytes.

3. Collect supernatant and centrifuge cells at 500 *g* for 8 min.

4. Resuspend cell pellet in 24 ml of double-distilled water for 30 sec to lyse remaining erythrocytes, then add 8 ml of 3.6% NaCl solution and centrifuge at 500 *g* for 8 min.

5. Resuspend cell pellet with 5 ml Hank's Balanced Salt Solution (HBSS), supplemented with 0.5 mg/ml of human serum albumin (HSA).

Protocol 2. *Continued*

6. Load leukocyte suspension carefully on 16 ml Percoll gradient and centrifuge in fixed angle rotor (Beckman R30) for 30 min at 20 000 *g*.

7. Collect the polymorphonuclear cell fraction, resuspend in 10 ml HBSS and centrifuge at 500 *g* for 8 min.

8. Resuspend cell pellet in 1 ml HBSS, determine cell number and viability, dilute to 3×10^7 neutrophils per millilitre, and use immediately in chemotaxis assay.

2.1 Preparation of test samples

The preparation of test samples can be done during the isolation of the cells used to measure chemotaxis. Since chemotactic substances, including naturally occurring cytokines, often have a high specific activity, samples should be tested at multiple dilutions (e.g. $0.5 \log_{10}$ steps) to obtain titration curves. Moreover, when chemotactic substances are applied at doses above their optimum, a reduction in migration can occur. A dose–response curve of a standard chemotactic preparation should be included in each assay, because the maximal migration distance obtained also depends on the origin and viability of the cells. The chemotactic agent formylmethionyl-leucylphenylalanine (fMLP) is usually used as a standard. If the optimal dose of fMLP (e.g. 10^{-7} M) in a given test system is reproducible, it might be preferable to test a single dose with multiple replicates (e.g. in every agarose plate). However, for testing cytokines (e.g. the neutrophil activating protein, IL-8), preference should be given to an internal standard preparation (purified if available) of the cytokine involved allowing expression of the potency of the test samples in units. The use of the corresponding cytokine standard might also be helpful when test samples are too weak in potency to be diluted. Indeed, if compared to an optimal dose of fMLP, such samples may be evaluated as negative, since fMLP often induces a more pronounced migration than cytokines. If single dilutions of the test compounds are used their chemotactic activity should be determined in triplicate, preferably in different agarose plates, in order to increase reproducibility. If dilution series are used, measurements in duplicate should be sufficient, especially when large series of fractions are to be tested. If testing column fractions obtained during cytokine purification, it is essential that these are made physiological before assay. This can be achieved either by dialysis or by sufficient dilution (if possible) in control buffer (HBSS).

2.2 Assay procedure

The configuration of cutting wells in an agarose plate (six series of three wells) allows testing of six individual samples (standard included) in each plate according to the procedure described in *Protocol 3*.

Protocol 3. Chemotaxis assay under agarose

Equipment and reagents

- Agarose plates (see *Protocol 1* and *Figure 1*)
- Non-chemotactic medium (e.g. HBSS + HSA)
- Humidified 5% CO_2 incubator
- Absolute methanol
- 37% formaldehyde
- May–Grünwald and Giemsa stains

Method

1. Fill the centre well of each series of three wells (*Figure 1*) with 10 μl of cells, e.g. 3×10^5 neutrophils.

2. Add 10 μl of control non-chemotactic medium (e.g. HBSS + HSA) and test sample dilution, to the inner and outer well, respectively.

3. Incubate agarose plates for 2 h at 37°C in a humidified CO_2 (5%) incubator.

4. Terminate the assay by adding absolute methanol (3 ml) to the agarose plates for 30 min at room temp. Decant carefully.

5. Fix cells with formaldehyde (37%) for 30 min, decant.

6. Carefully remove the agarose (without rotation of the gel) from the culture dish and stain the cells with May-Grünwald's and Giemsa's solutions.

7. Score potency of the samples by counting the number of migrated cells or by measuring the effective migration distance (*Figure 1*).

The optimal incubation time for maximal migration depends on the cell type tested, monocytes requiring a longer period (3 h) than neutrophils. During incubation the actual migration distance should be checked under the microscope so that the assay is stopped before the cells start to migrate into the well containing the test sample. Longer incubation reduces the sensitivity of the assay since further migration (toward test sample well) is stopped while spontaneous migration might still continue.

For quantification of the potency of the samples, microscopic measurement of the migration distance is preferred, since counting of the cells is more labour intensive. However, when single, supra-optimal doses are tested, the migration distance might be suboptimal, whereas the number of migrated cells remains maximal. In addition, microscopic counting of cells might become essential when cell isolation did not result in a completely pure population. In practice, the migration distance towards both the chemotactic sample (induced migration) and the control medium (random migration) are determined (*Figure 1*). The effective migration distance is calculated by subtraction of the random migration (Y) from the induced migration (X). The potency of a sample tested at a single dilution can be expressed as percentage of the maximal effective migration distance of the internal standard. Alternatively, the potency can be expressed as a stimulation index, which is obtained by dividing the effective migration distance by the random migration distance $[(X-Y)/Y]$. If samples are tested at multiple dilutions, a titration end-point can be calculated from the half-maximal effective migration distance. As a consequence the chemotactic potency of cytokine preparation can be expressed in units (U), 1 U/ml corresponding to the half-maximal effective migration distance obtained with an optimal dose of the internal cytokine standard.

3. Chemotaxis in micropore filters

This method of chemotaxis is based upon active migration of test cells through a filter with pores of a precise size. The filter can be placed in a

A

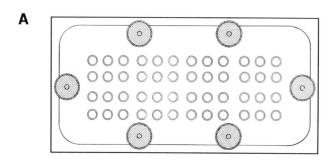

B

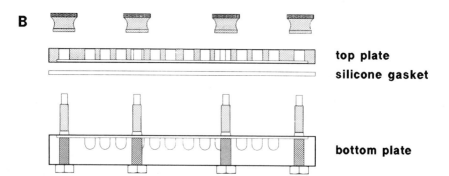

top plate
silicone gasket

bottom plate

Figure 2. Schematic representation of the Neuroprobe 48-well chemotaxis chamber. The microchamber consists of a top and bottom acrylic plate, sealed by a silicone gasket. The upper wells (containing cells) are separated from the lower wells (containing chemoattractant) by the micropore membrane, in which the test cells migrate toward the chemotactic gradient. **(A)** Top view; **(B)** side view.

chamber to create two compartments, as originally introduced by Boyden (5). Cells are added to the upper compartment, whereas the lower compartment is filled with the chemotactic substance. As a consequence a chemotactic gradient is created and cells penetrate through the pores of the filter to the lower compartment. The number of migrated cells serves as a parameter to determine the potency of a chemotactic substance.

This assay system demands more sophisticated equipment than the agarose assay, but multi-well microchamber assemblies allowing rapid and accurate measurements can be obtained (10). A commercially available device commonly used to detect chemotactic cytokines is the 48-well chemotaxis chamber (*Figure 2*) from Neuroprobe. In order to increase the reproducibility of the assay it is recommended that test samples should be assayed in triplicate within one chamber. The number of samples that can be tested per chamber therefore remains restricted, especially if dilution series are made.

The filter separating the two chamber compartments is the essential part of the test system. Depending on the cell type, different filter materials and pore sizes should be used. Cellulose ester filters (150 μm thick) allow to measure the migration depth of neutrophils (3–5 μm pores), monocytes (8 μm pores) or lymphocytes (8 μm pores) into the filter (11). Polycarbonate membranes are convenient when the number of migrated cells has to be determined (*Protocol 4*). For monocytes, polyvinyl pyrrolidone (PVP)-polycarbonate membranes with 5–8 μm pores are recommended, whereas for neutrophils PVP-free membranes (3–5 μm) should be used. To prevent migrating lymphocytes dropping to the bottom of the lower chamber, the lower surface of the PVP-free polycarbonate filter (5 μm) should be coated with collagen (type IV) (12).

Protocol 4. Microchamber chemotaxis assay for monocytes

Equipment and reagents

- Fresh anticoagulated human peripheral blood
- Ficoll-sodium metrizoate (Lymphoprep, Nyegaard)
- PBS
- Standard cooled laboratory centrifuge
- RPMI 1640
- HSA

- Trypan blue
- Standard chemoattractant (e.g. fMLP)
- Neuro Probe microchamber (*Figure 2*)
- PVP-polycarbonate membrane (5 μm pore size, Nuclepore Inc.)
- 5% CO_2 incubator
- 70% methanol
- Diff-Quick (Harleco)

Method

1. Isolate mononuclear cells from fresh anticoagulated human peripheral blood by centrifugation (400 *g* for 30 min, without the brake on) on Ficoll-sodium metrizoate.

2. Wash mononuclear cells twice with phosphate-buffered saline (PBS), centrifuge at 400 *g* for 10 min.

Protocol 4. *Continued*

3. Resuspend cells in RPMI 1640 plus 0.5 mg/ml HSA. Determine the cell number, viability and dilute to 1.5×10^6 cells/ml.

4. Prepare serial dilutions of test samples and standard chemoattractant (e.g. fMLP at 10^{-8} M) in RPMI 1640 supplemented with 0.5 mg/ml of HSA.

5. Add 27 µl of the test samples to the lower compartments of the Neuro Probe microchamber (*Figure 2*).

6. Put the PVP-polycarbonate membrane on the bottom plate and re-assemble the microchamber.

7. Add 50 µl of the mononuclear cell suspension (75×10^3 cells) to each well of the upper compartment.

8. Incubate chamber for 2 h at 37°C in a 5% CO_2 incubator to allow cell migration.

9. Dismount the microchamber unit, wet the non-migrated cell side of the membrane with PBS, and wipe the cells off this filter side.

10. Fix the cells which adhere to the lower surface of the membrane with 70% methanol, dry, and stain with Diff-Quick (Harleco, Gibbstown, New Jersey, USA).

11. Place the membrane on a microscope slide to dry and microscopically ($400 \times$ magnification) count monocytes in five oil immersion fields for each well.

If cellulose ester filters are used, the test samples can be evaluated by measuring the migration distance into the filter (leading-front assay), whereas with polycarbonate membranes the number of cells migrated through the pores has to be determined. In the latter case the potency of a test sample is calculated from the average cell counts of three wells. The chemotactic activity can be expressed as percentage of the maximal number of cells migrated to the reference chemoattractant, e.g. fMLP 10^{-8} M. Alternatively a chemotactic index can be calculated from the effective number of cells migrated to the test sample, divided by the number of cells migrated to the control medium. Preference should be given to use a purified standard preparation of the chemotactic cytokine in test as a reference. For the evaluation of the chemotactic potency of samples from stimulated cells, a supernatant from unstimulated cells should be run in parallel as a control.

4. *In vivo* chemotactic responses to cytokines

The biological relevance of chemotactic activities observed with cytokines *in vitro* can be demonstrated by their effect on cellular localization *in vivo*. This approach involves specific expertise such as histopathological techniques, radiolabelling of cells or implantation of chambers (13). Although such *in vivo* techniques also suffer from low specificity and sensitivity they might be complementary to *in vitro* assays in order to unravel the complexity of cytokine interactions. An *in vivo* test to measure increased microvascular permeability in response to intradermal injection in rabbit skin has been applied successfully to cytokine research (14).

The sensitivity of this test for measuring infiltration of neutrophils into the skin can be increased by injecting the cytokine in combination with a vaso-dilator substance (e.g. PGE_2), allowing detection of pmol amounts of IL-1, IL-8, and TNF. In addition, quantitation of this skin reactivity is facilitated by the use of radiolabelled neutrophils and albumin, resulting in the detection of both plasma protein extravasation and cell accumulation. Such oedema formation serves as an additional parameter for local inflammation. Plasma leakage induced by certain cytokines (e.g. IL-8) is dependent on neutrophil emigration since it is abolished when animals are made neutropenic (14). The specificity of the skin test can be enhanced by following the kinetics of the response to different cytokines. Indeed, it was observed that IL-8 elicited its skin reactivity faster than IL-1 (8). Similarly, upon intravenous injection IL-1 causes a slower increase in the number of circulating granulocytes than IL-8 (8, 15).

Although previously been reported to be chemotactic (16, 17), pure IL-1 does not exert *in vitro* chemotactic activity for granulocytes (8). This dis-crepancy between the *in vivo* and *in vitro* effects of IL-1 on granulocytes can in part be explained by the finding that IL-1 is a potent inducer of the granulocyte chemotactic protein (GCP) (*Table 1*), identical to IL-8, and a monocyte chemotactic protein (MCP). GCP/IL-8 and MCP have been iso-lated from cell cultures with the help of the chemotactic assays described here

Table 1. Induction of chemotactic factors by IL-1

Inducer		Production of chemotactic activity for [a]		
Type	**Dose**	**Monocytes**		**Granulocytes**
		Microchamber assay (%) [b]	**Agarose test (%)** [c]	**Agarose test (%)** [c]
IL1β	100	70	59	51
(U/ml)	10	54	37	75
	1	62	15	24
	0.1	31	0	0
Unstimulated	—	15	0	0

[a] Monolayers of human fibroblasts were stimulated for 48 h.
[b] Percentage of the maximal number of effectively migrated monocytes to pure MCP (*Protocol 3*).
[c] Percentage of effective maximal migration distance obtained with an optimal dose of pure MCP or GCP/IL-8 (*Protocols 1* and *2*).

(8, 9). However, in view of the low specificity of these chemotactic assays it is essential to use purified test cell population and specific antibody against each of these molecules in order to characterize the activity measured. The development of sensitive immunoassays in addition to bioassays is therefore crucial for the detection of these substances in body fluids.

Acknowledgements

The authors thank D. Brabants and C. Callebaut for excellent editorial help.

References

1. Wolpe, S. D. and Cerami, A. (1989). *The FASEB Journal*, **3**, 2565.
2. Oppenheim, J. J., Zachariae, C. O. C., Mukaida, N., and Matsushima, K. (1991). *Annual Review of Immunology*, **9**, 617.
3. Van Damme, J. (1994). In *Immunology and molecular biology of cytokines* (2nd edn) (ed. A. Thomson). Academic Press, London. (In press.)
4. Leonard, E. J. and Yoshimura, T. (1990). *Immunology Today*, **11**, 97.
5. Boyden, S. (1962). *Journal of Experimental Medicine*, **115**, 453.
6. Nelson, R. D., Quie, P. G., and Simmons, R. L. (1975). *Journal of Immunology*, **115**, 1650.
7. Nelson, R. D. and Herron, M. J. (1988). In *Methods in enzymology*, Vol. 162 (ed. G. Di Sabato), p. 50. Academic Press, San Diego.
8. Van Damme, J., Van Beeumen, J., Opdenakker, G., and Billiau, A. (1988). *Journal of Experimental Medicine*, **167**, 1364.
9. Van Damme, J., Decock, B., Lenaerts, J.-P. Conings, R., Bertini, R., Mantovani, A., and Billiau, A. (1989). *European Journal of Immunology*, **19**, 2367.
10. Falk, W., Goodwin, R. H. Jr., and Leonard, E. J. (1980). *Journal of Immunological Methods*, **33**, 239.
11. Wilkinson, P. C. (1988). In *Methods in enzymology*, Vol. 162 (ed. G. Di Sabato), p. 38. Academic Press, San Diego.
12. Larsen, C. G., Anderson, A. O., Appella, E., Oppenheim, J. J., and Matsushima, K. (1989). *Science*, **243**, 1464.
13. Boyle, M. D. P., Lawman, M. J. P., Gee, A. P., and Young, M. (1988). In *Methods in enzymology*, Vol. 162 (ed. G. Di Sabato), p. 101. Academic Press, San Diego.
14. Rampart, M., Van Damme, J., Zonnekeyn, L., and Herman, A. G. (1989). *American Journal of Pathology*, **135**, 21.
15. Van Damme, J., De Ley, M., Opdenakker, G., Billiau, A., and De Somer, P. (1985). *Nature*, **314**, 266.
16. Movat, H. Z. (1985). In *The inflammatory reaction* (ed. H. Z. Movat). Elsevier, Amsterdam.
17. Di Giovine, F. S. and Duff, G. W. (1990). *Immunology Today*, **11**, 13.

14

Cytokine regulation of endothelial cells

P. ALLAVENA, E. DEJANA, F. BUSSOLINO, A. VECCHI, and
A. MANTOVANI

1. Introduction

Endothelial cells (EC) have long been considered a 'passive' lining of blood vessels, endowed with negative properties, the most important one being that of representing a non-thrombogenic substrate for blood. As such, EC were viewed to participate in tissue reactions essentially as targets for injurious agents. The possibility of isolating and culturing EC from various tissues gave the tools for studying their complex reactions to a variety of activating stimuli. EC have, in this way, emerged as active participants in many physiological and pathological processes. It is now evident that haemostasis, inflammatory reactions, and immunity involve close interactions between immunocompetent cells and vascular endothelium. In particular, the ontogeny and function of white blood cells require an intimate relationship with vascular EC. Cytokines are mediators of these complex bidirectional interactions between leukocytes and vascular elements (for review see 1, 2).

EC represent both a source and a target of cytokines. Activation of EC by inflammatory stimuli or other modulatory peptides dramatically changes EC function and surface properties. Functional reprogramming of EC by cytokines follows discrete patterns with limited redundancy. Typically IL-1 and TNF induce a programme related to inflammation and immunity, whereas IFN-γ activates accessory cell function (2).

In addition to responding to a variety of cytokines, EC are important producers of several of these polypeptide mediators, including IL-1, IL-6, colony-stimulating factors (G, M, and GM), and chemotactic cytokines (IL-8 and monocyte chemotactic protein-1, MCP-1).

In this chapter we will discuss selected aspects of the methodology involved in the evaluation of the interaction of cytokines with vascular cells. In particular we will focus on methods and problems that are specifically related to the study of EC. These include the culture of EC, the measurement of procoagulant activity (PCA) and platelet activating factor (PAF), the evaluation of the adhesive properties and migratory capacity of EC. Other functions, such as measurement of production of cytokines, present no EC-related specific problems and the reader is referred to other chapters in this book and elsewhere. We will also discuss the use of the polyoma middle T (PmT) oncogene to generate mouse endothelial lines which retain responsiveness to cytokines.

2. Endothelial cells

2.1 HUVEC

Human umbilical venous endothelial cells (HUVEC) have frequently been used for studies on cytokines. The method for culturing HUVEC has been previously described in detail (3, 4). It includes collagenase digestion of the vessels to isolate the cells. Collagenase has the advantage over other enzymes of selectively digesting the subendothelial basement leaving the cell membrane and most of the integral membrane glycoproteins intact. Isolation of EC from different tissues (i.e. aorta, brain, kidney, skin) has been described in detail (for review see (4–8)).

Protocol 1. Culture of endothelial cells

Equipment and reagents

N.B. All culture reagents in this and the following protocols are from Gibco, unless otherwise stated

- Umbilical cords (at least 20 cm in length) collected in sterile plastic bags
- Ca^{2+}- and Mg^{2+}-free Hank's balanced salt solution (HBSS)
- Collagenase solution (0.1% in HBSS containing Ca^{2+} and Mg^{2+}) (*Cl. histolyticum*, CLS type I, Worthington Biochemical Corp.)
- Centrifuge

- M199 medium + 20% fetal bovine or human serum
- Culture flasks
- Trypsin (1.5 U/ml)–EDTA (0.02%)
- Endothelial Cell Growth Supplement (ECGS, Collaborative Research)
- Heparin (Sigma)
- 0.1% gelatin (Difco) coated culture flasks

Method

1. Collect umbilical cords (at least 20 cm in length) from normal deliveries or Caesarean sections in sterile plastic bags and excise any crushed area. In these conditions, cords can be maintained at 4°C up to one week before processing.

2. Perfuse the umbilical vein with Hank's balanced salt solution (HBSS) Ca^{2+}- and Mg^{2+}-free.

3. Remove HUVEC by short treatment at 37°C with the collagenase solution.

4. Flush out the contents of the vein and wash the lumen with HBSS.

5. Spin cell suspensions at 1200 r.p.m. for 10 min.

6. Resuspend the EC pellet in tissue culture medium (most commonly medium 199 (M199) with fetal bovine, or human serum at a 20% concentration).

7. Seed cells in culture flasks at a concentration not less than $20-40\times10^3$/cm^2. Cell counting is not easy, since cells detached from the vessel wall are not dispersed but are in sheets of 5–20 cells, the success of the culture being apparently related to the presence of such aggregates. In successful primary cultures the cells reach confluence in 7–10 days.

8. Passage the cells using trypsin–EDTA. After primary culture HUVEC can only be maintained for further passages in the presence of 50 µg/ml ECGS and 100 µg/ml heparin and should always be grown on 0.1% gelatine coated culture flasks.

2.1.1 Comments

Studies on the influence of maternal variables on the success of EC cultures showed that if the mother habitually smoked more than 15 cigarettes per day this negatively influenced the success of the culture, whereas age, parity, use of oxytocin, and pathologies such as diabetes and hypertension had no significant effect.

It is of interest to consider how variables related to culture conditions can affect HUVEC response to cytokines. When the cells are stimulated with cytokines, these peptides are added to complete culture medium in the presence of serum. During the time of cell activation by cytokines, serum can be substituted with 1% human or bovine serum albumin without any apparent change in cell response (9). The response of HUVEC to IL-1 and TNF declines with cell passage. This is particularly dramatic for PAF production which is almost abolished within 10 passages (10). However, other activities as prostacyclin (PGI$_2$), PCA or induction of membrane adhesive proteins for leukocytes gradually decrease with increased passage number. It is advisable therefore to test HUVEC response to cytokines within the fifth passage. Cytokine activities are reversible and are not cytotoxic.

2.2 Mouse PmT-transformed lines

Normal EC of human and murine origin are cumbersome to obtain and culture. EC lines have been generated sporadically, and, in our experience, at least some of them lack important functions of normal EC. The polyoma middle T (PmT) oncogene transforms mouse EC (11) and can be used to generate immortalized EC cell lines, possibly representative of microvascular elements. These lines retain many properties of normal EC including production of and responsiveness to cytokines (12–14).

Protocol 2. Generation of murine endothelioma cell-lines

Equipment and reagents

- 15 days' gestation fetuses
- 0.05% trypsin + 0.02% EDTA
- DMEM medium + 20% FCS (Hyclone labs) (complete medium)
- Centrifuge
- 6- and 12-well plates
- Retrovirus vector N-TKmT (produced by GgP+E cell line, obtained courtesy of Dr E. Wagner, Wien, Austria)
- G418
- Ca^{2+}- and Mg^{2+}-free PBS or saline

Method

All procedures must be performed with sterile material in aseptic conditions.

1. Remove fetuses or organs of interest from 6–8 fetuses of 15 days gestation.

2. Cut organ or fetus in small pieces and trypsinize (trypsin 0.05% + EDTA 0.02%, 20 min at 37°C)

3. Collect supernatants and add the same volume of DMEM medium with 20% FCS.

4. Centrifuge at 1200 r.p.m. for 10 min.

5. Resuspend the pellet in 2–5 ml of DMEM + 10% FCS (complete medium), count and bring the suspension to $0.5–1 \times 10^6$ cells/ml.

6. Distribute 2 ml of cell suspensions to each well of 12-well plates and incubate at 37°C, 5% CO_2.

7. After 24 h remove medium and add about 10^5 neo CFU of the retrovirus vector N-TKmT per well in 1 ml of complete medium. Virus was produced by the GgP + E cell-line obtained through the courtesy of Dr E. Wagner (Wien, Austria).

8. After 2 h remove medium and add fresh complete medium.

9. After 72 h select PmT infected-neomycin resistant cells with G418, 800 μg/ml.

10. Change the medium twice a week, keeping G418 at 800 μg/ml.

11. Check wells for G418 resistant cells. They usually are observed after 15–20 days.

12. When cells are confluent, wash the wells thoroughly two times with PBS Ca^{2+} and Mg^{2+} free or with saline, add 0.3 ml of trypsin 0.05% + EDTA 0.02% for 2–3 min at 37°C, resuspend the detached cells and add 1 ml of complete medium. Transfer all the suspension to a well of a 6-well plate and bring to the final volume of 3 ml, with G418 at the 800 μg/ml final concentration.

13. Check cells for the growth every day.

14. When confluent, passage the cells 1:3.

At this stage, cells should not be diluted too much, even if they are growing very well. Confluent monolayers can usually be kept 1–2 days without damage to the cells. On the other hand, if cells are diluted too much, they stop growing and can either remain quiescent for some time and eventually grow again, or die. Maintain selection with G418.

Following this protocol, we have obtained stable cell-lines from heart, brain and whole embryo of C57B1. Cell lines show a cobblestone morphology at confluency and maintain a monolayer structure without overgrowth. Cells are positive for CD31/PECAM-1 antigen, show rapid uptake of fluorescinated acetylated low-density lipoprotein, produce IL-6 constitutively and are negative or weakly positive for factor VIII-related antigen. Transmission electron microscopy revealed that they were uniformly negative for the presence of Weibel-Palade bodies.

Transformed cells maintain many characteristics of normal endothelial cells, for instance CD31-expression, modulation of adhesion molecules by cytokines and cytokine production (13). They do not constitutively express ICAM-1, VCAM-1, E- and P-selectin. These can however be induced, with the exception of ICAM-1, by exposure to TNF-α and LPS, but not IL-1.

Endothelioma cells produce IL-6 and MCP-1/JE, whose production can be increased by IL-1 exposure, and EDMF, an endothelial factor able to induce EC-migration, recently described (14).

2.2.1 Comments

Lines originated from embryo tissues infected *in vitro* with the PmT oncogene of the polyoma virus have been growing in this laboratory for the past 3 years. They represent an easy and reliable source of endothelial cells of murine origin, suitable for studies on EC biology (12, 15). These lines do not need exogeneous growth factors for proliferation.

All lines have been frozen, stored in liquid nitrogen and recovered without problems.

PmT murine EC-lines originated from haemangiomas (16, 17) have been used to generate mAbs against EC-specific antigens expressed constitutively, such as CD31 or those induced by cytokines, such as VCAM-1 and ELAM-1 (18).

EC lines from different organs can be useful to study the potential diversity of the microvasculature of different organs.

3. Chemotaxis

Chemotaxis is defined as the directional locomotion of cells sensing a gradient of the stimulus. Chemotaxis has been extensively studied with leukocytes that are 'professional migrants', but a variety of cell types including fibroblasts, melanoma cells, keratinocytes, and vascular endothelial cells exhibit directional locomotion *in vitro*. Migration as well as proliferation underlie the process of new blood vessel formation. Cytokines such as fibroblast growth factor (FGF), TNF, G-, and GM-CSF induce endothelial cell migration *in vitro* and angiogenesis *in vivo*.

Two main techniques have been used to measure EC-migration *in vitro*: repair of a wound inflicted on a cell monolayer and chemotaxis across porous membranes. While the former approach may more closely resemble the *in vivo* condition of endothelial monolayer lining blood vessels, the latter is easier to quantitate and allows analysis of directional versus random locomotion. We will therefore focus on the description of EC-migration through a porous membrane. Both a 'classic' modified Boyden chamber assay (19) and a micromethod (20, 21) will be described. Assays for chemotaxis and locomotion of polymorphonuclear leukocytes and monocytes are also discussed and described in Chapters 10 and 12.

3.1 Boyden chamber assay

Protocol 3. Boyden chamber assay

Equipment and reagents

- Boyden chamber (Neuroprobe, see Section 3.1.1)
- HUVEC
- 1.5 U/ml trypsin + 0.02% EDTA
- M199 medium containing 1% BSA
- Standard chemoattractant: human plasma fibrinogen (purified material demonstrating only Aα-, Bβ-, and γ-chains (19)) at 1 mg/ml

- Cytokines: basic FGF (5 ng/ml) (Farmitalia), GM-CSF (5 ng/ml), TNFα (500 U/ml)
- Humidified 5% CO_2 incubator
- Cotton wad
- Forceps
- Glass slides
- Diff-Quik (Harleco)

Method

1. Detach confluent HUVEC by brief exposure of the cells to trypsin (1.5 U/ml)–EDTA (0.02%).

2. Centrifuge the cells and resuspend at a concentration of 2×10^6/ml in medium 199 containing 1% fetal bovine serum (FBS) or 0.2% bovine serum albumin (BSA).

3. Seed 200 μl of stimulus (diluted in the same medium used for cell suspension) in the lower well of the chamber, carefully avoiding air-bubble formation.

4. Lay the filter on the surface of the stimulus with the opaque side up.

5. Screw tightly the upper part of the chamber so as to create an upper well in which 200–400 μl of cell suspension is seeded.

---

6. Incubate the chamber at 37°C in humidified air with 5% CO_2 for 6 h.

7. Absorb the liquid in the upper well with a cotton wad and clean with care the upper side of the filter. The remaining unmigrated cells should be removed.

8. Unscrew the upper part of the chamber, catch the filter with forceps, turn over the filter and put it on a glass slide so that the side that has contacted with the stimulus is now face-up.

9. Stain the filter with Diff-Quik.

3.1.1 Modified Boyden chambers and filters

These chambers (Neuroprobe, Maryland, USA) are composed of two wells: the stimulus is placed in the lower well while the cells are placed in the upper well. Polycarbonate filters (polyvinylpirrolidone-free), 13 mm diameter with 5 μm pore size and 10 μm thickness are obtained from Nucleopore, Pleasanton, California. The chemotactic activity of some cytokines is best observed after coating the filters with gelatin (or other extracellular matrix protein such as fibronectin). The filter should be soaked in 0.5 M acetic acid for 2 min, washed with PBS, incubated for 24 h in a 0.01% gelatine solution and air-dried.

3.1.2 Reading the assay

Chemotaxis is evaluated as number of cells that migrate across the filter and adhere to its lower surface. For this reason, cells in a certain number (usually ten) of high power fields (with oil immersion) are counted. The results are expressed as the mean number ($\pm$SE) of migrated cells in three replicates. In order to compare the results from different experiments, it is important to count the same number of fields and to use the same microscope and objective. Reading should be done after coding samples.

3.2 Micro-chemotaxis method

The micro-chemotaxis method has the advantage of utilizing a minimal quantity of cells (1×10^5 cells/well compared to $4–8 \times 10^5$/well) in the Boyden chamber, and permits evaluation of more samples in one experiment. The only difference from the Boyden chamber technique is that the micromethod uses a chamber of Plexiglass possessing 48 wells with volume of 25 μl. The cover piece, when it is mounted and screwed, forms 50 μl upper wells. One microchamber thus contains 48 replicates. The preparation of cells, filter, and chemoattractants is the same as described for the conventional method in *Protocol 3*.

Further discussion and description of this method can be found in Chapter 13.

Protocol 4. Micro-chemotaxis method

Equipment and reagents

- As *Protocol 3*
- Modified Boyden chamber (see above)
- Clamps (Neuroprobe)

Method

1. Aliquot 25 µl chemoattractant in each lower well. The 25 µl vol. may have some variations (2–3 µl more or less), depending on the micro-chamber used. It is advisable to calibrate in advance the lower wells, so that having seeded the chemoattractant, the liquid in the lower well forms a small convex surface that guarantees a perfect adhesion of the filter avoiding air bubble formation.

2. Put the filter (25 × 85 mm) on the lower compartment. To avoid confusing the order of the experimental groups in the filter, cut a small piece from one edge of the filter.

3. Mount the silicon trimming and cover piece. Press the cover piece tightly to avoid air-bubbles.

4. Seed 50 µl cell suspension (1×10^5 cells) in the upper well by leaning the pipette tip on the border of the well and quickly ejecting the cell suspension.

5. Incubate the chamber at 37°C in 5% CO_2 for 6 h.

6. Unscrew and turn over the chamber. Hold the upper compartment tightly and remove the lower compartment, keeping the silicon trimming and the filter adhered to the upper compartment of the chamber. At this point the migrated cells will be on the upper surface of the filter.

7. Lift the filter and hold it with a clamp on each end (the clamps are purchased from the manufacturer of the chamber, Neuroprobe, Maryland, USA).

8. Wash the opaque side of the filter, where the non-migrated cells remain, by passing this side over PBS. Do not immerse entire filter in PBS or the migrated cells will be lost.

9. Hold the filter with one of the clamps and clean the opaque side by scraping the filter against a special rubber policeman (purchased from the manufacturer) to remove all non-migrated cells.

10. Stain the filter with Diff-Quik. Read the results as described for conventional method.

3.3 Comments

The methodology described above, with variations in type of filter, incubation time, etc., could be utilized for the evaluation of the motility of haematopoietic cells, fibroblasts, smooth muscle cells, and tumour cells. For cells of non-haematopoietic origin, coating of the filter with extracellular matrix protein is often needed, although it increases background migration. Since the micro-method is technically more difficult, classical modified Boyden chambers may be a better choice for occasional use. It is important to define whether a cytokine that induces migration across filters, does so by random locomotion or chemotaxis. A checkerboard experimental design should be used whereby different concentrations of the cytokine are seeded in the upper and/or lower compartments of the chamber. For a chemotactic signal, maximal migration should occur in the presence of a positive concentration gradient between the lower and upper compartments, with little or no effect under negative gradient conditions (chemoattractant only in the upper well) or in the absence of a gradient (chemoattractant added to the upper and lower well) (19–21).

4. Procoagulant activity (PCA) or thromboplastin activity

Thromboplastin is a membrane glycoprotein that is constitutively expressed by several cell types. It can be also defined as a high-affinity cell-surface receptor and essential cofactor for the serine protease factor VII. The functional bimolecular complex possesses serine protease activity that mediates the initial proteolytic activation of the extrinsic coagulation cascade and thrombin formation. EC do not express PCA in control conditions but they can be strongly activated by endotoxin, IL-1 and TNF to do so. An easy, and relatively inexpensive way to measure PCA is through a biological assay which evaluates the time required by an EC lysate to promote clot formation (22). Alternatively, tissue factor can be evaluated using mAb and commercial ELISA.

Protocol 5. Procoagulant[a] activity assay

Equipment and reagents

N.B. All reagents should be endotoxin-free, as assessed by the Limulus assay (Sigma).

- HUVEC cells
- 4 cm^2 culture wells
- Medium (most commonly M199 + 20% FCS)
- PBS
- Cytokines
- Dry ice/methanol mixture
- Prewarmed transparent plastic tubes
- Citrated normal or coagulant-deficient plasma
- 0.025 M CaCl$_2$

Protocol 5. *Continued*

Method

1. Culture HUVEC cells in 4 cm² culture wells to confluence (2.5 to 3.5 × 10^5 cells/well) as described above.

2. Remove the culture medium [most commonly M199 with 20% fetal calf serum (FCS)] and wash once with 1 ml of phosphate buffer saline (PBS).

3. Incubate at 37 °C with 600 μl of culture medium in the presence or absence of cytokines.

4. At predeterminated intervals remove the supernatant and wash three times with 1 ml of PBS.

5. Fill wells with 300 μl of PBS and place at −20 °C. Disrupt the cells by freeze-thawing: place the cell plates on top of dry-ice/methanol mixture for freezing and thaw in a 37 °C waterbath. Repeat twice more. Standard cell culture plates do not crack, even after several freeze-thawing cycles.

6. Measure clotting time in prewarmed transparent plastic tubes.

    (a) Add in sequence 0.1 ml test sample, 0.1 ml citrated normal or coagulation-deficient plasma and 0.1 ml of 0.025 M $CaCl_2$.

    (b) Optically evaluate the time in seconds required for clot formation in a waterbath at 37 °C.

    (c) Express the results in arbitrary units/mg protein by comparison with a standard curve of clotting times produced by dilution of standard rabbit brain thromboplastin (1000 units of thromboplastin causes normal plasma to clot in 20 sec).

[a] The type of procoagulant activity can be identified using human plasmas selectively deficient in factors II, VIII, VII, IX, or X (Merz-Dade, Duding, Switzerland). Only substitution of normal plasma with factor VII deficient plasma can abolish PCA activity. Further characterization can be performed by use of known inhibitors of cell procoagulants, namely cysteine protease inhibitors $HgCl_2$ and iodoacetamide (Sigma Chemicals, St Louis, Missouri, USA).

4.1 Comments

IL-1, TNF and endotoxin are potent inducers of PCA in HUVEC. This activity requires protein synthesis and is fully inhibited by protein synthesis inhibitors. The time-course of PCA induction is identical for the two cytokines and endotoxin: the activity requires at least 1 h to be apparent, peaks at 4–6 h, and declines within 12 h. The active concentrations for endotoxin are 1–10 μg/ml (using *Salmonella enteritidis* lipopolysaccharide or *Escherichia coli*, Difco, Detroit, Michigan, USA) for IL-1 1–10 units/ml, and for TNF 1–10 units/ml.

For this assay it is of importance to use endotoxin-free reagents (media and sera) for culturing the cells before the assay due to very high sensitivity of HUVEC to endotoxin-induced PCA. Both IL-α and β are able to promote PCA in HUVEC without any apparent difference in activity. Lymphotoxin (TNF-β) is also active. Other cytokines we tested (IFNs -α, β, and γ, G-, M-, and GM-CSF, IL-6) were inactive (1).

5. Platelet-activating factor (PAF)

The term PAF defines a group of potent biologically active, ether-linked phosphocholines, the alkylacetylglycerophosphocholines (23). In addition to its activity on platelets, PAF possesses a broad range of biological activities at nanomolar concentrations. A number of cells and tissues produce PAF upon stimulation, through a deacylation–acetylation cycle that is catalysed by a phospholipase A2 and by a specific acetyl CoA acetyltransferase (23). The time-course of stimulation of this pathway can be rapid (few minutes) or prolonged (several hours). EC represent a good model for these two types of stimulation. For instance, they produce PAF in a few minutes when stimulated with thrombin, or in several hours when stimulated with IL-1/TNF (9, 24). A second, *de novo*, pathway has been described which is suggested to maintain a basal level of PAF in certain cells and tissues (23).

It is difficult to obtain highly accurate measurements of picomols or fento-mols of PAF produced by cells. Several approaches have been used to measure PAF. The most commonly used and sensitive method is based on activation of washed rabbit platelets. The incorporation of labelled acetate (25) can be employed to show the synthesis of PAF by stimulated cells, but the method does not give information on the true PAF concentration or on bioactivity. Mass spectrometry gives accurate results, permits the identifica-tion of different molecular species of PAF, but requires careful preparation of the samples and is also limited by the sample size.

More recently, polyclonal antibodies have been produced against PAF, but measurement by radioimmunoassay is not highly sensitive and the antibody can cross-react with other phosphoglycerides. We summarize here the current techniques routinely used in our laboratories to measure PAF in EC.

5.1 Cell cultures used for PAF assay

Several endothelial cell types (aortic and capillary bovine EC, human EC from veins, arteries or capillaries, aortic rabbit EC) are able to produce PAF, that remains in part (80%) associated to the cells, and in part released. EC should be grown to confluence in plastic Petri dishes (35 mm or 100 mm diameter purchased from Costar or Becton Dickinson) without any coating with exogenous proteins, as these can bind PAF. Endothelioma cell-lines deriving of different murine tissues and immortalized by mT antigen of polyoma virus represent a useful tool to study the synthesis of PAF by microvasculature (12). Stimulate cell in M199 or Iscove's medium containing 0.25% BSA (fraction V), as FCS can contain inhibitors of PAF synthesis (anti-proteinases) or a specific acetylhydrolase that destroys PAF. For short stimulations, maintain pH with 20 mM Hepes, in bicarbonate-free medium. For prolonged stimulations it is advisable to use a bicarbonate/carbonic anhydride buffer system.

5.2 Extraction of PAF

PAF in cells and medium is routinely extracted by a modification of the procedure of Pinckard (26), in which methanol contains 50 mM acetic acid or 1 M formic acid to improve the recovery of acetic lipids.

Protocol 6. PAF extraction from cells and medium

Equipment and reagents

- Endothelial cells
- Cold acidified methanol
- 35 mm diameter dishes
- [^{3}H]PAF (Amersham, 9.25 MBq, 250 μCi/ml)
- Rubber policeman
- Polypropylene tubes
- Chloroform

- 0.1 M sodium acetate
- Centrifuge
- Nitrogen gas stream
- TLC system (silica gel H, Merck)
- Chloroform:methanol:water (65:35:6, v/v and 1:2:0.8, v/v)

Method

1. To extract PAF from adherent EC, rapidly remove the medium and add 1 ml of cold acidified methanol to a 35 mm diameter dish. Add a small amount of ^{3}H-PAF (50 000 c.p.m.) (Amersham) to monitor recovery.

2. Scrape the cells off with a rubber policeman and wash the plastic dish twice with 1 ml of acidified methanol.

3. To 3 ml of acidified methanol collected in a polypropylene tube (a glass tube is not advised since its surface can bind PAF) add 1.2 ml of water and 1.5 ml of chloroform.

4. Mix the monophasic mixture and leave at room temp. for 30 min.

5. Add 1.5 ml of sodium acetate 0.1 M and 1.5 ml chloroform.

6. Mix thoroughly by shaking and centrifuge at 200 *g* for 5 min. Collect the bottom organic phase and extract the aqueous phase twice with 2 ml of chloroform.

7. Dry the sample under a nitrogen stream.

8. To extract PAF from 1 ml of medium, add 2.4 ml of acidified methanol and 1.2 ml of chloroform and then process the mixture as above.

---

9. PAF is isolated from other lipids by thin-layer chromatography (TLC) (silica gel H, Merck) by using chloroform:methanol:water (65:35:6, v/v) as solvent system. The lipid material, with an RF from 0.18 to 0.22, is extracted by incubating the silica for 20 min at room temp. with 3 ml of chloroform:methanol:water (1:2:0.8, v/v). This procedure is repeated three times. The extracted lipid is used for characterization and biological assay.

5.3 Identification of PAF

There are a number of well-defined criteria for positive identification of PAF.

(a) The test lipid should migrate with authentic PAF on TCL with an RF of 0.21 between lyso-phosphatidylcholine (RF 0.11) and phosphatidylcholine (RF 0.31) (solvent system: chloroform:methanol:water, 65:35:6, v/v).

(b) In the HPLC system, the separation is carried out on a 300×4.5 mm microporasil column (water) eluted with chloroform:methanol:water (60:55:5) at a flow rate of 1 ml per min. PAF shows a retention time of 21 min, between phosphatidylcholine (retention time 10 min) and lyso-phosphatidylcholine (retention time 26 min).

(c) PAF activity should be destroyed by base-catalysed methanolysis (3 min incubation in 1 ml of NaOH 0.05 N dissolved in methanol) and by phospholipase A2 treatment [0.03 mg of phospholipase A2 from pig pancreas (Sigma), in 1 ml of Tris-buffered saline containing 10 mM $CaCl_2$, 1 h incubation at 37 °C], that both remove the acetyl group at sn-2 position.

(d) Removal of the phosphocholine by treatment with phospholipase C destroys the PAF activity [0.05 mg of phospholipase C from *Bacillus cereus* (Sigma), in 1 ml of 0.1 M Tris–HCl buffer, pH 7.4, containing 1 mM $CaCl_2$, 6 h incubation at 22 °C].

(e) Treatment of the sample with lipase A1 [0.2 mg lipase A1 from *Rhizopus arrhizus*, (Sigma), 1 ml of 0.1 M borate buffer, pH 6.5 containing 10 mM $CaCl_2$, 1 mM deoxycholate and 0.4% BSA, 6 h of incubation at 22 °C], acid (0.03 N HCl, 3 h at 22 °C), and weak base (28% ammonium hydroxide for 30 min at room temp.) does not abolish the biological activity of PAF. After the mentioned treatments, the lipids are extracted as described and compared to untreated controls.

5.4 Quantitation of PAF by bioassay

Both aggregation and degranulation of washed rabbit platelets are commonly used to measure PAF activity in the sample. Rabbit platelets respond readily at concentrations from 10 pM to 1 nM. The response to PAF of platelets prepared from the blood of adult or old New Zealand white rabbits is better in terms of sensitivity and time to response than that of platelets from young rabbits. The preparation of platelets is carried out at room temperature by using polypropylene materials.

Protocol 7. Quantitation of PAF by platelet aggregation

Equipment and reagents

- Rabbit blood (see step 1)
- EDTA
- Centrifuge
- Ca^{2+}-free gelatine Tyrode's buffer: 137 mM NaCl, 1 mM $MgCl_2$, 2.6 mM KCl, 12 mM $NaHCO_3$, 5 mM glucose, 0.25% Difco gelatine, pH 6.5

- Gelatine Tyrode's buffer (as above) containing 1 mM $CaCl_2$, pH 7.4
- Elvi 540 aggregometer + cuvettes
- Standard PAF (Bachem)
- Saline containing 0.25% BSA

Method

1. Collect blood from the central ear artery of a rabbit in 5 mM EDTA (final concentration) and centrifuge at 375 *g* for 20 min.

2. Collect the top two-thirds of platelet-rich plasma (PRP) and centrifuge at 1450 *g* for 15 min.

3. Wash platelets twice in 50 ml of Ca^{2+}-free gelatine Tyrode's buffer and resuspend in the same buffer at a concentration of 6×10^8/ml.

4. For aggregation, dilute platelets 1 to 6 in gelatine Tyrode's buffer (Ca^{2+}-free gelatine-Tyrode's buffer containing 1 mM $CaCl_2$, pH 7.4) in the cuvette (0.5 ml vol.) of an Elvi 540 aggregometer. Test samples and standard PAF (16:0 PAF, Bachem) at the desired dilutions (4 points in the linear range should be used) in saline containing 0.25% BSA. The PAF content of the samples is quantitated by testing several dilutions of the same sample and linearly correlating the aggregation with that of the standard curve.

Protocol 8. Degranulation assay for PAF

Equipment and reagents

- PRP
- [^{14}C] serotonin (Amersham, 9.25 MBq, 1 μCi/ml)
- Ca^{2+}-free Gelatine Tyrode's buffer (see *Protocol 7*)

- Plastic tubes
- 1.5 M formaldehyde
- Scintillation counter
- 2.5% Triton X-100
- Saline containing 0.25% BSA

Method

1. Incubate PRP with [^{14}C]serotonin (Amersham, 1 μCi/ml) for 15 min at 37°C. The labelled platelets are then treated as described above.

2. Add 5–10 μl of the dilutions to 0.2 ml of diluted platelets in gelatine Tyrode's buffer in plastic tube, and after 1 min of incubation at 37°C, stop the reaction by 0.05 ml of 1.5 M formaldehyde.

3. Centrifuge the samples at 12 000 *g* for 30 sec, and count released radioactivity. Calculate the total amount of label in the platelets, lysing with 0.01 ml 2.5% Triton X-100. The background release is determined by incubation of platelets with 0.01 ml of saline containing BSA.

4. Plot the results of the assays as % of [^{14}C] serotonin release. This rises linearly with the concentration of standard up to 30 to 50% release. Comparison of the amounts of the sample required to give the same activity as in a standard calibration curve provides a measure of the moles of PAF present.

5.5 Quantification of PAF by HPLC-tandem mass spectrometry

A newly developed technique based on HPLC-tandem mass spectrometry (HPLC-MS/MS) has been developed by our colleagues and it permits chemical analysis and quantitation of PAF without any derivatization (27, 28).

Protocol 9. Mass spectrometry assay of PAF

Equipment and reagents

- API III (Perkin-Elmer Sciex) mass spectrometer with an ion spray articulated source, interfaced to a syringe pump (HPLC, Italy), and a Perkin-Elmer ISS-101 autosampler
- TLC purified samples
- Mobile phase: methanol:2-propanol:hexane:0.1 M aqueous ammonium acetate (100:10:2:5, v/v)

- HPLC equipment (Waters, Millipore) equipped with a reverse-phase column (Phase separations, Spherisorb C18, 5 mm, 100 × 1 mm i.d.)
- Argon

Method

1. Resuspend TLC-purified samples in 50 μl of mobile phase (methanol-2-propanol-hexane 0.1 M aqueous ammonium acetate; 100:10:2:5, v/v) and inject 20 μl of sample into HPLC (Waters, Millipore) equipped with a reverse-phase column (Phase separations, Spherisorb C18, 5 mm, 100 × 1 mm I.D.). Elute the sample with the mobile phase at a flow rate of 50 μl/min.

2. Perform mass spectrometry analyses under MS/MS conditions by parent ions scan or by multiple reactions monitoring (MRM). Fragmentation is obtained by collision with argon at a collision gas thickness of 2.7 × 10^{12} atoms/cm^3 and at an impact energy of 70 eV. Parent ions spectra, positive mode, are obtained from daughter ion with m/z 184,

Protocol 9. *Continued*

corresponding to phosphocholine fragment; the scanning range is m/z 100–600.

3. In the MRM analyses, acquired in positive mode, the study of different PAF molecular species is done by using the following reactions (parent ions → daughter ions): alkyl-PAF C16:0, 524 → 183.8; alkyl-PAF C18:0, 552 → 183.8; alkyl-lyso-PAF C16:0, 482 → 183.8; acyl-PAF C18:1, 550 → 183.8; acyl-PAF C16:0, 538 → 183.8.

4. The quantitative data are obtained by comparing the peak areas, at the retention time(s) characteristic of each compound, in the unknown samples to those from standard containing known amount of each analyte extracted and processed as the real samples.

5.6 Comments

The specificity of the two bioassays can be checked by the use of a PAF receptor antagonist (5 to 15 μM, CV-3988 from Takeda, Japan, BN52021 from IHB, France, SRI-6341 from Sandoz, USA), that block completely the aggregation and degranulation of platelets by authentic PAF. Addition of indomethacin (10 μM) and a creatine phosphate (0.3 mg/ml)-creatine-phosphokinase (0.15 mg/ml) enzymatic system minimizes the possible influence of arachidonic acid and ADP on the assays and increases the specificity for PAF.

In our experience, the aggregation of washed rabbit platelets is the easiest and least expensive method of quantifying PAF produced by EC, in particular when cytokines are used as stimulus. It is possible to measure 20 to 40 TLC-purified samples per day, with 80–100 ml of blood. It is important to ascertain that the procedures give a good recovery and are yielding reliable results by using appropriate standards. The standard curve should be linear, and controlled during the aggregation test each hour, to be sure of the sensitivity of the platelets. Many authors quantitate PAF by a radiometric method based on the evaluation of labelled acetate incorporation into a phospholipid having the same mobility properties on TLC systems of synthetic PAF (29). In EC stimulated with cytokines, this is not an adequate method. In EC stimulated with IL-1 and TNF, which trigger PAF synthesis after 4–6 h, the radiometric method is not adequate. Comparison of HPLC-MS/MS and bioassay with the incorporation of [^{3}H]acetate into the newly synthesized PAF indicates that the radiometric technique is suitable for stimuli that trigger a rapid synthesis of PAF, such as thrombin and elastase, but not for stimuli that require a long period of incubation, such as cytokines. The moles of [^{3}H]acetate incorporated in PAF newly synthesized after stimulation with TNF-α and IL-1α did not correspond to the moles of PAF detected by biological assay or HPLC-MS/MS. In fact resting EC (500 000 cells) contain about 0.1–0.3 pmoles PAF, that cannot be measured by radiometric assay. However, when elastase was used as stimulus the results obtained as acetate incorporation were comparable with that obtained with the biological assay and HPLC-MS/MS (F. Bussolino, L. Silvestro, and G. Camussi, in preparation). Moreover by radiometric technique one can quantitate the moles of acetate incorporated into newly synthesized PAF, but not the moles of PAF produced because there is no knowledge of the size of endogenous acetate pool of EC. This may prove to be relevant in diluting and competing with labelled acetate especially in long-term experiments. Furthermore exchange acetate reactions between lipids, proteins and carbohydrates could be relevant in these conditions.

6. Leukocyte adhesion and transmigration

The emigration of leukocytes from blood to tissues is essential to mediate immune surveillance and to mount inflammatory responses. The interaction of leukocytes with EC can be divided into four sequential steps: tethering, triggering, strong adhesion, and migration. The selectin family of adhesion molecules mediates tethering; strong adhesion is mediated by the integrin family, which need to be activated (triggering), and finally migration is induced by local promigratory factors including some cytokines and chemokines (30, 31).

We have studied the adhesive properties and transendothelial migration of various leukocyte subsets, but our interest has particularly focused on NK cells. These methods may also apply for investigation of other cell types, for instance tumour cells.

Protocol 10. Adhesion assay

Equipment and reagents

- Normal blood
- Complete medium: RPMI 1640 + 10% FBS
- ^{51}Cr (Amersham, 37 MBq, 1 mCi)
- Endothelial cells
- 96-well plates
- IL-1
- 0.05% NaOH + 1% SDS
- Gamma counter windowed for ^{51}Cr

Method

1. Separate PMN, monocytes, NK cells, or lymphocytes from buffy-coats of normal blood donors, as described (32, 33 and Chapters 10, 11, and 12).

2. Resuspend cells at 10^7 cells/ml in complete medium and label by incubation with 100 μCi ^{51}Cr for 1 h at 37°C.

3. After labelling wash extensively and resuspend in complete medium.

4. Culture EC in 96-well plates (1 × 10^4/well) in order to reach a confluent monolayer in 36–48 h. Stimulate designated wells with IL-1 (10 ng/ml) during the last 18 h of culture.

Protocol 10. *Continued*

5. Incubate EC with 100 μl of ^{51}Cr-labelled cells resuspended at 10^6 cells/ml and incubate at 37°C for 30 min.

6. Carefully remove the supernatant and wash the cells twice to remove non-adherent cells.

7. Incubate the adherent cells with 100 μl of 0.05% NaOH +1% SDS for 5 min and count radioactivity using a gamma-counter. Express cell adhesion as % of input cells.

6.1 Comments

The spontaneous adhesion of resting leukocytes to unstimulated EC varies for different subsets. For instance, the adhesion of NK cells is usually 5–15%, a value intermediate between that of monocytes (20–40%) and the very low value of T cells and PMN (< 5%). With monocytes and NK cells there is usually a high degree of variability among different donors.

The adhesive capacity of leukocytes to EC can be modulated by various signals. When EC are stimulated with IL-1, leukocyte adhesion increases, as EC express new adhesion molecules. Studies with specific mAb have demonstrated that the interaction of NK cells and monocytes with resting EC is mediated by the LFA-1/ICAM-1,2 pathway, while that on IL-1-activated EC involves both LFA-1/ICAM-1 and VLA-4/VCAM-1 (33).

Cytokines can also affect leukocytes directly. IL-2 and IL-12, for instance, increase NK cell adhesion to EC, while IL-4 has inhibitory activity.

Protocol 11. Transmigration assay

Equipment and reagents
- See Reference 34
- EC cells
- Gelatine precoated PVP-free polycarbonate filters (5 μm pore)
- 24-well plates
- Boyden chambers
- M199 complete medium
- Uncoated nitrocellulose filter (5 μm pore)
- Leukocytes
- Cotton buds
- Vials
- Gamma counter windowed for ^{51}Cr

Method

A schematic representation of the radioisotopic transmigration assay is shown in *Figure 1* and described in ref. 34.

1. Subculture EC to confluence on gelatine precoated PVP-free polycarbonate filters (5 μm pore) in 24-well-plates (8–10 × 10^4 EC per well reach confluence in 5 to 6 days).

2. Mount the filters with confluent EC monolayers on Boyden chambers and cover with 0.15 ml of M199 complete medium. The lower compart-

ment contains 0.2 ml of complete medium overlayered by an uncoated nitrocellulose filter (5 μm pore).

3. Seed leukocytes 3–6 × 10⁵ in 0.15 ml of complete medium in the upper compartment of the chamber and incubate at 37°C for 60 min.

4. After incubation, collect the medium with unattached cells and wash the EC monolayers with 0.5 ml warm medium. These two fractions are pooled (non-adherent fraction). The intact EC monolayers together with adherent leukocytes are gently collected with cotton buds (bound fraction). Transfer the double-filter system to vials together with the medium of the lower compartment (migrated fraction). Measure radioactivity in the three fractions. The recovery of radioactivity is more than 95%.

With NK cells, only a proportion (usually 30%) of adherent cells are able to transmigrate during the assay. When EC are activated with IL-1 a greater number of cells adhere, but usually the same proportion transmigrate. It should be noted that IL-1 does not change the state of confluence of the monolayer, as determined by staining.

Activation of NK cells with IL-2 increases their transmigration. As IL-2-activated NK cells are potent killers, and EC cells are a sensitive target, these experiments must be performed with a shorter incubation time of 30 min.

As for adhesion, the process of migration also is mediated via the interaction of LFA-1 and VLA-4 with their respective ligands on EC.

The transmigration assay can be performed also in Transwell plates, which are easier to use, but expensive.

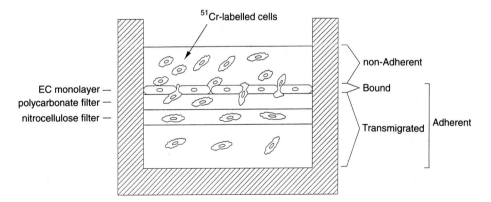

Figure 1. Cross-section of the compartments of a transmigration apparatus; a modified Boyden chamber was used.

Acknowledgements

This work was supported by CNR, Project ACRO, by Istituto Superiore di Sanità (VII Progetto AIDS); by Associazione Italiana per la Ricerca sul Cancro.

References

1. Mantovani, A. and Dejana, E. (1989). *Immunology Today*, **10**, 370.
2. Mantovani, A., Bussolino, F., and Dejana, E. (1992). *FASEB Journal*, **6**, 2591.
3. Gimbrone, M. A., Cotran, R. S., and Folkman, J. (1974). *Journal of Cell Biology*, **60**, 673.
4. Balconi, G. and Dejana, E. (1986). *Medical Biology*, **64**, 231.
5. Rayan, U. S., Ryan, W., Whitaker, C., and Chin, A. (1978) *Tissue Cell*, **8**, 125.
6. Schwartz, S. M. (1978). *In vitro*, **14**, 966.
7. Sherer, G. K., Fitzharris, T. P., Faulk, W. P., and LeRoy, E. C. (1980). *In vitro*, **16**, 675.
8. Spatz, M., Bembry, J., Dodson, R. F., Hervconen, H., and Murray, M. R. (1980). *Brain Research*, **1941**, 577.
9. Bussolino, F., Breviario, F., Tetta, C., Aglietta, M., Mantovani, A., and Dejana, E. (1986). *Journal of Clinical Investigations*, **77**, 2027.
10. Breviario, F., Bertocchi, F., Dejana, E., and Bussolino, F. (1988). *Journal of Immunology*, **141**, 3391.
11. Williams, R. L., Courtneidge, S. A., and Wagner, E. F. (1988). *Cell*, **52**, 121.
12. Bussolino, F., De Rossi, M., Sica, A., Colotta, F., Wang, J. M., Bocchietto, E., Padura, I. M., Bosia, A., Dejana, E., and Mantovani, A. (1991). *Journal of Immunology*, **147**, 2122.
13. Garlanda, C., Parravicini, C., Sironi, M., De Rossi, M., Wainstok de Calmanovici, R., Carozzi, F., Bussolino, F., Colotta, F., Mantovani, A., and Vecchi, A. (1994). *Proc. Natl. Acad. Sci. USA*, **91**, 7291.
14. Taraboletti, G., Belotti, D., Dejana, E., Mantovani, A., and Giavazzi, R. (1993). *Cancer Research*, **53**, 3812.
15. Bocchietto, E., Guglielmetti, A., Silvagno, F., Taraboletti, G., Pescarmona, G. P., Mantovani, A., and Bussolino, F. (1993). *Journal of Cell Physiology*, **155**, 89.
16. Vecchi, A., Garlanda, C., Lampugnani, M. G., Resnati, M., Matteucci, C., Stoppacciaro, A., Ruco, L., Mantovani, A., and Dejana, E. (1994). *European Journal of Cell Biology*, **53**, 247.
17. Piali, L., Albelda, S. M., Baldwin, H. S., Hammel, P., Gisler, R. H., and Imhof, B. A. (1993). *European Journal of Immunology*, **23**, 2464.
18. Hahne, M., Jager, U., Isenmann, S., Hallmann, R., and Vestweber, D. (1993). *Journal of Cell Biology*, **121**, 655.
19. Dejana, E., Languino, L. R., Polentarutti, N., Balconi, G., Ryckewaert, J. J., Larrieu, M. J., Donati, M. B., Mantovani, A., and Margurie, G. (1985). *Journal of Clinical Investigations*, **75**, 11.
20. Falk, W., Goodwin, R. H., Jr., and Leionard, E. J. (1980). *Journal of Immunological Methods*, **33**, 239.

21. Bussolino, F., Wang, J. M., Defilippi, P., Turrini, F., Sanavio, F., Edgell, C.-J., S., Aglietta, M., Arese, P., and Mantovani, A. (1989). *Nature*, **337**, 471.
22. Colucci, M., Balconi, G., Lorenzet, R., Pietra, A., Locati, M., Donati, M. B., and Semeraro, N. (1983). *Journal of Clinical Investigations*, **71**, 1893.
23. Prescott, S. M., Zimmermann, G. A., and McIntyre, T. M. (1990). *Journal of Biological Chemistry*, **268**, 17381.
24. Kuijpers, T. W., Hakkert, B. C., Hart, M. H. L., and Ross, D. (1992). *Journal of Cell Biology*, **117**, 565.
25. Prescott, S. M., Zimmermann, G. A., and McIntyre, T. M. (1984). *Proc. Natl. Acad. Sci. USA*, **81**, 3534.
26. Pinckard, R. N., Farr, R. S., and Hanahan, D. J. (1979). *Journal of Immunology*, **119**, 2185.
27. Silvestro, L., Da Col, R., Scapaticci, E., Libertucci, D., Biancone, L., and Camussi, G. (1993). *Journal of Chromatography*, **647**, 261.
28. Montrucchio, G., Bergerone, S., Bussolino, F., Aloatti, G., Silvestro, L., Lupia, E., Cravetto, A., DiLeo, M., Emanuelli, G., and Camussi, G. (1993). *Circulation*, **88**, 1476.
29. McIntyre, T. M., Zimmermann, G. A., Setoh, K., and Prescott, S. M. (1985). *Journal of Clinical Investigations*, **76**, 271.
30. Springer, T. A. (1994). *Cell*, **76**, 301.
31. Adams, D. H. and Shaw, S. (1994). *Lancet*, **343**, 831.
32. Sozzani, S., Luini, W., Molino, M., Jilek, P., Bottazzi, B., Cerletti, C., Matsushima, K., and Mantovani, A. (1991). *Journal of Immunology*, **147**, 2215.
33. Allavena, P., Paganin, C., Martin Padura, I., Peri, G., Gaboli, M., Dejana, E., Marchisio, P. C., and Mantovani, A. (1991). *Journal of Experimental Medicine*, **173**, 439.
34. Bianchi, G., Sironi, M., Ghibaudi, E., Selvaggini, C., Elices, M., Allavena, P., and Mantovani, A. (1993). *Journal of Immunology*, **151**, 5135.

Biological assays for haemopoietic growth factors

N. G. TESTA, C. M. HEYWORTH, B. I. LORD, and
E. A. DE WYNTER

1. Introduction

The aim of this chapter is to summarize *in vitro* standard assays used to investigate the myeloid growth factors interleukin-3 (IL-3), granulocyte-macrophage colony-stimulating factor (GM-CSF), granulocyte colony-stimulating factor (G-CSF), macrophage colony-stimulating factor (M-CSF), and erythropoietin (Epo). Such methods are illustrated by experimental protocols using murine cells, and with selected examples using human bone-marrow cells.

Two primary approaches are used for the *in vitro* study of these factors. The first uses the clonal assays that first detected the CSFs, and then monitored their purification (1, 2). In these assays, in which cell proliferation is absolutely dependent on the continuous presence of CSFs, the progeny of single cells (colony-forming cells, CFC) are observed as colonies composed of haemopoietic cells (1). The second approach uses growth factor-dependent cell-lines which provide a quick method for their titration. Some of these are described here and other cell-line assays are described in Chapter 2.

The advantages of the clonal assays are:

(a) They are performed with freshly isolated haemopoietic cells, which are the physiological targets of these growth factors (1, 2).

(b) The CFC range from primitive multipotential cells, with extended proliferation capacity (and some degree of self-renewal), to later cells committed to differentiation into one cell lineage, and which show a more restricted capacity for proliferation (1, 3, 4). Thus, the action of growth factors on a wide range of different populations of progenitor cells can be investigated (5, 6).

(c) They allow the study of the mature cells in the colonies, and thus the influence of a given factor on the determination of the lineage along which CFC differentiate (1, 5, 6, 7).

(d) Purified subpopulations of CFC, and serum-free cultures, allow the distinction between direct and indirect biological effects of the growth factor in primary cultures of haemopoietic cells (5, 8, 9). This type of information is particularly important for the understanding of the complex networks that regulate their proliferation and differentiation.

Factor-dependent cell-lines provide quicker results than clonal assays for screening purposes (10). In addition, they offer the possibility of studying the mechanisms of action of the growth factor (11) in

(a) Promoting cell survival.

(b) Stimulating proliferation.

(c) Inducing differentiation.

(d) Inducing the functional activity of the mature cells.

Similar studies like those performed with cell-lines are at present being undertaken using purified populations of CFC (9).

2. Preparation of cell suspensions

2.1 Murine bone-marrow

Femora are the usual source of bone-marrow cells. For standard plating experiments, three femora are used to prepare a cell suspension as shown in *Protocol 1*.

Protocol 1. Preparation of a cell suspension from murine bone-marrow

Equipment and reagents

- Sterile gauze
- Mouse femora
- Clean scissors
- Syringe fitted with 24-gauge needle
- Culture medium
- Microscope and haemocytometer

Method

1. Clean dissected femora of muscle tissue by wiping with sterile gauze.

2. Cut the epyphysis with clean scissors.

3. Insert a needle (23-gauge) attached to a syringe (containing 1 ml of the culture medium, preferably the same as to be used for culture) into the bone cavity.

4. Wash the bone cavity two or three times gently but thoroughly by flushing the medium up and down, into a sterile container.

5. Count the cells. About $1.5-2.0 \times 10^7$ cells per femur are usually obtained.

2.1.1 Comments

For plating experiments, the cell suspension may be diluted to 5×10^5–10^6 cells per millilitre, or $10 \times$ the cell concentration desired for plating. Although plating is usually performed as soon as the cells are ready, the cell suspensions can be kept on ice for 4–5 h without loss of cell viability of the CFC (3).

2.2 Human bone-marrow

Protocol 2. Preparation of cell suspensions from human bone-marrow

Equipment and reagents

- Human bone-marrow aspirates
- Sterile tubes
- Iscove's modified Dulbecco's medium (IMDM)
- Preservative-free heparin
- Centrifuge (MSE 3000i)
- FCS
- Pipettes
- Microscope and haemocytometer

Method

1. Collect bone-marrow aspirates in sterile tubes containing 5–10 ml of Iscove's modified Dulbecco's medium (IMDM) plus 400 units of preservative-free heparin.

2. Centrifuge the cells (600–1000 g) for 10 min, discard the supernatant and resuspend the cells in IMDM plus 2% fetal calf serum (FCS), and mix gently by pipetting.

3. Count the cells in a haemocytometer and adjust concentration as needed (see *Protocol 1*). As the cellularity of bone-marrow aspirates varies with the pathology of the patient, and also with the aspiration technique, some judgement must be exercised about the volume used to resuspend the cells. Hypocellular marrow samples are resuspended in 1–2 ml, but usually 5–10 ml are used.

4. Keep the cell suspensions on ice until plating.

2.2.1 Comments

Sometimes the source of bone-marrow is not a diagnostic aspirate, but marrow obtained from sections of ribs removed for access during surgery. Protocols for processing these samples may be found in ref. 12.

It is important to realize that *all* human bone-marrow samples are considered to be of potential risk (hepatitis virus, HIV, and other possible infective agents) and need to be handled with care. The use of appropriate laminar flow cabinets, centrifuges which minimize aerosol contaminations, the careful use (or avoidance when possible) of sharp instruments, the use of appropriate containers in transport, are essential precautions that **must** be undertaken. The legal requirements for safety at work are available in all laboratories. Careful reading of the required regulations and adequate instruction of the operators are essential requirements for working with clinical samples (3, 12).

2.3 Separation of excess red cells from human bone-marrow samples

This step is needed when it is judged that there is a large contamination with peripheral blood in the bone-marrow aspirate, or to separate the red cells in samples of peripheral blood.

Protocol 3. Separation of red cells from human bone-marrow

Equipment and reagents

- Human bone-marrow aspirate
- Ficoll-Hypaque (d = 1.077 g/ml)
- Centrifuge
- Pipettes
- FCS

Method

1. Layer the bone-marrow aspirate on a volume ratio of 4:3 slowly down the side of the tube (to avoid mixing) on top of Ficoll-Hypaque preparation (density 1.077 g/ml).

2. Spin at 400 g for 30 min. There should be a red pellet at the bottom of the tube, a clear layer of medium, a cloudy interface layer (with the cells to be harvested) and a plasma/medium layer on top.

3. Remove the interphase layer aspirating with a pipette (**never** mouth pipette, see Section 2.2.1) and discard the rest.

4. Suspend the cells in 10 ml of IMDM plus 2% FCS, and proceed as described in Section 2.2.

---

3. Standard clonal assays

3.1 Murine bone-marrow

As an example, *Protocol 4* details the steps for the assay of Mix-CFC from murine bone-marrow. This assay allows the growth of mixed colonies, which may contain several myeloid lineages (neutrophils, eosinophils, basophils, erythroid cells, monocytes-macrophages, and megakaryocytes) resulting from clonal proliferation of Mix-CFC. In addition, pure erythroid colonies (progeny of BFU-E), as well as colonies of neutrophilic granulocytes and macrophages (derived from GM-CFC), pure macrophage colonies (from M-CFC) or pure colonies of neutrophilic granulocytes (from G-CFC), eosinophils (from Eo-CFC) or megakaryocytes (from Meg-CFC) also grow. For further information, see ref. 13.

Protocol 4. Mix-CFC assay for murine bone-marrow

Equipment and reagents

- Murine bone-marrow cells (see *Protocol 1*)
- 3 cm diameter Petri dishes (Falcon)
- BSA (10% stock solution)
- Growth factor or conditioned medium
- FCS (pre-tested, Flow or Gibco)
- Erythropoietin (2 U)
- IMDM (see *Protocol 2*)

- Agar (3.3%)
- Fully humidified gas incubator (5% CO_2 in air, or 5% CO_2 plus 5% O_2 in nitrogen)
- Microscope with zoom lens or an inverted microscope
- Pasteur pipettes
- Cytospin

Method

1. Plate three aliquots of 1 ml containing 5×10^4 cells each in Petri dishes of 3 cm diameter (Falcon).

2. Make up the plating mixture as follows, to a total value of 3.3 ml (to allow 0.3 ml for waste).

	Volume (%)	For 3.3 ml
Cell suspension ($10 \times$ desired final cell concentration)	10	0.33
BSA (10% stock solution, see Sections 4.2 and 3.2.2)	10	0.33
Growth factor or conditioned medium	10	0.33
FCS (pre-tested, Flow or Gibco)	20	0.66
Erythropoietin (2 units)	2	0.066
IMDM	48	1.32

3. Put agar (3.3%) into a boiling-waterbath to melt.

4. (Optional). While the agar is melting, warm the plating mixture to 37 °C in a water bath to prevent the agar from setting too quickly when added.

5. Add 0.30 ml of agar, mix thoroughly but gently, and plate 1 ml per dish.

6. Place dishes on a tray and allow to set. They can be placed in a refrigerator for about 2 min to speed the process.

7. Check that the agar has set (it will not set into a smooth gel at 37 °C) and put the plates into a fully humidified gas incubator at 37 °C. The gas mixture may be 5% CO_2 in air or 5% CO_2 plus 5% O_2 in nitrogen. Low O_2 concentration may improve growth (3, 14).

Protocol 4. *Continued*

   8. Incubate for 10 days.

   9. Score the colonies under about 40 × magnification using a micro-scope with a zoom lens, or an inverted microscope.

   10. Individual colonies can be picked up for cytological examination using a Pasteur pipette. They are suspended in 0.1 ml of medium (plus a source of protein, either 1% serum or 0.1% BSA) for standard cytospin preparations.

The conditions can be varied to select for the growth of discrete populations of murine CFC (13).

(a) Omit Epo if erythroid growth is not desired.

(b) BSA is not essential for the growth of GM-CFC, G-CFC, and M-CFC.

(c) Granulocytic colonies are best scored after 7 days.

(d) Colonies derived from relatively mature erythroid progenitors (CFU-E) are scored after two days of incubation. They disappear at later times (3). These colonies only need 0.2 units of Epo per millilitre.

3.2 Human bone-marrow or peripheral blood

Basically similar conditions to those for murine bone-marrow are used, but 7 days of incubation are required for CFU-E, 9–14 days for GM-CFC, and 14 days for BFU-E and Mix-CFC. The main modification that may be needed is when the cells are recovered for cytogenetic examination (14, 15). Good chromosome preparations have not been obtained when the matrix used for colony growth is agar. Therefore, methylcellulose (viscosity 4000 c.p.s., Dow Chemicals) is used as a 1.6–2.7% stock solution in IMDM (different batches may vary in viscosity, and concentrations should be tested for every batch). A detailed protocol for its preparation may be found in ref. 12. Volumes of the plating mixture in *Protocol 4* are adjusted to allow 50–50 mix of the stock methylcellulose preparation and the rest of the culture components.

3.3 General comments

3.3.1 Factors and CFC

The CFU-E are the progenitor cell populations most sensitive to Epo, and the choice population of CFC to study its effects (17). Earlier erythroid progenitors (BFU-E) need other factors (IL-3 or GM-CSF) to proliferate before Epo becomes essential for further progression and maturation along the erythroid lineage. Multipotential Mix-CFC will proliferate in response to IL-3 and, to a lesser extent, GM-CSF, while the bipotential GM-CFC will respond to GM-CSF. The progenitors with the most restricted differentiation potential, G-CFC and M-CFC, will respond preferentially to their respective lineage specific stimulators, G or M-CSF (1, 2). These factors also interact producing additive or potentiating effects on colony growth (5, 6, 18). Other factors not covered in this chapter (like stem cell factor, SCF, IL-1, IL-6, cMlp ligand) also have additive or potentiating effects (12, 13).

3.3.2 Technical comments and possible problems

Possible problems may be encountered at each step in a culture protocol.

(a) Concentrated cell suspensions, or those with numerous red cells may contain cell aggregates. To plate single-cell suspensions, those aggregates must be dispersed by gentle pipetting.

(b) Two main precautions should be taken with synthetic culture medium. Even after preparing stock according to manufacturer's instructions, it should be checked that the osmolarity is 280–300 m osmol for murine cells (19) and 300 for human cells (3). Freshness of the medium is also essential. Storage at 4°C for more than a fortnight is discouraged (20).

(c) Every batch of serum should be pre-tested before use (see Section 4.1). Sometimes screening of several batches is necessary to find one that supports adequate growth. Serum may be kept frozen at −20°C for about a year. Thawed aliquots should be used within a fortnight.

(d) BSA (Grade V, Sigma) also needs to be pretested. Deionized solutions should be used. Detailed protocols are found in ref. 3. Stock solutions may be kept frozen at −20°C for about a year, aliquoted in convenient volumes to avoid repeated thawing and freezing.

(e) The main problems encountered in incubators are drying of cultures and temperature gradients. An increase in temperature of 0.1°C over 37°C may be critical, but decreases are better tolerated. Excessive opening of the incubator or excessive gas flow may contribute to dryness, resulting in poor growth.

4. Serum-free cultures

4.1 General comments

The assay of murine myeloid progenitor cells is usually performed in soft gel media supplemented with either FCS or horse serum (EqS). The variation in plating efficiencies achieved with different batches of sera can be enormous, and it is necessary to test each batch prior to purchase of large quantities to maintain a consistent standard. This variation in batches of serum indicates that they contain varying levels of agents capable of modulating myeloid cell development. The development of a serum-free medium which supports colony formation overcomes the difficulties that the additions of unknown factors to colony-forming assays might cause, and negates the need to test these batches of serum. The use of serum-free media introduces a greater degree of certainty into experimental procedures employed to determine the relative effects of cytokines on myeloid progenitor cell development (12, 13).

4.2 Preparation of the reagents

The so-called serum-free medium (8) is generally prepared by the addition of a number of constituents to IMDM. A fully detailed protocol of the method is described elsewhere (13).

4.3 Preparation of serum-free medium

The solutions described in *Table 1* can be employed as substitutes for serum in the mixed colony-forming assay described above (*Protocol 4*). In these assays, serum-free medium constitutes 1 ml of the total volume of 3.3 ml.

Further additions of other reagents such as haemin (final concentration 200 μM) can be made by adjusting the volume of Iscove's medium added to make the total volume of 1.0 ml.

Table 1. Supplements for serum-free cultures

Reagent	Volume (μl)	Final concentration
BSA	330	10 mg/ml
Soya bean lipids	42	25 μg/ml
Cholesterol	33	7.8 μg/ml
Linoleic-acid	33	5.6 μg/ml
Sodium pyruvate	33	1 mM
Glutamine	33	2 mM
α-thioglycerol	33	100 μM
Transferrin	33	300 μg/ml
Iscove's medium	430	

4.4 Comments

Serum-free colony-forming assays must use highly purified natural or recombinant growth factors as the crude conditioned medium preparations of haemopoietic growth factors that are often employed in such assays will often contain serum or other impurities. Dose-response relationships for growth factor-stimulated colony-formation are sometimes different from those seen in serum containing assays, which must be accounted for in experimental design. Although colony formation may take slightly longer in serum-free cultures, the maximal plating efficiency achieved is generally equal to, if not greater than that seen in serum-containing cultures.

5. Growth factor-dependent cell-lines

A number of haemopoietic growth factor-dependent cell-lines exist which can be employed to give an indication of the concentration and type of growth

---

Table 2. Examples of cell-lines dependent on haemopoietic growth factors for proliferation

Cell-line	Growth factor-dependence
FDCP-1	IL-3, GM-CSF
FDCP-2	IL-3
32DCL23	IL-3, G-CSF
BAC1.2F5	M-CSF, GM-CSF
DA1	GM-CSF
NFS-60	GM-CSF
M1	G-CSF
J774	M-CSF

---

factor present in biological preparations (10). Some of those frequently used are listed in *Table 2*. Others are described in Chapter 2. It should, however, be noted that antibody-based assays are often available to give unequivocal identification of the growth factors present.

To establish whether a specific growth factor is present in a solution, and determine its approximate concentration, proceed as follows:

Protocol 5. Biological assay for growth factors

Equipment and reagents

- Cells in suspension
- Benchtop centrifuge with swing-out rotor
- Culture medium
- Fischer's medium (Gibco)
- Horse serum
- CO_2 incubator
- Trypan blue (Northumbria Biologicals)
- [^{3}H]thymidine (37 kBq)
- GF/C filters (Whatman)
- Millipore cell harvester or similar apparatus
- Trichloroacetic acid (10% w/v)
- Liquid scintillation counter

Method

1. Culture cells in suspension until they reach an exponential growth rate.

2. Harvest by spinning at 800 *g* for 5 min on a bench centrifuge using a swing-out rotor.

3. Resuspend the cell pellet in the appropriate medium, centrifuge again and resuspend a further two times.

4. Resuspend cells at a concentration of 1–2 × 10^6/ml in Fischer's medium (Gibco-plus glutamine) plus horse serum (10% v/v) and plate out in a total volume of 100 μl at trial concentrations of 1–2 × 10^5 cells/ml with either dilutions of the preparation under test or dilutions of a standard preparation of the appropriate growth factor. The final concentration of horse serum (or if the cells are normally cultured in it, fetal calf serum) should be 10–20% (v/v). Incubate the cells in a gassed CO_2 incubator at a temperature of 37°C.

Protocol 5. *Continued*

5. After 24 or 48 h the effects of the growth factors on survival and proliferation can be assessed using the Trypan blue assay in which the cell suspension is mixed 1:1 with Trypan blue solution and those (viable) cells excluding dye counted.

6. A second method for assessment of proliferation (DNA synthesis) involves the measurement of [^{3}H]thymidine incorporation. After 24 or 48 h [^{3}H]thymidine (37 kBq) is added to each well and the incubation continued for 4 h. The cells are then removed from the incubator and serially transferred to GF/C filters on a Millipore cell harvester or similar apparatus. The cells are given three washes with trichloroacetic acid (10% w/v, 5 ml) and the TCA-precipitable material retained on the filter counted on a liquid scintillation counter.

7. The relative growth promoting activities of the standard and the diluents of the preparation under test are compared to quantify the growth promoting activity in the sample.

6. Enrichment of progenitor cell populations from murine bone-marrow

6.1 General comments

The progenitor cells exist in very small numbers, the pluripotent cells representing about 0.4% of the total bone-marrow cells and the more restricted committed progenitors being perhaps 2–3% of the population. To study direct effects of growth factors on progenitor cells, therefore, it is desirable to enrich the concentration of the appropriate cells. Historically cells are separated on physical parameters—density and size—and these properties are incorporated into the modern techniques which employ electronic sorting of cells tagged with fluorescent molecules or a process of continuous flow centrifugal elutriation. It is these two processes that will be outlined. Numbers of animals and/or cells used will be stated merely as a guide to convenient and practical handling of a sort. They may of necessity vary with specific practical requirements.

6.2 Murine cells

6.2.1 FACS sorting for *in vivo* multipotent and *in vitro* committed CFC

The essence of this method is to label the appropriate cells with a fluorescent marker, to carry out a preliminary density selection, then to sort the cells using a flow cytometer on the basis of their size, cellular heterogeneity and fluorescence intensity. The fluorescent markers generally used are fluoresceinated wheat-germ agglutinin (WGA-FITC) which binds selectively mainly to multipotent and some committed progenitors and rhodamine-123 (Rh-123) which is incorporated by the more mature elements of the WGA selected cells. Hence the more primitive (stem) progenitors are designated as WGA$^+$, Rh-123 dull cells.

Protocol 6. FACS sorting of murine bone-marrow cells

Equipment and reagents

- Stock solutions of metrizamide (Nygaard, Oslo)
High-density *Solution A* ($\rho_A \sim 1.10$ g/ml) and low-density *Solution B* ($\rho_B \sim 1.06$ g/ml) are made by dissolving 21 and 11 g metrizamide powder respectively in 42 and 64 ml Hank's balanced salt solution, buffered with Hepes (pH = 6.7). In each case, 1 g BSA is added and the volumes made up to 100 ml with double-distilled water

Measurements and adjustments
  i. pH to 6.5 ± 0.1
  ii. Osmolarity to 300 ± 10 mOsm
  iii. Measure density (density meter or bottle) at 4°C to 3 decimal places
  iv. Sterilize by filtration through 0.45 μm filtration units
  v. Store at −20°C in 10–20 ml aliquots
- Metrizamide for 1-step density cut. Mix solutions A and B to give a density of 1.080 ± 0.001 g/ml, maintaining all solutions at 4°C

To 10 ml B metrizamide $x = \dfrac{1.080 - \rho_B}{\rho_A - 1.080} \times 10$

add x ml A metrizamide

- Fluoresceinated wheat-germ agglutinin (Polysciences Inc., Pennsylvania, USA). WGA-FITC. Store at −20°C in 15 μl aliquots (15 μg) aliquots
- Rhodamine-123—made up at 1 mg/ml in double distilled water. Use fresh
- PBS and FHS (Fischer's medium with added glutamine (0.7 ml, 200 mM per 100 ml Fischer's and horse serum at 10%)
- 0.2 M *N*-acetyl-D-glucosamine (NADG). Store frozen
- Murine femora (see *Protocol 1*)
- Laminar flow hood
- 15 ml tubes
- 5 ml *glass* tubes
- Needles
- Centrifuge
- Pipettes or syringes and needles
- FACS system

Method

All solutions and cell suspensions should be maintained at 4°C.

1. Remove and clean femora from 20 mice. Operating under a laminar flow hood, flush the marrow into 7 ml of metrizamide B. (Yield ⩾ 4 × 10^8 cells.)

2. Add 15 μl WGA-FITC and mix thoroughly.

3. Place 3 ml metrizamide ($\rho = 1.080$ g/ml) in a 15 ml tube. Holding the tube at 45°, carefully layer 1 ml of cell suspension (max ~ 5 × 10^7 cells) on to the surface of the metrizamide by trickling it down the wall of the tube. Run six tubes in this way.

4. Break the sharp partition between cells and metrizamide by stirring gently with a needle point.

5. Centrifuge 15 min at 1000 *g*, 4°C. Use minimum braking force on the centrifuge to minimize gradient disturbance.

Protocol 6. *Continued*

6. Collect the band of low density ($\rho \leqslant 1.080$ g/ml) cells from the interface using a pipette or syringe and needle.

7. Wash once in Fischer's medium, centrifuging for 10 min at 800 *g*, 4°C.

8. Resuspend the cell pellet in 4 ml PBS and count the cells. (Yield $\sim 2 \times 10^7$ cells.)

9. Transfer to FACS. The detailed running of this instrument is beyond the scope of this chapter. It is recommended to have a tame operator available.

10. Set the pre-sterilized FACS to record forward-angle light-scatter (FALS: proportional to cell size) and right angle light-scatter (RALS: proportional to heterogeneity of cellular contents). Run a few thousand cells through the instrument to establish their distribution as illustrated in *Figure 1a* and *1b*. Set FALS windows as shown to exclude small cells (channels 1–120, mainly lymphocytes and residual erythrocytes) and very large cells (channels 200–255). Set RALS windows to exclude cells with high scatter (channels 95–255, mainly maturer granulocytic cells and monocyte/macrophagic cells).

11. Switch FACS to record fluorescence in the cells selected by step 10 above. *Figure 1c* shows the distribution, together with a further window set to detect cells showing the greatest affinity for WGA.

12. Put 4 ml FHS in a 5 ml *glass* tube, wetting the inside of the tube to the top. (This is to prevent drying of the film of cells directed into the tube.) Place the tube below one of the deflection plates of the sorter, start flow (~ 3000 cells/sec) and collection of the cells in the window defined.

13. On completion of the sort (~ 2–3 h, yielding $\sim 10^6$ cells), centrifuge the cells (10 min, 800 *g*, 4°C) and resuspend the pellet in 1 ml PBS.

14. Resort the cells at 300 cells/sec using the same selection parameters. The final yield ($\sim 10^5$ cells) can contain better than 90% *in vivo* CFC and are now ready for assay. WGA-FITC is not toxic to either the multipotent cells assayed by the spleen colony technique (21) or to committed progenitor cells assayed as *in vitro* colony-forming cells (see above). For use in growth-factor assays, however, its presence may complicate interpretation and it can be removed as follows.

15. Centrifuge the collected cells and resuspend the pellet in 5 ml NADG. Incubate 15 min at 37°C.

16. Centrifuge and resuspend as required for assays or further sorting. *Note:* For selection of the more primitive WGA$^+$, Rh-123 dull progenitors, it is necessary at step 11 to open the collection window to receive all WGA positive cells.

17. Mix 0.1 ml Rh-123 solution with 1 ml cell suspension (at 10^7 cells/ml).

18. Incubate for 20 min at 37°C then centrifuge (10 min at 800 *g*).

19. Resuspend the cell pellet in fresh medium and incubate for 15 min at 37°C.

20. Wash twice by centrifugation in PBS and resuspend for further FACS analysis.

21. On the FACS instrument, set the fluorescence window to exclude the Rh-123 positive cells and collect those with dull labelling only.

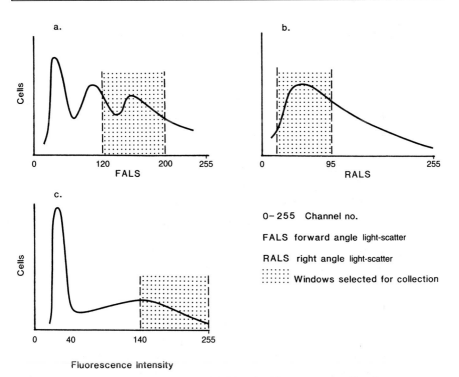

Figure 1. Diagrammatic presentation of FACS-sorted bone-marrow. Numbers represent channel numbers.

For further reading on this technique, including extended sorting with rhodamine-123, see refs 9 and 21–26. It is also possible to use immunological markers, essential for work on human marrow, see Section 6.3, to select for early progenitors (27–29).

6.2.2 Centrifugal elutriation for *in vitro* committed CFC

The protocol outlined below is developed from that reported by Williams *et al.* (30) who obtained highly enriched populations of GM-CFC containing little contamination with other committed progenitors (BFU-E, Mix-CFC, Meg-CFC). A preliminary density selection is followed by elutriation according to cell size. A Beckman centrifuge with the JE-6B elutriation rotor is used. The reagents and density separation methods are as described in the FACS *Protocol 6* above.

Protocol 7. Centrifugal elutriation

Equipment and reagents

Reagents and density separation system as *Protocol 6*

- Mice (adult, ♂ or ♀)
- Cyclophosphamide dissolved in physiological saline (200 mg/kg dose)
- Beckman centrifuge with a JE-6B elutriation rotor, chamber, and pumping system
- 70% ethanol

Method

1. Inject 20 mice i.p. with 200 mg/kg cyclophosphamide dissolved in physiological saline.

2. 4th day (e.g. Monday to Thursday) remove femora and suspend marrow in 10 ml Fischer's medium with added glutamine. Centrifuge 10 min, 900 g, 4°C.

3. Remove 9 ml of supernatant and resuspend the cell pellet in the remaining supernatant.

4. Add 1 ml of low density (B solution) metrizamide and mix.

5. Layer cells on two tubes of metrizamide at $\rho = 1.080$ g/ml at 4°C. (For density separation procedure see under FACS *Protocol 6*, stages 3–8, above.) Collect low-density cells from the interface and wash once in Fischer's medium. Centrifuge 10 min, 800 g, 4°C.

6. Resuspend the cell pellet in 3 ml Fischer's medium. (Starting with 20 mice this should yield ~ 2×10^7 cells.) *Note*: For the elutriation process from this stage, use FHS containing 5% EqS which should be at 4°C at the start of elutriation. The elutriation rotor, chamber and pumping system should be set up as described in the manual, sterilized with 70% ethyl alcohol and thoroughly washed with sterile Fischer's medium prior to starting stage 2. After use it should again be cleaned and both the rotor and chamber dismantled and dried.

7. Load the 3 ml of cells directly into the 50 ml loading chamber (see manual for elutriator operating procedure) and flush the cells into the elutriator rotor with FHS at a flow rate of 10 ml/min. Collect 10 ml effluent and switch flow of FHS to bypass the loading chamber.

8. Increase flow rate sequentially to 14, 16.5, 19.5 ml/min collecting 100 ml for each fraction. (These fractions may be collected together and discarded.)

9. Collect further 100 ml fractions, sequentially increasing the flow rate to 22, 25.5, 29, 32 ml/min. These may again be collected cumulatively and together represent the GM-CFC-rich fraction. Together this should yield $2–6 \times 10^6$ cells with a 7–14 day colony plating efficiency in agar with IL-3, CSF-1, or GM-CSF of 20–50%. *Note:* collecting the fractions sequentially and pooling gives a cleaner sort than collecting all four together at the highest flow rate.

10. Finally, maintaining the flow rate at 32 ml/min, switch off the centrifuge and collect cells sedimented in the chamber. If sorting for GM-CFC, these also may be discarded.

6.2.3 Use of elutriator for *in vivo* CFC

Multipotent progenitor cells may also be enriched using the elutriator (31). Animals must first be primed by thiamphenicol and bleeding after which the spleen becomes highly enriched in these progenitors. The elutriation procedure is carried out as above, collecting fractions in the range of 12–16 ml/min flow rates. Low-density cells ($\leqslant 1.079$ g/cm) subsequently separated from this fraction contain spleen CFC at 50–100% purity. This method is particularly useful for high cell-number yields, but the CFC obtained are less primitive than those obtained by FACS analysis.

6.3 Human cells

6.3.1 General comments

Like the murine stem cells, human haemopoietic progenitor cells exist in very small numbers. The CD34 antigen is expressed on haemopoietic colony-forming and stem cells, but is absent on more mature blood cells. These $CD34^+$ progenitor cells occur at a frequency of 0.5–4% in bone-marrow, 0.05–0.2% in steady-state peripheral blood and 0.1–2% in cord blood. A number of techniques have been developed to enrich or purify the CD34 population utilizing either physical characteristics (centrifugal elutriation) or immunological characteristics (immunoadsorption). See *Table 3*. Performance varies in terms of purity, yield and enrichment of CFC. FACS-sorting permits separation of any population or sub-population of cells by combining both physical and immunological parameters. As cells labelled with a fluorescent marker pass through a laser beam, they are examined individually and can be characterized based on fluorescence intensity, size, cell density and nucleo-cytoplasmic ratio. Highly purified populations are obtained with this method.

6.3.2 Immunoadsorption methods

These methods are commonly used as an alternative to FACS sorting. Mononuclear cell (MNC) fractions are labelled with a CD34 antibody and the target

Table 3. Commercially available systems used for enrichment of CD34 cells

1. Fluorescence activated cell sorting (FACS)
2. Systems based on immunoadsorption
    (a) Panning (AIS)
    (b) Affinity column chromatography (CELLPro)
    (c) Adsorption on ferromagnetic beads
        (i) Large beads (DYNAL and BAXTER)
        (ii) Middle-size beads (IMMUNOTECH)
        (iii) Microbeads (MILTENYI)

Table 4. CD 34^+ cell enrichment using different separation techniques

Technique	Total MNC recovery (%)	Purity of CD34$^+$ separated (%)	Yield of CD34$^+$ cells (%)	CFC enrichment (× fold)
FACS ($n = 23$)	1.4	71.2	38.3	30
CELLector ($n = 11$)	1.4	32.5	23.0	17
DYNABEADS ($n = 11$)	0.6	28.4	4.9	14
CELL-PRO ($n = 5$)	1.1	76.8	36.8	45
MiniMACS ($n = 12$)	1.4	76.6	62.6	104

n = number of experiments.

cells then adsorbed to antibody coated beads, flasks or retained on columns. Unwanted cells are removed and target cells detached, collected and assessed for purity. Results obtained in our laboratory using five of these methods are shown in *Table 4*.

Detailed comments on the methods of separation are beyond the scope of this chapter, but for further reading on the use of these systems see ref. 32.

6.3.3. Comments on advantages and disadvantages of the systems

These are all procedures relying on the CD34 monoclonal antibody–antigen interaction. Different CD34 monoclonal antibodies are used in the different methods. With each system it is important to follow the manufacturer's directions particularly concerning sample preparation. The preparation method may vary if human umbilical cord blood or cells from leukapheresis harvests are used.

(a) CELLector flasks (AIS uses clone ICH-3). Some progenitor cells are lost in the first stage of this process on the soyabean agglutinin flasks. However, the flasks are sterile tissue culture flasks and are easily manipulated. The capacity of the system is restricted to 2×10^7 cells/T25 flask unless the large scale T150 flasks are used. In our hands MNC from cord blood do not perform well in this system.

(b) Immunomagnetic CD34 conjugated beads (DYNAL uses clone 9C5). Again sterility can be maintained in this system but it does require some expertise to obtain good results. There are a number of problems with detaching the cells from the beads unless enzymic methods are used. Do not start with less than 5×10^7 MNC.

(c) Immunomagnetic beads (MILTENYI—Kit uses QBEND 10). Good capacity and up to 4×10^8 cells can be processed on the MiniMACS column. Care must be exercised when preparing the sample as the column may become blocked if the sample contains large numbers of red cells or platelets. Aim for a good single-cell suspension. This system is easy to use, yields good enrichment of CFC with high purity of CD34$^+$ cells (after two cycles). Highly purified populations of CD34 cells can be obtained in one hour.

(d) Immunoaffinity columns (CELLPro—Kit uses clone 12.8). Excellent capacity and up to 5×10^8 cells can be processed with each column. However, the performance of this system is reduced if cord blood cells are used, probably due to the large numbers of red cells even after density centrifugation. In addition, there may be problems associated with the biotinylated antibody which in our hands loses affinity for the target cells after storage at 4°C.

Note which antibody is used in the separation procedure as the same one *cannot* be used in the analysis. This is of particular importance if enzymic methods have been used for detaching cells, as the antigen may be destroyed (see ref. 32).

6.3.4 FACS sorting of CD34 positive cells

Mononuclear cells are labelled with a CD34 antibody directly conjugated to a fluorochrome and the cells sorted by flow cytometry. Generally sorting speeds used are 4–5000 cells/sec. Nevertheless, if 5×10^7 cells are to be sorted, this can take up to 4 h. To reduce the time taken to sort cells, any of the systems outlined above may be used as a pre-enrichment step. We routinely use the MiniMACS columns which yield enriched populations in one hour.

Protocol 8. Labelling cells for FACS sorting with CD34 antibody

Equipment and reagents

- IMDM/2% FCS (Iscove's modified Dulbecco's medium plus 2% fetal calf serum)
- PBS/1% BSA
- CD34 monoclonal antibody conjugated to FITC/PE (HPCA-2, Becton-Dickenson)
- 5 ml sterile tubes
- Screw top 15 ml centrifuge tubes—conical bottom
- Avoid capping of the antibody on the cell surface by working quickly and maintaining all solutions at 4°C

Method

1. Cell suspensions are obtained from bone marrow or other sources using either *Protocol 2* or *3*.

2. Decide total number of cells in the suspension to be labelled with the CD34 antibody.

3. Suspend these cells in PBS/1% BSA at a concentration of 2×10^8 cells/ml.

4. Incubate cells with the antibody for 20 min at 4 to 12°C. (Quantity of antibody to be used is predetermined by titration.) See ref. 32.

5. Wash cells once with 10–15 ml PBS/1% BSA and resuspend in IMDM/2% FCS at a concentration not exceeding 5×10^6 cells/ml.

6. Prepare two 5 ml tubes containing IMDM/2% FCS for collection of the sorted cells.

7. Maintain on ice ready for cell sorting.

6.3.5 Comments

(a) A portion of the mononuclear cell suspension may be removed prior to labelling with the CD34 antibody and used as a negative control. If required these cells can be labelled with an isotype matched control antibody conjugated with the same fluorochrome.

(b) If sub-populations are required, eg. CD34/38/33, it is possible to label with more than one antibody simultaneously, providing they are all *directly* conjugated to different fluorochromes.

6.3.6 Analysis and sorting of CD34⁺ cells on FACS

The set-up for analysis of CD34⁺ cells will vary according to the instrument used and it is essential to have the co-operation of a trained operator. A number of methods have been previously described for analysis and the reader is referred to ref. 32 for detailed discussion.

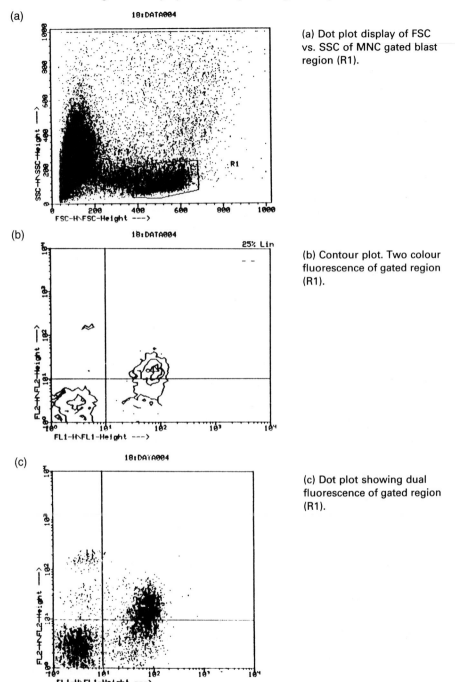

(a) Dot plot display of FSC vs. SSC of MNC gated blast region (R1).

(b) Contour plot. Two colour fluorescence of gated region (R1).

(c) Dot plot showing dual fluorescence of gated region (R1).

Figure 2. FACS analysis of CD34 positive labelled cells. See *Protocol 9*.

The following procedure is used in our laboratory on the FACS Vantage:

Protocol 9. Analysis of CD34$^+$ cells by FACS

Equipment and reagents

- CD34-FITC labelled cells
- FACS Vantage system
- Unlabelled 10 μm beads

- FITC- and PE-labelled Calibrite beads (Becton-Dickinson)
- Control samples labelled with isotype matched monoclonal antibodies

Method

1. Following instructions provided by the manufacturer, unlabelled 10 micron Calibrite beads are used to optimize instrument settings and compensation parameters set using FITC and PE labelled beads.

2. Cells are passed through the instrument and at least 20 000 events collected and examined for correct distribution of the populations, i.e. Forward Scatter vs. Side Scatter (FSC vs. SSC), see *Figure 2a*.

3. A 'gate' is set around blasts and small lymphocytes excluding granulocytes, monocytes and nucleated red cells and debris (*Figure 2a*).

4. Control samples labelled with isotype matched monoclonal antibodies are then examined and adjusted so that all cells fall within the first decade (log scale FL1). Negative cells fall in lower left quadrant. Percentages and other statistics are automatically recorded.

5. CD34-FITC labelled cells are then passed through the machine and again 20 000 events collected.

6. Using the same 'gate' as for control samples, cells are examined for fluorescence and all cells outside the first decade considered to be positive (lower right and upper right quadrant). Again, percentages and other statistics are recorded (*Figure 2b* and *c*).

By subtracting the values for the control sample from that of the CD34 labelled samples, the percentage of CD34$^+$ cells can be measured.

CD34 positive cells can be expressed either as a percentage of the lymphocyte population or as a percentage of the mononuclear cells. If the latter figure is required, omit step 3 as there is no need to gate on the lymphocyte population. Values thus obtained for the ungated population correspond to the CD34$^+$ cells in the total mononuclear fraction.

For extensive details on FACS sorting the reader is referred to ref. 33.

References

1. Metcalf, D. (1988). *The molecular control of blood cells.* Harvard University Press, Cambridge, Mass.
2. Testa, N. G. and Dexter, T. M. (1990). In *Clinical endocrinology and metabolism*, Vol. 4 (ed. S. M. Shalet), p. 177. Bailliere Tindall, London.
3. Testa, N. G. (1985). In *Cell clones* (ed. C. S. Potten and J. H. Hendry), p. 37. Churchill Livingstone, Edinburgh.
4. Humphries, R. K., Evans, A. C., and Eaves, C. J. (1981). *Proceedings of the National Academy of Sciences, USA*, **78**, 3629.
5. Heyworth, C. M., Pontin, I. L. O., and Dexter, T. M. (1988). *Journal of Cell Science*, **91**, 239.
6. Cook, N., Dexter, T. M., Lord, B. I., Cragoe, E. J., and Whetton, A. D. (1989). *EMBO Journal*, **8**, 239.
7. Metcalf, D. (1980). *Proceedings of the National Academy of Sciences, USA*, **77**, 5327.
8. Cormier, F., Ponting, I. L. O., Heyworth, C. M., and Dexter, T. M. (1991). *Growth Factors*, **4**, 157–64.
9. Lord, B. I. and Spooncer, E. (1986). *Lymphokine Research*, **5**, 59.
10. Whetton, A. D. and Dexter, T. M. (1990). *Biochimica Biophysica Acta*, **989**, 111.
11. Whetton, A. D. and Dexter, T. M. (1985). *Nature*, **303**, 639.
12. Coutinho, L. H., Gilleece, M. H., De Wynter, E. A., Will, A., and Testa, N. G. (1993). In *Haemopoiesis: a practical approach* (ed N. G. Testa and G. Molineux), pp. 75–106. IRL Press, Oxford University Press.
13. Heyworth, C. M. and Spooncer, E. (1993). In *Haemopoiesis: a practical approach* (ed. N. G. Testa and G. Molineux), pp. 37–53. IRL Press, Oxford University Press.
14. Bradley, T. R., Hodgson, G. S., and Rosendaal, M. (1978). *Journal of Cell Physiology*, **97**, 517.
15. Duke, I. D., Eaves, C. J., Kalousek, D. K., and Eaves, A. C. (1981). *Cancer Genetics and Cytogenetics*, **4**, 157.
16. Coutinho, L. H., Geary, C. G., Chang, J., Harrison, C., and Testa, N. G. (1990). *British Journal of Haematology*, **75**, 16.
17. Testa, N. G. (1989). In *Critical reviews in hematology and oncology*, Vol. 9, p. 17. CRC Press, Boca Raton, Florida.
18. Williams, D. E., Straneva, J. E., Cooper, S., Shadduck, R. K., Waheed, A., Gillis, S., Urdal, D., and Broxmeyer, H. E. (1987). *Experimental Hematology*, **15**, 1007.
19. Iscove, N. N., Guilbert, L. S., and Wyman, C. (1980). *Experimental Cell Research*, **126**, 121.
20. Testa, N. G. and Molineux, G. (ed.) (1993). Haemopoiesis: a practical approach, IRL Press, Oxford University Press.
21. Till, J. E. and McCulloch, E. A. (1961). *Radiation Research*, **14**, 213.
22. Visser, J. W. M., Bauman, J. G. J., Mulder, A. H., Eliason, J. F., and Leeuw, A. M. (1984). *Journal of Experimental Medicine*, **59**, 1576.
23. Bertoncello, I., Hodgson, G. S., and Bradley, T. R. (1985). *Experimental Hematology*, **13**, 999.
24. Mulder, A. H. and Visser, J. W. M. (1987). *Experimental Hematology*, **15**, 99.
25. Ploemacher, R. E. and Brons, N. H. C. (1988). *Journal of Cell Physiology*, **136**, 531.

26. Visser, J. W. M. and van Bekkum, D. W. (1990). *Experimental Hematology*, **18**, 248.
27. Spangrude, G. J., Heimfeld, S. and Weissman, I. L. (1988). *Science*, **241**, 58.
28. Spangrude, G. J. (1989). *Immunology Today*, **10**, 344.
29. Baines, P., Mayani, H., Bains, M., Fisher, J., Hoy, T., and Jacobs, A. (1988). *Experimental Hematology*, **16**, 785.
30. Williams, D. E., Straneva, J. E., Shen, R.-N., and Broxmeyer, H. E. (1987). *Experimental Hematology*, **15**, 243.
31. Nijhof, W. and Wierenga, P. K. (1984). *Experimental Cell Research*, **155**, 583.
32. de Wynter, E. A., Countinho, L. H., and Testa, N. G. (1993). In *Hematopoietic stem cells—the Mulhouse manual* (ed E. Wunder, H. Sovolat, P. R. Henon, and S. Serke). Alpha Med Press, Dayton, Ohio.
33. Hoy, T. G. (1990). In *Flow cytometry: a practical approach* (ed. M. G. Ormerod), pp. 125–136. IRL Press, Oxford University Press.

Assays for macrophage activation by cytokines

ANTHONY DOYLE, MICHAEL STEIN, SATISH KESHAV, and SIAMON GORDON

1. Introduction

The circulating monocyte may undergo a wide spectrum of functional changes during further differentiation. In the absence of inflammatory stimuli monocytes enter the tissues constitutively to become resident macrophages (1). These cells are relatively quiescent with respect to secretory or microbicidal activity and may have as yet uncharacterized trophic functions. In contrast, inflammatory recruitment causes monocytes to rapidly induce their secretory activity. These cells produce a large variety of enzymes, growth factors and cytokines when appropriately triggered. Lymphokines produced at the site of inflammation regulate the fate of elicited macrophages determining whether they are further stimulated to a heightened state of activation with maximal host-defence function, e.g. by interferon-γ (IFN-γ) or whether their secretory and defence activity is held in check (e.g. by IL-4 and IL-10).

Previously, macrophage activation has been defined in terms of an increased killing capacity for micro-organisms or tumour cells and respiratory burst activity has often been used as a measure of cytocidal activation. As our understanding of the repertoire and regulation of macrophage activity expands, the concepts of priming and activation have become highly complex. Nevertheless, by applying a small number of carefully selected assays it is possible to distinguish macrophages at a number of points on the activation spectrum (2). Our purpose here is to describe a panel of well-characterized assays that do not involve interactions between macrophages and a living target cell, but nevertheless readily discriminate between resident and elicited macrophages, and immunologically activated macrophages with enhanced microbicidal and tumouricidal potential. All the assays have been used in our laboratory on murine macrophage cultures and can be applied to human cells with the caveat that detection of NO is highly variable.

In vitro culture of monocytes and macrophages modulates the activity of the cells even in the most defined and controlled culture systems. Therefore, the assays described herein are best performed on freshly isolated primary cells, although in our experience the assays are also useful in comparing various longer term stimuli under the same culture conditions. Monocytes and macrophages adhere to tissue culture plastic, glass and bacterial plastic in the presence of 10% fetal calf serum (FCS). This makes isolation of almost pure monolayers a simple procedure when using elicited or resident cells where most other contaminating immune cell populations are non-adherent in the presence of FCS.

2. Assay for respiratory burst activity

Superoxide anion release assay measures the change in colour of cytochrome C when reduced by superoxide anion released from the stimulated macrophage (3).

Protocol 1. Superoxide anion release assay

Equipment and reagents

Reaction mixture:
- Hank's buffered saline solution (HBSS, phenol-red-free)
- 80 µM Ferricytochrome C (Sigma, type IV)
- 2 mM Sodium azide (this is a cytochrome oxidase inhibitor and prevents re-oxidation of cytochrome C)
- Macrophages

- RPMI 1640 + 10% FCS
- 24-well plates
- Superoxide dismutase (Sigma type I)
- Phorbol myristic acid (PMA)
- Zymosan
- PBS
- Centrifuge (Beckman G 5–6R)

Method

1. Plate macrophages at $1–5 \times 10^5$ well in RPMI 1640, 10% FCS in a 24-well tissue culture plate. This density ensures that oxygen anion release is proportional to cell number.

2. Set-up negative and positive controls:

   (a) a cell blank as a negative control

   (b) wells in which 30 µg/ml superoxide dismutase (Sigma type I) should be added as a negative control

   (c) elicited macrophages stimulated with a known trigger of respiratory burst such as phorbol myristic acid (PMA) at 10–100 ng/ml or zymosan at 100 µg/ml

3. Wash adherent macrophages once with PBS and preincubate with 0.45 ml of the reaction mixture for 5 min at 37°C.

4. Add 0.05 ml of HBSS and stimulant.

5. After 60 min dilute a portion of the reaction mixture three-fold with cold HBSS. Remove cell debris by centrifugation at 200 *g* for 5 min.

---

6. The cytochrome C becomes a darker red in colour after reduction.

7. Calculate superoxide release from the difference in absorbance at 550 nm in the absence and presence of superoxide dismutase, using an extinction coefficient of 21.1/mM/cm (reduced minus oxidized).

Comments: This assay is easy to perform and provides a relatively direct estimate of microbicidal and cytocidal potential. There are several other assays of respiratory burst activity that have been described elsewhere. These include H_2O_2 release (4) and chemiluminescence (5) assays for monolayers and nitrobluetetrazolium (NBT) reduction assay for single cell analysis (6). In assays where a cytokine or other modulatory compound being tested may also alter cell numbers or mass, it is best to determine total cell protein, e.g. Lowry or BCA method (Pierce) and express results per mg cell protein.

3. Assay for secretion of nitric oxide

Nitric oxide is an extremely reactive molecule which mediates cytotoxic effects on microbes and tumour cells (7). Because it is rapidly converted to NO_2^- in the presence of oxygen, NO secretory activity of cells in culture can be estimated by determining NO_2^- concentrations by the colorimetric Griess reaction.

Protocol 2. NO_2^- accumulation assay

Equipment and reagents

- 24-well plates
- Peritoneal macrophages
- IFN-γ
- LPS (e.g. from *E. coli* 055:B5, Difco)
- N^G-monomethylarginine (NMMA, Sigma)

- Griess reagent: within 12 h of use, mix 1% (w/v) naphthylethylenediamine dihydrochloride with an equal volume of 1% (w/v) sulphanilamide in 5% (v/v) H_3PO_4 (all reagents from Sigma). Component solutions can be stored for up to 2 months at 4°C

Method

1. Plate macrophages in 24-well plate as per *Protocol 1*. Sensitivity of the assay is enhanced by culturing the cells in a relatively small volume of medium, e.g. 300–500 μl/well.

2. Prime macrophages for NO burst by adding IFN-γ (50–100 U/ml final concentration). Incubate at least 2 h.

3. Add LPS (20 ng/ml) to trigger NO burst. Incubate 16 h. For each different cell type or stimulus it is necessary to include a negative control in which NO synthesis is inhibited by N^G-monomethylarginine. This is added at a final concentration of 250 μg/ml at the same time as the LPS.

Protocol 2. *Continued*

4. Remove supernatant from culture, mix with an equal volume of Griess reagent, incubate at room temp. for 10 min and read absorbance at 550 nm.

5. Based on an expected yield of 0–50 nmol $NO_2^-/10^6$ peritoneal macrophages prepare an appropriate series of NO_2^- standards diluted in culture medium, perform Griess reaction and construct a standard curve for calculation of NO_2^- concentration in test samples. Express results as NMMA-inhibitable accumulation of $NO_2^-/10^6$ cells (or per mg cell protein).

Comments: This assay will reflect the cytocidal activation state of the cells. It can be modified to determine the effect of other compounds or cytokines on activation by pretreating the cells with these compounds.

4. MHC class II expression

Resident tissue macrophages not exposed to immunogen and specific T-cell cytokines express low levels of MHC class II(2). Expression is upregulated by T-cell derived cytokines such as IFN-γ and IL-4 (8, 9). Flow cytometric analysis of fluorescently stained cells is the method of choice for comparison of class II expression in different cell populations (standardization of cell numbers is easily achieved). Immunocytochemical analysis for class II expression has the additional advantage of single cell analysis.

Protocol 3. Assay for MHC class II expression

Equipment and reagents

- Bacterial plastic Petri dishes
- Macrophages
- PBS
- EDTA
- Narrow-bore pipette (e.g. 5 ml)
- PBA: PBS, 0.1% (w/v) BSA, 10 mM NaN$_3$
- PBA containing 2% serum (from same species as secondary antibody)
- Eppendorf tubes
- Microcentrifuge
- Anti-MHC class II antibody or isotype-matched negative control antibody
- Fluorochrome-conjugated secondary antibody
- Flow-cytometry system

Method

1. Culture macrophages in bacterial plastic Petri dishes.

2. De-adhere cells and harvest: wash monolayer 3 times with PBS to remove tissue culture medium, add PBS containing 5 mM EDTA, incubate at room temperature for 5 min, ensure cells have rounded up and flush cells from surface using a narrow bore pipette (e.g. 5 ml).

3. Pellet cells, resuspend in PBA, and transfer to Eppendorf tube. Wash cells 3 times with PBA to remove EDTA. For each wash, pellet cells by centrifugation, 20 sec pulse at low speed (approximately 6500 r.p.m.) on microcentrifuge, aspirate supernatant, and resuspend.

4. Resuspend cells in anti-MHC class II antibody or isotype-matched negative control antibody diluted in PBA containing 2% serum (from same species as secondary antibody) to saturate Fc receptors. Approximately 1 μg antibody in a 100 μl PBA per 10^6 cells is usually appropriate. Incubate on ice for 30–60 min. Gently mix cells occasionally.

5. Wash cells 3 times in PBA to remove excess primary antibody.

6. Resuspend pellet in fluorochrome-conjugated secondary antibody. Incubate on ice for 30–60 min, protected from light to avoid quenching of fluorochrome. Shake occasionally.

7. Wash cells once to remove free secondary antibody, resuspend cells in PBA and analyse by flow cytometry. Relative antigen expression can be calculated:

$$\frac{\text{geometric mean (sample 1)-geo. mean (isotype control 1)}}{\text{geometric mean (sample 2)-geo. mean (isotype control 2)}} \times 100$$

Comments: It is essential that antibodies are present at saturating concentrations and reach equilibrium. In general, a concentration of 1 μg/10^6 cells is optimal. The second antibody must be a F(ab′)$_2$ since macrophages have abundant Fc receptors. Cells can be fixed before immunostaining by resuspending in a freshly made solution of 2% (v/v) paraformaldehyde/PBS (do not use glutaraldehyde as a fixation as this is autofluorescent). Incubate for 15 min on ice and wash away excess paraformaldehyde with PBS (3 times). Incubate cells in 10% FCS/PBS to block remaining fixation sites, and wash 3 times with PBS. The advantage of fixation is that cells can be analysed up to 24 h later.

5. Macrophage mannosyl receptor (MMR) assays

MMR is a cell surface receptor that binds mannosylated or fucosylated proteins (10). MMR activity is downregulated on exposure to IFN-γ. We have found the degradation assay described below to be a very reliable and reproducible assay to discriminate between resident or elicited macrophages (high MMR activity) and IFN-γ activated cells (low MMR activity) (11).

Mannosylated-BSA (23 mol mannose/mol BSA; EY Labs. San Mateo, CA, USA), a glycoconjugate of mannose-bovine serum albumin, serves as a good ligand for the MMR. It is trace-labelled with Na[^{125}I] by a modified chloramine T method.

Protocol 4. Labelling of mannosylated BSA

Equipment and reagents

- Mannosylated BSA
- Na[^{125}I]iodide (Amersham IMS 30, 100 mCi/ml)
- Chloramine T
- Sodium metabisulfite
- Potassium iodide
- 0.1 M sodium phosphate buffer, pH 7.6
- PD-10 column (Sephadex G-10, Pharmacia)
- DMEM + 5% FCS
- TCA (10% v/v)

Method

1. Incubate 50 μl of mannosylated BSA at 2 mg/ml with 1 mCi Na[^{125}I] and 30 μl chloramine T at 10 mg/ml on ice for 10 min.

2. Add 436 μg sodium metabisulphite (190 μl) and 1.9 mg potassium iodide (190 μl) to terminate the reaction.

3. Suspend all reagents in 0.1 M sodium phosphate buffer pH 7.6.

4. Pass the reaction products down a PD-10 column (Sephadex G-10; Pharmacia), pre-washed with 0.1 M phosphate buffer, pH 7.6.

5. Collect 300 μl aliquots, pool the first labelled peak and dilute in 5 ml of DMEM + 5% FCS.

6. Test the ligand for trichloroacetic acid (TCA) precipitability (10% v/v) and screen by uptake or degradation assay using macrophages which express MMR activity (resident peritoneal macrophages or elicited but not IFN-γ-activated macrophages).

Protocol 5. Binding and uptake of mannose-specific ligands

These assays are less time consuming than the degradation assays (described below) but not as sensitive, with potentially higher backgrounds. Binding and uptake are assayed at saturating concentrations of ligand using trace labelled mannosylated BSA (120 ng/ml ligand/5 × 10^5 macrophages) or β-2 glucuronidase (300 μg/ml/5 × 10^5 macrophages) in the presence or absence of mannan. Binding is assayed at 4°C, uptake at 37°C. The reaction mixture contains FCS which does not inhibit mannose-specific binding. MMR-ligand binding is cation dependent.

Equipment and reagents

- Macrophages/monocytes
- 24-well plates
- PBS
- DMEM
- FBS
- Hepes buffer, pH 7.0
- [^{125}I] mannosylated BSA (as prepared in *Protocol 4*—3 × 10^6 cpm/μg, 1.7 Ci/μg)
- Mannan (Sigma)
- 10 mM sodium azide
- 1 M NaOH
- Gamma counter

Method

1. Adhere macrophages/monocytes (5×10^5) for at least 1 h in 24-well plates.

2. Wash in PBS and incubate in 300 μl DMEM + 5% FBS, with Hepes buffer, pH 7.0 and 40 ng [^{125}I]mannosylated BSA (3×10^6 c.p.m./μg), with or without 1.25–2.5 mg/ml of mannan (Sigma).

3. Incubate cells in duplicate for 60 min at 4°C or for 20–30 min at 37°C.

4. Wash three times in ice-cold PBS with 10 mM sodium azide. Add 200 μl of 1 M NaOH to dissolve the cells, and measure cell-associated radio-activity in a gamma-counter. Express results as nanograms of manno-sylated BSA specifically bound or taken up per 5×10^5 macrophages plated or per mg cell protein.

5. Degradation of [^{125}I]mannosylated BSA by macrophages can be measured by the appearance of TCA-soluble labelled material in the medium. Degradation of [^{125}I]mannosylated BSA is detectable after ~40 min incubation at 37°C and continues at a linear rate for several days if macrophages are maintained in the continuous presence of ligand. We have found overnight culture convenient.

Protocol 6. [^{125}I]mannosylated-BSA degradation assay

Equipment and reagents

- Adherent macrophage monolayers
- Sterile labelled ligand
- 96- or 24-well plates
- Culture medium + FCS
- Mannan (Sigma)
- TCA

- 30% (v/v) H_2O_2
- 4 M KI
- Chloroform
- Vortex mixer
- Microcentrifuge
- Gamma counter

Method

1. Add trace amounts of sterile ligand (approximately 10^6 c.p.m. in 10 μl) to monolayers of adherent macrophage populations (5×10^5–1×10^6 macrophages/well) in 96- or 24-well plates, in culture media (0.2–1.0 ml) with FCS, in the absence or presence of mannan (1–2 mg/ml).

2. Include cell-free blanks, with and without mannan.

3. Incubate for 16 h to 2 days and then remove an aliquot of medium. Determine the TCA-soluble radioactivity.

4. If the supernatant following TCA precipitation is not clear or there is a high background, contaminating protein and/or free iodine can be removed by incubating a 0.25 ml aliquot of the supernatant with 10 μl

Protocol 6. *Continued*

of 30% v/v H_2O_2 followed by excess potassium iodide (5 μl of 4 M KI). Leave the mixture to react for 10 min at room temp.. Thereafter, add 0.7 ml chloroform (to extract protein), vortex the tube vigorously for 20 sec, centrifuge (microcentrifuge, high speed, 5 min), and measure radioactivity in an aliquot of the clear aqueous phase in a gamma-counter. Cell-dependent, mannan-inhibitable degradation per unit time is calculated as a function of macrophage number or protein.

5.1 Single cell assays

Uptake of $[^{125}I]$-mannose-BSA or $[^{125}I]$-β-glucuronidase by macrophages can be detected at the single-cell level by autoradiography. To increase the cell-associated radioactivity a higher specific activity ligand should be used (5 × 10^6 c.p.m./μg) at saturation. Incubate coverslips with adherent macrophages as above for 20–30 min at 37°C, wash well and fix. Radioactivity associated with the coverslips can be determined directly before processing.

5.1.1 Comments

Optimal culture conditions are essential. pH variations in crowded cultures may perturb MMR activity. In prolonged degradation assays, cell viability should be monitored by phase-contrast microscopy. β-glucuronidase is a more stable ligand than mannosylated BSA.

Kinetic parameters of specific binding and uptake of ligand can be determined. The degradation assay is not saturable, but is sensitive and linear for prolonged periods of incubation. Specificity is established by using unlabelled competitive inhibitors of MMR binding such as mannan or unlabelled ligand. Cell-free blanks are essential and provide a measure of background activity in binding, uptake and degradation assays.

Theoretically, biotinylated ligand may be substituted for iodine label in single cell analyses. A streptavidin immunocytochemical detection system is then used instead of autoradiography.

6. Macrophage cytokine assays

Following exposure to bacterial products such as LPS, macrophages are a major source of cytokines including IL-1, IL-6, and TNF-α. Secreted cytokine levels reflect the state of macrophage activation, e.g. resident peritoneal macrophages produce about 100-fold less TNF-α following LPS challenge than do macrophages primed by lymphokines (e.g. IFN-γ) and phagocytic stimuli such as uptake of the yeast cell wall product zymosan (12). Quantitation of protein is usually more amenable than measurement of mRNA. Secreted protein can be measured by bioassay or more directly by ELISA (see Chap-

Table 1. Activation profiles of resident, elicited, and activated macrophage populations

	Macrophage population:		
	Resident	**Elicited**	**IFN-γ activated**
Respiratory burst	low	low/high [a]	high
NO burst	low	low	high
MHC class II	low	low	high
MMR	high	high	low
TNF-α secretion	low	high	high

[a] Thioglycollate broth elicited peritoneal MØ can produce high levels of superoxide anion, but low H_2O_2 is detected, possibly as a result of scavengers in the cells.

ters 22 and 23). Many ELISA kits or pairs of monoclonal antibodies suitable for ELISA are now commercially available (e.g. Genzyme, Pharmingen).

7. Conclusions

The markers described here provide a basis for correlation of activation state with functional phenotype of macrophages recovered from experimental animals. From *Table 1* it is evident that activated macrophages, which express high levels of MHC class II and produce large amounts of reactive oxygen and nitrogen intermediates and cytokines, are the opposite extreme in the activation spectrum to resident macrophages with their low secretory capacity, low MHC class II and high MMR expression. Elicited populations are best distinguished by high MMR-expression and significant cytokine secretion. It is important to emphasize that the macrophage population exposed to cytokines produced by antigen-responsive lymphocytes at the site of an immune response will not necessarily become activated. In contrast to IFN-γ, the cytokines IL-10, IL-4, and IL-13 are known to down-grade or temper macrophage activation, such that if these are the dominant cytokines produced at the site, macrophages are likely to be maintained in a non-activated state (high MMR, reduced pro-inflammatory cytokine secretory ability) (13). Indeed, IL-4 has been shown to greatly enhance MMR activity in contrast to IFN-γ (14). Further studies concentrating on *in vivo* events, the cytokine network and control of gene expression will hopefully enhance our understanding and characterization of macrophage activation.

References

1. Gordon, S. (1986). *J. Cell Sci.*, Suppl. **4**, 267.
2. Gordon, S., Starkey, P., Hume, D., Ezekowitz, A., Hirsh, S., and Austyn, J. (1985). In *Handbook of experimental immunology*, 4th edn (ed. D. M. Weir, L. A. Herzenberg), pp. 43.1–43.14. Blackwell Scientific, Oxford.

3. Babior, B. M., Kipnes, R. S., and Curnutte, J. T. (1973). *J. Clin. Invest.*, **52**, 741.
4. Nathan, C. and Root, R. K. (1980). *J. Exp. Med.*, **146**, 1648.
5. Trush, M. A., Wilson, M. E., and Van Dyke, K. (1978). *Meth. Enzym.*, **57**, 462.
6. Murray, H. W. and Cohn, Z. A. (1980). *J. Exp. Med.*, **152**, 1596.
7. Ding, A. H., Nathan, C. F., and Stuehr, D. J. (1988). *J. Immunol.*, **141**, 2407.
8. Basham, T. Y. and Merigan T. C. (1983). *J. Immunol.*, **130**, 1492.
9. Crawford, R. M., Finbloom, D. M., Ohara, J., Paul, W. E., and Meltzer, M. S. (1987). *J. Immunol.*, **139**, 135.
10. Ezekowitz, R. A. B. and Gordon, S. (1984). In *Contemporary Topics in Immunobiology* (ed. D. O. Adams and M. G. Hanna Jr), **Vol. 18**, pp. 33–56. Plenum, New York.
11. Ezekowitz, R. A. B., Austyn, J. M., Stahl, P. D., and Gordon, S. (1981). *J. Exp. Med.*, **154**, 60.
12. Stein, M. L. and Gordon, S. (1990). *Eur. J. Immunol.*, **21**, 431.
13. Doyle, A. G., Herbein, G., Montaner, L. J., Minty, A. J., Caput, D., Ferrara, P., and Gordon, S. (1994). *Eur. J. Immunol.*, **24**, 1441.
14. Stein, M., Keshav, S., Harris, N., and Gordon, S. (1992). *J. Exp. Med.*, **176**, 287.

Measurement of proliferative, cytolytic, and cytostatic activity of cytokines

FRANCES BURKE, ENRIQUE ROZENGURT, and
FRANCES R. BALKWILL

1. Introduction

Most cytokines have proliferative, cytostatic, and/or cytolytic activity which is often dependent on the context in which they are acting. IFN-α, β, and γ, TNF, Lymphotoxin, IL-1, IL-4, have all been shown to have proliferative, cytostatic and cytolytic activity depending on the environment (1). Synergistic interactions between cytokines, can often produce powerful antiproliferative effects (2). There is some experimental evidence to suggest that growth inhibitory cytokines such as IFN-α may act *in vivo* as autocrine or paracrine growth inhibitors (3). Some of the methods used to assess the above parameters will be described in the following chapter.

2. Effects of cytokines on cell growth *in vitro*

The protocols described in this chapter all require basic tissue culture experience and sterile technique. It is recommended that cell lines are grown under pyrogen-free conditions when studying effects of cytokines. For this sterile plastic pipettes and tubes should be used and glassware Pasteur pipettes triple-baked. Fetal calf serum should be chosen for its minimal endotoxin content.

2.1 Direct cell counting

Direct cell counting is valuable for assessing cell viability and doubling time. However to assess whether a cytokine induces proliferation or is cytostatic or cytotoxic at specific stages during the cell cycle, more extensive analysis is required.

Two methods will be described for direct cell counting. The simplest method is to count a suspension of cells on a haemocytometer. This is very time-consuming especially when comparing the effects of different concentrations and/or combinations of cytokines. A more accurate method involves the use of a Coulter counter. The disadvantage of this method is that it does not distinguish viable cells from dead cells. This is not so important for cells growing in monolayers, since dead cells become detached and float off, thus being removed by the washing procedure prior to trypsinization.

2.1.1 Direct cell counting using an improved Neubauer haemocytometer

This procedure is described in *Protocol 1*.

Protocol 1. Direct cell counting using an improved Neubauer haemocytometer

Equipment and reagents
- Neubauer haemocytometer with coverslip
- Phase-contrast microsope
- Pasteur pipette

Method

1. Wash haemocytometer in detergent and water followed by an alcohol rinse.

2. Moisten edge and position coverslip until Newtons rings are obtained. Touch side of coverslip to check that Newton's rings do not move. Using a Pasteur pipette add one drop of cell suspension under the coverslip to fill the haemocytometer. Be careful not to flood the area, the channels should not be filled.

3. The central 25 squares of the grid form a 1 mm² area. Count all the cells in this area under phase contrast. Viable cells will appear bright and refractive whilst dead cells will appear dull. Alternatively, five squares can be counted and the result multiplied by five. Count at least 100 cells to ensure accuracy.

4. Calculate the cell count/ml using the following equation:
   Cell count/ml = number in central 1 mm² area × 10^4 × dilution factor.

2.1.2 Direct cell counting using a Coulter counter

This method makes use of an electronic means of cell counting. An electric current is passed through an 100 μm aperture in a probe which is subsequently placed into a single-cell suspension using isoton as a buffering solution. The cells are drawn through the probe by a mercury manometer controlled pump which alters resistance to the current flow and produces a series of pulses which are counted and displayed on a digital readout. The change in resistance produced is thus proportional to the volume of the cell passing through the probe. Using the manufacturer's instructions a threshold is set so that background counts from cell debris is excluded.

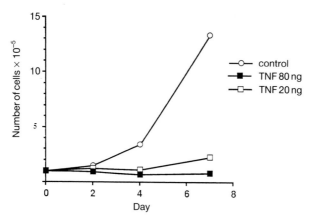

Figure 1. The growth inhibitory effect of tumour necrosis factor on the MCF-7 breast cancer cell line.

This procedure is routinely used in our laboratory where 6-well plates (3.5 × 1.0 cm) are used, each with 10^5 cells per plate. All parameters are assayed in triplicate and counts are made at various time points to obtain a cell growth curve as shown in *Figure 1*.

Protocol 2. Direct cell counting using a Coulter counter

Equipment and reagents

- Accuvettes (Coulter electronics Ltd.)
- Coulter counter (Coulter electronics Ltd.)
- Isoton
- Syringe fitted with 19-gauge needle
- (a) Trypsin 0.25% in Tris saline pH 7.7— filtered sterilized and stored at −20°C. For one litre:

Trypsin	2.5 g
NaCl	8 g
KCl (19% solution w/v in distilled water)	2 ml
Na$_2$HPO$_4$	0.1 g
Dextrose	1 g
Trizma base	3 g
Phenol red (1% solution w/v in distilled water)	1.5 ml

Penicillin	1 × 10^5 units
Streptomycin	0.1 g

Adjust pH to 7.7 with 1 M HCl

- Versene in PBS pH 7.2 (autoclave and store at 4°C). For one litre

Diaminoethane tetraacetic acid disodium salt (EDTA)	0.2 g
NaCl	8 g
KCl	0.2 g
Na$_2$HPO$_4$	1.15 g
KH$_2$PO$_4$	0.2 g
Phenol red (1% solution w/v in distilled water)	1.5 ml

- Use 1:4 trypsin : versene

Method

1. Trypsinize cells if adherent. Wash the wells with trypsin/versene. Use a 0.5 ml solution to disperse the cells and stop the action of trypsin/versene with 1.5 ml of isoton/2% fetal calf serum. If cells grow in suspension, spin down and resuspend in 2 ml of isoton/2% foetal calf serum.

Protocol 2. *Continued*

2. Obtain a single-cell suspension by dispersing the cells with a 19-gauge needle and wash the dish with up to 8 ml of isoton. Transfer the final volume which will be 10 ml into an accuvette.

3. Allow the Coulter counter to warm up and prepare for use following manufacturer's instructions.

4. Count the contents of each accuvette and multiply by 20.[a]

[a] This can accurately measure up to 10 000 cells. The accuracy will decrease if the total number exceeds this as a result of coincident passage of cells through the probe aperture. In these cases a conversion chart provided by the manufacturers must be referred to. The mean of the triplicates should be calculated and growth curves plotted.

3. Determination of effects on growth of cytokines using indirect methods

Two protocols will be outlined. One method uses a fluorogenic assay, whilst the other is a kit adaptation of the conventional MTT assay as also described in Chapter 21, *Protocol 6*.

3.1 The alkaline phosphatase assay

Protocol 3 is based on a fluorogenic enzyme assay which measures alkaline phosphatase activity, a membrane bound enzyme present in most human cell types. This enzyme is found in three main forms; in the placenta, in the intestine and liver, and in the bone or kidney. The conditions of the assay allow the non-fluorogenic substrate 4-methylumbelliferyl phosphate to react with all forms of the enzyme to produce free phosphate and the fluorogenic compound 4-methyl umbelliferylerone. The fluorescence is then measured on a multiscan plate reader where the total enzyme activity given is directly proportional to the the cell number. The activity of alkaline phosphatase increases proportionately with cell number and is not affected by time in culture.

This assay has been reported to accurately assess as few as 10^4 cells. The cells are plated in 96-well plates in the range of 10^4 to 10^5 per well. Cells are grown as monolayers and analysed at defined time points.

Protocol 3. Alkaline phosphatase assay

Equipment and reagents

- Dynatech Fluor Tm reader (Dynatech products)
- Assay solution (0.2 M Boric acid, 1 mM Mg Cl2, 1.2 mM 4-methyl umbelliferyl phosphate)
- Phosphate buffered saline (PBS)
- 37°C humidified 7% CO_2 incubator

Method

1. Wash the cells twice in PBS to remove any dead cells.

2. Add 200 μl of freshly prepared assay solution to each well.

3. Incubate at 37°C for 3 h.

4. Measure the relative fluorescence using a Dynatech Fluor Tm reader (Dynatech products).

3.2 The MTT-type assay

The conventional MTT assay has been described elsewhere (Chapter 21, *Protocol 6*). We have used a Promega kit which is a modified version of the conventional MTT assay. The data from the Promega cell titre ^{96}TM non-radioactive cell proliferation/cytotoxicity assay is analysed at 570 nm on a standard plate reader. An example of this assay is shown in Figure 2.

4. Measurement of cell lysis

We have used two methods for cell lysis in our laboratory. An easy and convenient method uses crystal violet. This is routinely used to measure TNF levels. More recently we have made use of a Promega kit which is based on the release of the lactate dehydrogenase enzyme from cells.

4.1 Measurement of cell lysis using crystal violet

The cytotoxic effect of cytokines can be simply measured using a microtitre plate and staining the viable cells using crystal violet. This procedure can be used in various biological assays which are available to look for the presence of cytokines in serum and in cell lines. The WEHI 164 cell line assay described here can be used for the measurement of TNF and LT. In this case the neutralizing antibody is added to the TNF or LT and incubated at room temperature for two hours prior to adding to the cells. The WEHI assay using cell lysis and crystal violet is outlined in *Protocol 4*.

Protocol 4. WEHI assay for TNF production

Equipment and reagents

- Dynatech Fluor Tm reader (Dynatech products)
- Methanol
- Aqueous crystal violet (1%)
- 96-well Costar plate
- RPMI 1640
- FCS
- 37°C humidified 7% CO_2 incubator
- TNF stock solution
- Paper towels
- Multiscan plate reader

Protocol 4. *Continued*

Method

1. To a 96-well Costar plate add 5×10^4 cells/100 μl RPMI 10% FCS/well. Leave 2–3 h in a 37°C incubator as described earlier.

2. Dilute TNF stock solution in RPMI only (no FCS), using ten-fold dilutions so that the final concentrations are from 100 ng/ml to 1 pg/ml or lower if required.[a]

3. Add 100 μl of the above to the plate in duplicate.

4. Add 100 μl RPMI to 4 control wells.

5. Add test samples at a 1:5 dilution to duplicate wells.

6. Transfer plate to 37°C incubator and leave for 18–24 h.

7. Flick out medium and fix the cells by filling the wells with methanol for 30 sec.

8. Remove the methanol, add 100 μl of 1% aqueous crystal violet to each well and leave for 5–20 min.

9. Flick out the stain and wash with distilled water. Invert plate over a paper towel for a few minutes to allow to dry.

10. Measure the optical density 620 nm on multiscan plate reader, following the manufacturer's instructions.

[a] Remember the final concentration will be half that put in.

4.2 Measurement of cell lysis using the lactate dehydrogenase assay

We have also investigated another technique for measuring the cytotoxic action of cytokines and cytokine combinations. This is based on the release of lactate dehydrogenase (LDH) from dying cells. LDH is a stable cytosolic enzyme and is released from lysed cells in a similar way to ^{51}Cr release from pre-labelled target cells (4). An LDH kit (CytoTox 96™) is available from Promega and has been useful for assaying cytotoxic lymphocyte killing non-isotopically (see Chapter 12). Our experience using this assay on the OVCAR-3 ovarian cancer cell-line is detailed below:

(a) The spontaneous release is approx. 10% and does not change through the culture period. The majority of 'spontaneous' LDH-release occurs during the first 24 h after plating-out.

(b) At 4–7 days culture we can detect significant and specific release (up to 97% of total release) with high-dose cytokine combinations.

(c) Cytostasis can also be measured because the total release value obtained on lysis of all cells with detergent falls, as compared to control wells.

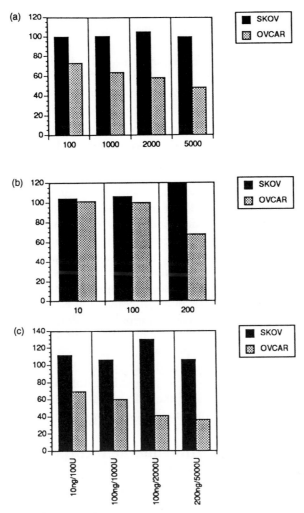

Figure 2. Four day MTT-type assay of ovarian cancer cell lines SKOV-3 and OVCAR-3 treated with (a) 100–5000 U/ml IFN-γ, (b) 10–200 ng/ml TNF, and (c) A combination of TNF and IFN-γ.

(d) Linear assay results were only obtained between 2×10^4 and 5×10^4 cells/well.

However, time-lapse cinematography studies on the same cell line revealed that cytokine induced cell death was apoptotic in nature and that the contents of many dying cells, particularly in the first few days of culture, were rapidly ingested by the surrounding viable cells. Hence, in spite of quite extensive cell death, little LDH was released into the tissue culture medium during the first few days of culture. We conclude that LDH-release is a simple and non-invasive method of investigating cell lysis, but results should be interpreted with caution when apoptotic cell death is occurring. Release will only be detected when a significant proportion of the cell population is dead or dying.

5. Cell cycle studies

As described in Section 2, cell counting provides simple growth curves which can be used to assess cell viability and doubling time. However, the effect of a cytokine on specific stages during the cell cycle must be studied by more complex techniques.

De novo synthesis of DNA can be quantified by measuring the incorporation of tritiated thymidine (present in the culture medium) during S phase. Incorporated label is subsequently detected by autoradiography or liquid scintillation counting. Incubating cells with the cytokine for a range of times followed by pulsing with tritiated thymidine provides information on the effect of the cytokine on early phases of the cell cycle to the S phase. It is not, however, an indication of quiescence. Another way to examine increases or decreases in DNA synthesis in a cell population is to use cytofluorimetry staining the nuclei with dyes such as propidium iodide. These protocols are outlined below. Quiescent cells can be obtained from fibroblasts and primary cultures. Fibroblasts, for example Swiss 3T3 cells, are plated and incubated for a further 4–7 days before use in the assay described in *Protocol 6*. At this time the cells are arrested in G_1 as judged by cytofluorometric analysis and by the fact that less than 1% of the cells are autoradiographically labelled after a 40 h exposure to [^{3}H]thymidine (see *Protocol 5*).

5.1 Measurement of [^{3}H]thymidine incorporation into acid-insoluble material

This section refers particularly to the effects of murine cytokines or cross-species reactive cytokines on murine fibroblast cell lines. Other medium and conditions will be required for different cell lines. The incorporation of [^{3}H]thymidine into acid-insoluble material has been widely used as a convenient method of assessing the stimulatory and inhibitory effects on the growth of many different cells. However, this technique not only assesses effects on DNA synthesis, but can also reflect alterations in the transport of thymidine across the plasma membrane, and in the phosphorylation of this nucleoside by thymidine kinase which occurs before it can be incorporated into DNA. In some systems, agents previously thought to have antiproliferative activity have subsequently been shown merely to inhibit thymidine transport. An easy way to circumvent any effects on thymidine transport is to vary the concentration of thymidine used in the experiment (e.g. from 1 mM to 10 mM). If the apparent inhibition of DNA synthesis is entirely due to a decrease in thymidine transport, then the effect should be overcome by increasing the concentration of this nucleoside, because at high concentrations, the nucleoside permeates the cell membrane by simple diffusion rather by carrier-mediated transport. Thus, this technique provides a convenient measurement of changes in DNA synthesis, provided that appropriate controls are included. The assay described in *Protocol 5* is carried out on Swiss 3T3 cell in the absence of serum in DME medium supplemented with Waymouth's medium which contains additional amino acids and vitamins present in serum but lacking in DME (5). The protocol should be adapted for other cells as appropriate.

Protocol 5. Measurement of [^{3}H]thymidine incorporation into acid-insoluble material

Equipment and reagents

- Swess 3T3 cells
- Scintillation counter
- DME medium
- Waymouth's medium (MB 752/1 powdered formula plus 1.6 mM FeSO$_4$)
- [^{3}H]thymidine (1 μCi/ml, 1μM)
- Phosphate-buffered saline (0.15 M NaCl in 0.1 M potassium phosphate buffer, pH 7.4)

- 5% trichloroacetic acid (TCA)
- 0.1 M NaOH
- 2% Na$_2$CO$_3$
- 1% sodium dodecyl sulphate (SDS)
- Picoflor

Method

1. Wash confluent and quiescent cultures of Swiss 3T3 twice with DME medium at 37°C to remove residual serum.

2. Incubate cultures in 2 ml DME/Waymouth's medium (1:1) containing [^{3}H]thymidine (1 mCi/ml; 1 mM).

3. After 40 h at 37°C wash cultures twice with ice-cold phosphate-buffered saline (0.15 M NaCl in 0.1 M potassium phosphate buffer, pH 7.4) and remove acid-soluble radioactivity by a 2 min treatment with 5% trichloroacetic acid (TCA) at 4°C.

4. Wash cultures twice with ethanol and solubilize cells by a 30 min incubation in 1 ml of 0.1 M NaOH containing 2% Na$_2$CO$_3$ with 1% sodium dodecyl sulphate (SDS).

5. Determine the radioactivity incorporated into acid-insoluble material by liquid scintillation counting in Picoflor.

6. After 40 h, [^{3}H]thymidine incorporation reaches saturation and thus ensures that the maximum response of the cell population is being determined rather than variations in the rate of DNA synthesis.

Protocol 5. *Continued*

7. In contrast, to determine the rate of DNA synthesis the cultures should be pulse-labelled with [^{3}H]thymidine (e.g. 5–15 min) but it should be borne in mind that this measurement is meaningful only if the specific radioactivity of the precursor pool is taken into consideration. (This is outside the scope of this section.)

5.2 Autoradiography of labelled nuclei

To substantiate further that neither the transport nor the phosphorylation of thymidine is affected by a cytokine, the proportion of cells actually synthesizing DNA can be measured by autoradiographic techniques. This method is much less sensitive to changes in the specific radioactivity of the precursor pool. Furthermore, it is possible to stimulate DNA synthesis in quiescent mouse fibroblasts by the addition of growth factors without completion of the cell cycle, i.e. arrest occurs after completion of S phase when highly radioactive [^{3}H]thymidine is present (1 mM, 1 mCi/ml^{-1}). Mitotic cells do not accumulate under these conditions because of radiation damage due to incorporation of [^{3}H]thymidine into DNA. The cells progress to G_2, remain arrested there due to radiation effects, and very few cells detach. Thus, with highly radioactive [^{3}H]thymidine generally used in our protocols (1 mM, 1 mCi/ml^{-1}), most cells never reach mitosis and they remain firmly attached to the dish after one round of DNA replication. To determine the length of the G_1 period and the rate at which the cell population starts DNA synthesis (rate of entry into S phase) the percentage of labelled nuclei should be determined after various times of growth factor stimulation in the continuous presence of the labelled precursor. *Protocol 6* describes the autoradiography of labelled nuclei quiescent cultures of 3T3 cells. The protocol should be adapted for other cell lines.

Protocol 6. Autoradiography of labelled nuclei

Equipment and reagents

- Swiss 3T3 cells
- Scintillation counter
- Waymouth's medium (MB 752/1 powdered formula plus 1.6 mM FeSO$_4$)
- [^{3}H]thymidine (5 μCi/ml, 1 μM)
- Isotonic saline
- 5% TCA
- Ethanol

- Chrome alum solution: 5 g gelatin heated in 40 ml glass distilled water to dissolve; 0.5 g chrome alum is dissolved separately in 400 ml of water. Mix solutions when cool, and make up to 1 litre
- Kodak AR10 stripping film
- Kodak D19 developer
- Hypam fixer (Ilford, Basildon, Essex, UK)
- Giemsa stain

Method

1. Wash confluent and quiescent cultures of 3T3 cells and incubate as described in *Protocol 6* except that [^{3}H]thymidine is added to a concentration of 1 mM and 5 mCi/ml.

Protocol 6. *Continued*

2. After a 40 h incubation, wash cultures twice with isotonic saline, extract with 5% TCA twice for 5 min, wash three times with ethanol, and dry.

3. Coat dishes with chrome alum and leave to dry.

4. Lay Kodak AR10 stripping film on dishes and store them in the dark for 1–3 weeks.

5. Develop film with Kodak D19 developer (4 min) and fix for 5 min with Hypam fixer (Ilford) diluted 1:4.

6. Stain cells with Giemsa stain.

Following this protocol, the nuclei engaged in DNA synthesis become intensively labelled. A typical result is shown in *Figure 3*. Labelled and unlabelled cells of several microscopic fields are counted and the results expressed as follows:

$$\% \text{ labelled nuclei} \times \frac{\text{No. of labelled cells} \times 100}{\text{Total No. of cells}}$$

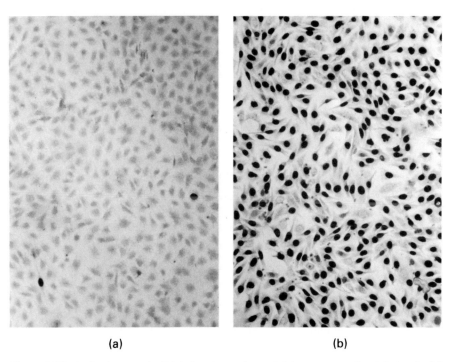

(a)   (b)

Figure 3 Typical microscopic field showing quiescent (a) or serum-stimulated cells (b), labelled with [³H]-thymidine and processed according to *Protocol 6*.

5.3 Cytofluorimetry

To confirm that the [³H]thymidine incorporation method is an accurate reflection of the state of DNA synthesis an independent technique, such as cytofluorimetry can be used, which interrupts the progression of the cells through the cell cycle in mitosis. In this manner, the proportion of cells that move from G_1 to M after treatment can be precisely assessed. Although colchicine does stimulate the incorporation of [³H]thymidine into DNA in 3T3 cells, it has been shown that its mitogenic activity is lost if it is added to the cells once S phase has been initiated, i.e. 20 h after stimulation with other factors (6). By adding colchicine, which blocks the cells in mitosis but does not potentiate the stimulate of DNA synthesis if added 20 h after the factors, the effect of growth factors and cytokines on the movement of the cells through the cycle from G_1 through S phase and into G_2 can be readily assessed.

Protocol 7. Cytofluorimetry

Equipment and reagents

- Fluorescence activated cell sorter (Becton Dickinson FACS-1)
- Colchicine, 1 mM
- Lysis buffer (0.5% Triton X-100, 4 mM MgCl₂, 0.6 M sucrose, 10 mM Tris–HCl, pH 7.5)
- Nuclei buffer (0.25 M sucrose, 5 mM MgCl₂, and 20 mM Tris–HCl, pH 7.4)
- Propidium iodide (0.05 mg/ml in 0.1% trisodium citrate)

- Quiescent Swiss 3T3 cells
- Trypsin (see *Protocol 2*)
- Syringe fitted with 27-gauge needle
- Ribonuclease A
- Centrifuge
- Cold phosphate-buffered saline (see *Protocol 5*)

Method

1. Treat cultures of quiescent cells with various agents as indicated and incubate for 40 h.

2. In order to ensure that cells which were stimulated to synthesize DNA are arrested in G_2, add the microtubule disrupting agent, colchicine at 1 mM, after 20 h of incubation. Colchicine has no mitogenic effect when added so late after the commencement of the experiment.

3. After the 40 h incubation, trypsinize cells from the monolayer, wash, pass through a 27-gauge needle to ensure a single-cell suspension, and incubate in lysis buffer for 3 min at room temp.

4. Incubate nuclei at 37°C for 30 min in nuclei buffer containing 0.5 mg/ml ribonuclease A.

5. Stain with propidium iodide for 20 min.

6. Spin down stained nuclei at 2000 r.p.m. at 4°C, resuspend in cold phosphate-buffered saline, and measure DNA content immediately on a fluorescence activated cell sorter at 488 nm excitation.

5.4 Measurement of bromodeoxyuridine incorporation to assess cell cycle changes *in vivo*

A useful method to assess cell proliferation *in vivo* in experimental animals involves labelling cells with bromodeoxyuridine (BrDu), a thymidine analogue which is incorporated into DNA during the S phase of the cell cycle. Subsequently, immunohistochemistry can be performed on excised tissue to detect the BrDu. Both growth inhibitory and stimulatory cytokines can be investigated. We have used this method to look at the effects of IFN-γ on the growth of solid tumours in nude mice (Burke *et al.*, in preparation). This is outlined in *Protocol 8*. The DNA needs to be denatured so the BrDu can be exposed to the specific antibody. This denaturation which is usually with acid can result in poor morphology of the tissues. The incubation period for the denaturation procedure is critical. There are many routine methods for immunohistochemistry. The method outlined below is used routinely in our laboratory and was originally obtained from Mount Vernon Hospital, Northwood, Middlesex.

Protocol 8. *In vivo* labelling of tumours/tissues with bromodeoxyuridine and subsequent immunohistochemistry

Equipment and reagents

- Bromodeoxyuridine made up in saline immediately before use.
- Anti-BrDu (Sigma)
- Reagents for immunohistochemistry are widely available. Our reagents were obtained from DAKO
- Mice with solid tumours
- Formol saline
- Poly-L-lysine coated slides
- CNP 30
- Alcohols: 100%, 95%, 50%
- Trypsin (see *Protocol 2*)
- 1 M HCl

- TBS, pH 7.6 (0.05 M Tris–HCl, pH 7.6, 0.15 M NaCl)
- 1% human AB serum diluted in TBS
- Dako biotinylated rabbit-anti-mouse
- ABC complex (Dako)
- DAB solution (CAUTION: DAB is a suspected carcinogen and may cause severe eye and skin irritation). DAB is available in tablet form. Weighing out the powder is not recommended. Wear gloves and make up in a fume hood. DAB is light sensitive
- Mayer's haematoxylin
- 1% HCl/alcohol
- DPX mountant

Method

1. Inject 50 mg/kg of BrDu in 0.1 ml into mice. Leave 45–60 min.
2. Sacrifice mice and remove tumour in formol saline.
3. Cut sections on to poly-L-lysine coated slides. Then treat slides as follows:
4. CNP 30 for 10 min.
5. 95% alcohol for 2 min.
6. 50% alcohol for 2 min.

Protocol 8. *Continued*

7. Tap-water for 2 min.

8. Block endogenous peroxidase with 0.1% hydrogen peroxide in methanol at room temp.

9. Wash in tap-water.

10. Wash in distilled water at 37 °C.

11. Transfer to freshly prepared trypsin which has been preheated to 37 °C (pH 7.8).

12. Rinse with cold water for 5 min.

13. Denature the DNA by immersing slides in 10% solution of 1 M HCl for 10 min.

14. Wash in tap-water for 5 min.

15. Flood slides with TBS pH 7.6 for 5 min. Tip off and wipe around section.

16. Incubate with anti-BrDu diluted 1/20 in 1% human AB serum diluted in TBS. Leave for 1.5–2 h.

17. Wash × 3 in TBS.

18. Add Dako biotinylated rabbit anti-mouse diluted 1/300 in 1% human AB serum diluted in TBS.

19. Wash × 3 in TBS.

20. Add ABC complex (Dako).

21. Wash × 3 in TBS.

22. Add DAB solution made up as per manufacturer's instructions. Leave on for 10 min.

23. Rinse in TBS.

24. Wash well in running tap-water for 10 min.

25. Counterstain in Mayer's haematoxylin. Time is dependent on age and frequency of use of reagent in range of 10 sec–2 min.

26. Wash in running tap-water for 3 min.

27. Fix in 1% HCl/alcohol in and out.

28. Wash in running tap-water for 5 min.

29. Dehydrate through 50% (2 min), 95% (2 min) and absolute alcohol (in and out).

30. CNP 30 for 2 min.

31. DPX mount.

5.5 Mitotic index

This technique looks at the rate of entry into mitosis and thus completes the picture of the cell cycle and the effects cytokines may have. The main disadvantage of this procedure is that mitoses occur rapidly usually only lasting 45 minutes and therefore frequent time-points have to be studied. The procedure involves gentle lysis of cells. Any resulting 'mitotic spreads' are fixed and stained and examined under a low-power microscope.

Protocol 9. Determination of the mitotic index

Equipment and reagents

- Tri-sodium citrate 0.684% in distilled water
- Carnoys fixative: one part glacial acetic acid plus three parts absolute alcohol
- 30 mm tissue culture dishes
- Orcein acetate (filtered)
- Absolute alcohol
- Petri dishes

Method

1. Aspirate off medium.
2. Add 1.9 ml of tri-sodium citrate to the Petri dishes followed by 0.4 ml of distilled water drop-wise.
3. Mix the contents by very gently swirling the dish.
4. Leave at room temp. for 10 min.
5. Add 2.3 ml of Carnoys fixative, pour off and repeat.
6. Leave at room temp. for 10 min.
7. Pour off the fixative and air-dry.
8. When dry, stain with filtered orcein acetate and leave for 10 min.
9. Wash off the stain with absolute alcohol and air-dry.
10. Observe under low-power magnification. Mitotic nuclei should be visible.
11. Count 1000 cell nuclei (200 per field of view) per dish and calculate the percentage of mitoses.

---

6. DNA fragmentation

Duke *et al.* (1983) have shown that during target cell lysis by sensitized T cells, fragmentation of target cell DNA may occur within 10 min of exposure (7). This fragmentation occurs in a specific manner resulting in the formation of 200 base-pair or multiples of 200 base-pairs indicating that the DNA is being cleaved at susceptible points between condensed regions of chromatin.

The DNA fragmentation observed in the Duke system has been shown to act via the activation of an endonuclease and has since been observed in lytic events mediated by lymphotoxin and TNF (5). Analysis of DNA fragmentation may provide useful information on mechanisms of cytokine mediated killing.

The assay described in *Protocol 10* is a modification of that described by Duke *et al.* (7).

Protocol 10. DNA fragmentation

Equipment and reactions

- Sensitized T cells
- Medium
- 24-well plates
- [³H]thymidine (5 µCi/ml)
- Cytokine of interest

- 25 mM sodium acetate buffer, pH 6.6
- Syringe fitted with 19-gauge needle
- Ultra centrifuge
- Beta scintillation counter

Method

1. Seed the target cells at a density of 1×10^5 cells in 2 ml of medium in 24-well plates.
2. Leave to attach for 6 h.
3. Incubate for 18 h with 5uCi/ml [³H]thymidine (925 GBq/mmol).
4. Wash the cells 3 times with culture medium.
5. Incubate for 30 min on ice in culture medium.
6. Remove unincorporated [³H]thymidine by further washing.
7. Culture at a range of time-points with the cytokine of interest.
8. At each time-point remove and retain culture medium.
9. Treat the cultures for 1 h with 25 mM sodium acetate buffer, pH 6.6.
10. Using a 19-gauge needle, syringe the contents of the dishes to detach the cells and centrifuge for 15 min at 27 000 *g*. This will separate the intact chromatin (which will form a pellet) from the fragmented DNA.
11. Count the radioactivities of the target cell DNA present in the culture medium, in the 27 000 × *g* supernatant, and in the 27 000 × *g* pellet using a beta scintillation counter.
12. Calculate the specific fragmentation using the formula below:

    % DNA fragmentation =

$$\frac{\text{c.p.m. DNA fragments}^a - \text{c.p.m. spontaneous DNA fragments}^b}{\text{total c.p.m.}^c - \text{c.p.m. spontaneous DNA fragments}^b}$$

[a] c.p.m. culture medium + c.p.m. of 27 000 *g* supernatant of cytokine treated cells retained at point 8 in protocol.
[b] c.p.m. culture medium + c.p.m. 27 000 *g* supernatant of parallel control cells.
[c] c.p.m. DNA fragments 1 + c.p.m. 27 000 *g* pellet of the cytokine treated cells.

Specific labelling of DNA fragmentation can also be investigated *in vivo* using
immunohistochemisty. *Protocol 11* is based on a paper published by Gavrieli *et al.* (8). The effects of cytokines such as IFN-γ can be investigated in this way. Sections were prepared for staining as described in the *Protocol 8* up to point 9.

Protocol 11. Specific labelling of DNA fragmentation *in vivo*

Equipment and reagents

- Slide box
- Proteinase k (20 µg/ml)
- Terminal transferase (Boehringer Mann-heim)
- Biotin-16-dUTP (Boehringer Mannheim)
- Bovine serum albumin (BSA)

- TDT buffer (39 mM trizma base, pH 7.2, 140 mM sodium cacodylate, 1 mM cobalt chloride)
- TB buffer (300 mM sodium chloride, 300 mM sodium citrate)
- General immunohistochemistry reagents (see *Protocol 8*)

Method

1. Follow *Protocol 8* to step 9.
2. Incubate tissue sections with 20 µg/ml proteinase k for 15 min.
3. Wash × 4 in distilled water.
4. Inactivate endogenous peroxidase activity by incubation with 0.1% hydrogen peroxide for 15 min.
5. Rinse sections in distilled water and immerse in TDT buffer.
6. Add TDT and biotinylated dUTP (in TDT buffer) to sections and incubate for 60 min.
7. Terminate reaction by immersing slides in TB buffer for 15 min at room temp. Rinse sections with distilled water.
8. Add 2% solution of BSA for 10 min at room temp.
9. Rinse in distilled water.
10. Immerse in PBS for 5 min, room temp.
11. Follow steps 18–31 from *Protocol 9* substituting PBS for TBS.

References

1. Burke, F., Naylor, M. S., Davies, B., and Balkwill, F. R. (1993). *Imm. Tod.*, **14** 165.
2. Dealtry, G. B., Naylor, M. S., and Balkwill, F. R. (1987). *Eur. J. Immunol.*, **17** 689.
3. Resnitzky, D., Yarden, A., Zipori, D., and Kimchi, A. (1986). *Cell*, **46**, 31.
4. Korzeniewski, C. and Callewaert, D. M. (1983). *J. Imm. Meths.*, **64**, 313.

5. Mierzejewski, K. and Rozengurt, E. (1977). *Exp. Cell Res.*, **106**, 394.
6. Wang, Z. W. and Rozengurt, E. (1983). *J. Cell Biol.*, **96**, 1743.
7. Duke, R. C., Chervenak, R., and Cohen, J. J. (1983). *Proc. Natl Acad. Sci. USA*, **80**, 6361.
8. Gavrieli, Y., Sherman, Y., and Ben-Sasson, S. A. (1992). *J. Cell Biol.*, **119**, 493.

Production of cytokine transgenic and knockout mice

MANOLIS PASPARAKIS and GEORGE KOLLIAS

1. Introduction

Transgenic mice have served for many years as fine tools in the dissection and understanding of complex biological phenomena. Studies on gene function in multifactorial and multicellular systems have most benefited, because they often require experimental settings which faithfully simulate the *in vivo* situation. Additional technological advances which become accessible to a growing number of investigators, offer the opportunity to inactivate or modify endogenous genes in the mouse and open up new possibilities for research. In general, perturbation of gene expression in transgenic systems often leads to measurable alterations in the physiology of the animal and important functional implications come to light.

Within the limitations of space in this chapter, we have sought to provide a technical guide to the production of transgenic and knockout mice for use by those who have limited experience in this field, taking care to include technical tips and details which have not been published elsewhere. Detailed laboratory manuals on transgenic technologies may also be found in refs 1, 2.

2. Production of transgenic mice by pronuclear injection of DNA

2.1 Equipment

Specialized equipment is needed for pronuclear injections. Inverted microscopes are more convenient to use for microinjection and they should be equipped with Nomarski differential interference contrast optics (DIC), which allow better visualization of the pronuclear membranes. A pipette puller able to generate good injection needles is an absolute requirement for successful injections. Two sets of micromanipulators are necessary for the microinjection and two stereo-microscopes and a cold light source will be needed for the surgical transfer of the injected eggs into the oviducts of the pseudopregnant females. In our laboratory we use a Diaphot TMD microscope (Nikon Ltd) with Nomarski optics and a set of Leitz micromanipulators (Leitz Instruments Ltd). A Kopf needle puller, model 750 (David Kopf Instruments), is used for needle-pulling and two Nikon stereo-microscopes and a Nikon fiber optic light source are used for the egg transfers. Other commercial suppliers of suitable equipment are Leitz, Carl Zeiss and Olympus for the inverted and the stereo-microscopes, Narishige Co. for the micromanipulators and Campden Instruments for the needle puller.

2.2 Animals

2.2.1 Mouse strains

Fertilized oocytes obtained from several strains of mice have been used for pronuclear injections. Zygotes produced by hybrid F_1 females are found to give better results than most inbred strains, possibly through a maternal or egg cytoplasmic effect that gives increased recovery of the eggs after microinjection. When inbred genetic backgrounds (e.g. specific MHC haplotypes, disease resistance/susceptibility) are essential, inbred instead of F_1 zygotes may be used albeit with lower overall efficiency. At least one inbred strain, the FVB/N, shows efficiencies comparable to F_1 hybrids (3). Good F_1 females are obtained when C57Bl/6 females are crossed with CBA, C3H or SJL/J males. We use (CBA × C57Bl/6) F_1 females both for the derivation of the eggs and as pseudopregnant foster mothers.

2.2.2 Scale

The levels of transgene expression may vary significantly between different mouse lines generated with the same DNA, due to the variable number of integrated copies and the unpredictable influences of neighbouring chromatin at the transgene integration site. When 'good' expression of a transgene is required, production of 4–6 transgenic mouse lines is usually sufficient, although in particular cases more mouse lines may be needed (e.g. when the generation of a phenotype is expected to be associated with the level of transgene expression).

To obtain four to six transgenic founders, at least 150 injected eggs should be transferred into pseudopregnant mothers. For this, 10–15 superovulated females will be needed. Depending on the supply of mice and the skills of the operator, this is usually accomplished in two to three days.

2.3 Preparation of DNA for microinjection

Any cloned fragment of DNA can be used for injection. Linear DNA fragments with sticky ends show much higher integration efficiencies than blunt-ended linear or supercoiled DNA (4). Prokaryotic vector sequences have been found to interfere with the expression of the transgene (5) and they should be removed before microinjection. For that reason, unique restriction sites on both sides of the insert should be considered when designing a construct. Since the presence of introns is shown to facilitate transcription of the transgene, intron containing genes are preferred over cDNA constructs (6).

2.3.1 Purification of DNA fragments

DNA purity is a very important factor for the successful generation of transgenic mice, since impurities such as traces of agarose or organic solvents can severely decrease the viability of the injected egg. Several methods are available for the isolation of DNA fragments (7). Depending on the size of the construct different approaches may be followed. For large DNA fragments (> 100 kb), such as those cloned in Yeast Artificial Chromosomes, specific successful protocols have only recently been described (8). For medium fragment sizes, e.g. cosmid inserts or ligated cosmid inserts (9), sodium chloride gradients or preparative pulse field gel electrophoresis may be used. For DNA fragments less than 25 kb in length, we routinely use preparative agarose gel electrophoresis followed by elution of the DNA fragment, either by glass powder adsorption after sodium iodide dissolution (Geneclean, BIO 101) or by digestion with β-Agarase I (New England Biolabs) according to the instructions of the manufacturers.

2.3.2 The concentration of DNA for microinjection

The concentration of the DNA used for microinjection is usually adjusted to 1–2 ng/μl. This concentration reflects the original descriptions referring to successful rates of integration when 500–1000 copies of a 3.5 kb linear DNA fragment were injected in each zygote (4). However, when low integration frequencies are encountered we have found that for DNA constructs ranging between 4–12 kb in size, the concentration can safely be increased to 4–6 ng/μl. For new DNA constructs prepared for microinjection, concentration is adjusted by comparison to an older fragment of similar size which has been used successfully in previous experiments. The new DNA is diluted until equal volume of solution gives equal band intensity with the control DNA on an ethidium bromide stained agarose gel.

2.4 Preparation of the pipettes used for pronuclear injections

2.4.1 The injection pipette

Needles for microinjection are made from glass capillaries (borosilicate glass capillaries, thin wall, with inner filament) using a pipette puller. The shape of the needle is the most important factor for egg survival after injection. With an optimal needle, up to 90% of the eggs should survive injection. Using the Kopf needle puller and glass capillaries purchased from Intracel Ltd, we get good needles using the following settings:

Heat 1	15.5
Heat 2	0
Sol	4
Delay	0
Sol	0.05

The two Sol are parameters related to the pulling force and pulling time (described in Kopf puller manual). The pulling time which should be around 5.5 sec can be controlled by adjusting the proximity of the heating elements.

2.4.2 The holding pipette

The holding pipette is a blunt, heat-polished pipette through which suction is applied to position and hold the egg for microinjection. Its external diameter should be between 80–120 μm with an opening of approximately 20–30 μm. It is easier to prepare optimal holding pipettes by using a micro-forge instrument (e.g. Micro Instruments) to control the size and the shape of the opening. However, good pipettes can also be drawn and flame-polished by hand using a stereoscope and the flame of a microburner.

2.5 Recovery of oocytes

Fertilized oocytes used for microinjection are usually obtained from matings between F_1 males and females. To control the timing of ovulation it is best that animals are kept in a constant light–dark cycle. Since ovulation occurs 3–5 h after the onset of the dark period and egg pro-nuclei are suitable for microinjection at 15–18 h after ovulation, a convenient light–dark cycle can be worked out according to the needs of the operator. For example, in a 7 pm–7 am dark, 7 am–7 pm light cycle, microinjection is best performed at 15:00–18:00 h.

Since natural matings produce low numbers of zygotes it is preferable that females are induced to superovulate. For superovulation, 3 to 5 weeks old females are injected i.p. on day one with 5 units of pregnant mare's serum (PMS) which mimics the effects of follicle stimulating hormone (FSH). This is done 5–6 h before the start of the dark period of the light cycle. On day three, 46–48 h after the PMS administration, mice are injected i.p. with 5 units of human chorionic gonadotropin (hCG) and placed with individually caged F_1 stud males (i.e. one female and one male per cage). To maximize the fertilizing efficiency of sperm, it is best to alternate the group of males so that each male gets one female every 2–3 days. The following morning, females are checked for vaginal plugs. Those that have mated are sacrificed and their oviducts are dissected out and placed in a 35 mm tissue culture dish containing standard M2 medium equilibrated at room temperature. Zygotes are collected in M2 medium containing 300 μg/ml hyaluronidase (Sigma) to remove the sticky cumulus cells. This step should not exceed 10–15 min as hyaluronidase may affect egg viability. Following this, eggs are washed several times in M2 medium to remove hyaluronidase, transferred in CO_2 buffered M16 medium and stored at 37°C, 5% CO_2 in micro-drop cultures overlaid with paraffin oil. M2 and M16 media can be prepared in the lab using analytical grade chemicals and pyrogen-free water according to published protocols (1, 2). We find it more convenient to buy both M2 and M16 media ready made, in powdered form (Sigma) and prepare them according to the manufacturer's instructions.

2.6 Microinjection procedure

Pronuclear injections are carried out in a paraffin-oil covered drop of M2 medium, set on a siliconized glass depression slide. The glass slide should be washed in a mild detergent and thoroughly rinsed in pyrogen-free water. The holding pipette which is filled with Fluorinert FC77 (Sigma) is controlled by a micrometer syringe which is connected to the pipette through a paraffin-oil-filled Tygon tubing. The injection needle is connected through air-filled tubing to an air-filled 50 ml glass syringe.

Protocol 1. Microinjection procedure

Equipment and reagents

- Holding and injection pipettes (see Sections 2.4.1, 2.4.2, and 2.6)
- Paraffin-covered drop of M2 medium (see above)
- M2 medium
- M16 medium
- Cytochalazin D

- Oocytes (see Section 2.5)
- Inverted microscope with Nomarski optics (see Section 2.1)
- One pair of micromanipulators (see Section 2.1)
- 50 ml air-filled syringe
- CO_2 incubator

Method

1. First lower the holding and injection pipettes in the drop of M2 medium.

2. Transfer a small number of eggs, usually 15–30 depending on the experience of the injector, into the drop.

3. To check that the injection needle is open move it close to one of the eggs and apply pressure observing the movement of the egg. If the needle is closed, try to brake its very tip by letting it carefully touch the holding pipette.

4. At low magnification use the holding pipette to pick an egg and observe at higher magnification for visible pro-nuclei. Select the most easily accessible pro-nucleus and use the holding and injection pipettes to bring the egg at the best position for injection. Depending on the structure of the opening of the holding pipette, it is usually safer for the egg to be held from the area of the zona pellucida next to the polar bodies.

Protocol 1. *Continued*

5. Move the injection needle close to the egg and use the micromanipulator adjuster to focus both the tip of the needle and the pro-nucleus at the same level.

6. Insert the needle through the zona pellucida and the egg membrane into the pro-nucleus. Apply some pressure to the 50 ml air-filled syringe. If enough pressure is applied and the needle has not yet penetrated the egg and pro-nuclear membranes, a 'bubble'-like structure should appear at the tip of the needle. Jab the nucleus with the needle and inject again. Visible swelling of the pro-nucleus indicates a successful injection and the needle must be carefully withdrawn (Figure 1). After injection the pronucleus tends to shrink back to its original size.

7. Move the injected oocyte to one side of the optical field using low magnification and proceed with the next one.

8. After having injected the first group of eggs, transfer them in M16 medium and store in the CO_2 incubator. Continue with the rest of the groups.

   An experienced injector should expect that approximately 70–90% of the eggs will survive injection. However, we and others have found that ease of injections and survival rates are increased when cyto-chalazin D (at 1 μg/ml) is included in the drop of M2 medium where injections are taking place. Cytochalazin D reversibly depolymerizes the cytoskeleton of the egg making the cell membranes more distortable and less susceptible to lysis. Injected eggs should be washed exten-sively in M2 medium to remove cytochalazin D.

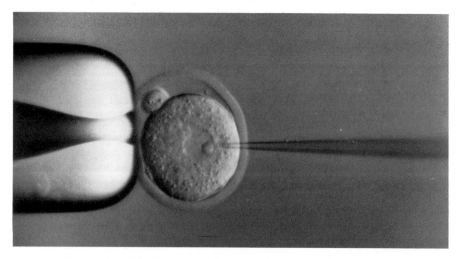

Figure 1. Injection of DNA solution into one of the pro-nuclei of a mouse zygote.

Protocol 2. Egg transfer into the oviducts of pseudo-pregnant
females

Equipment and reagents

- Anaesthetic (Hypnorm/Hypnovel/ddH$_2$O mix 1:1:6
- Blunt and fine forceps
- Surgical wound clips (e.g. Disposable Skin Stapler, Precise DS-25 from 3M)
- Binocular dissecting microscope and cold light source
- 0.1% adrenaline
- Mouth-driven glass transfer pipette
- Heat source for post-anaesthetic recovery

Mature (> 7 weeks old) F$_1$ females, mated with vasectomized males are
used as pseudopregnant recipients for the injected eggs. Egg transfer may
take place on either the same day of injection or the next day, when the
eggs will be at the two-cell stage. In either case, day 0.5 pseudo-pregnant
females should be used as recipients.

Method

1. Anaesthetize the recipient female by intraperitoneal injection of 0.3 ml
   of Hypnorm/Hypnovel/ddH$_2$O mix (10).
2. Swab the back of the mouse with ethanol and make a small mid-line incision in the skin at approximately the level of the last rib. Locate the position of the ovary indicated by a pink structure seen through the body wall.
3. Make a small incision through the body wall and use blunt forceps to
   grasp and gently pull out the fat pad which is attached to the ovary
   exposing the uterus.
4. Place a clip on the fat pad to hold the oviduct in place and move the
   mouse under a binocular dissecting microscope.
5. Expose the opening of the oviduct, which is found adjacent to the
   ovary, by tearing the bursa with fine forceps. Rupture of the bursa often
   causes bleeding which can obscure vision and lead to operational difficulties for the unexperienced. To avoid this, a drop of a 0.1% solution
   of adrenaline may be applied directly on to the bursa just before tearing
   it. This approach has no apparent adverse effects and prevents all
   bleeding from the ruptured bursa (G. Schmidt and J. O'Sullivan, personal
   communication).
6. Expel eggs into the oviduct using a mouth driven glass transfer pipette.
   Transfer approximately 15 eggs at the one-cell stage (or 10 eggs if at
   the two-cell stage) to each oviduct.
7. When the operation is completed, staple together the edges of the
   skin where the incision was made using one or two wound clips.

It is important to keep the mouse warm after the surgery until it has
recovered from the anaesthesia. Two to three recipient females are caged
together and they give birth usually 19–20 days after the transfer. Caging
of the females together lowers the possibility of loosing some new-borns
due to small litter size.

2.7 Integration and expression of injected genes

A few hundred copies of the gene of interest are usually injected into the pro-nucleus of the mouse zygote. The factors affecting subsequent integration into the mouse genome have not yet been defined. In general, it is accepted that integration occurs at a random site in the mouse genome and transgenes are found inserted usually at a single chromosomal locus, either as single copies or, more often, in tandem, head-to-tail arrays which are inherited as Mendelian traits. More rarely, transgene integration can occur at two or more positions in the mouse genome. In addition, it has been observed that in as many as 30–40% of the transgenic founder mice only a subset of cells carry the injected DNA and that these mice are therefore mosaic for the transgene. Such mosaicism could result from delayed DNA integration at the two-cell or even later stages of development. Mosaicism is not always a problem, since it can contribute to the survival of mice carrying an otherwise lethal or pathology-inducing transgene. This may be a common situation, in transgenic studies of cytokine function. For example, we have been able to analyse the pathology induced by human TNF in transgenic mice by studying the development of an early lethal phenotype in the progeny of an unaffected mosaic transgenic founder (11).

The transcriptional efficiency of transgenes is almost always influenced by the activation status of the neighbouring chromatin at the insertion site. It is generally observed that both the specificity and the levels of transgene expression may be influenced. However, in most cases, it has been shown that tissue specificity can be correctly conferred upon expression of the transgene, by inclusion in the gene construct of a few hundred base pairs of flanking, cis-acting DNA sequences (e.g. promoters and enhancers). Influences from the neighbouring chromatin do not affect expression of such transgenes in the 'correct' tissues although often they can cause expression in 'wrong' tissues. At present, position independent expression of transgenes can only be obtained by using specific cis-acting DNA elements called Locus Control Regions (LCRs) which are able to drive position independent, tissue specific and copy number dependent expression of the associated transgenes (12, 13).

2.8 Identification of transgenic progeny

Identification of transgenic mice is generally performed by standard Southern or dot/slot blot hybridization techniques on genomic DNA prepared from mouse tail fragments. Alternatively, transgene specific PCR amplification may also be applied allowing for the use of smaller samples of tissue, such as tail tips, ear pieces or blood samples.

Protocol 3. Preparation of DNA from tail fragments of mice

Equipment and reagents

- Tail buffer (50 mM Tris–HCl pH 8.0, 0.1 M EDTA, 0.1 M NaCl, 1% SDS)
- Proteinase K (10 mg/ml) in 50 mM Tris–HCl pH 8.0
- 25:24:1 Phenol/chloroform/isoamyl alcohol saturated in 100 mM Tris–HCl pH 8.0
- RNase A (10 mg/ml) in 0.3 M NaCl, 0.03 M Na_3-citrate, inactivate DNase impurities by a 5 min incubation in a boiling waterbath
- 1.5 ml safe-lock polypropylene tubes (e.g. Eppendorf)
- Vortex mixer
- 24:1 Chloroform/isoamyl alcohol
- Isopropanol
- 70% Ethanol
- TE buffer (10 mM Tris–HCl pH 8.0, 1 mM EDTA)
- Microcentrifuge
- Pipette tips with cut ends
- Sealed Pasteur pipettes
- Bench-top shaker
- UV Spectrophotometer

Method

1. Cut 0.5–1 cm of the tail into a 1.5 ml safe-lock polypropylene tube containing 0.6 ml of tail buffer.
2. Add 20 μl of 10 mg/ml Proteinase K solution, mix and incubate overnight at 55°C.
3. The next morning add 1 μl of 10 mg/ml RNase A and incubate for one hour at 37°C.
4. Add 0.6 ml phenol/chloroform/isoamyl alcohol to each tube and vortex for 5–10 min.
5. Centrifuge each tube for 10 min at 13000 *g* in a microcentrifuge at room temp. Remove the supernatant to a fresh tube using tips with cut ends to avoid any carry-overs from the interface.[a]
6. Repeat steps 4 and 5.
7. Add 0.5 ml chloroform/isoamyl alcohol and vortex for 5 min.
8. Centrifuge for 5 min at 13000 *g* in a microcentrifuge and remove the aqueous phase to a fresh tube.
9. Proceed with one tube at a time. Add 0.6 vol. isopropanol to each tube, mix several times by inversion until a white DNA mass is visible. Hook out the precipitated DNA using a sealed Pasteur pipette.
10. Immerse the DNA briefly in 70% ethanol, let it dry for a few minutes and place it in a microcentrifuge tube containing 100 μl TE buffer. Allow to stand for 10 min and then carefully remove and discard the Pasteur pipette.
11. Shake the tubes on a bench-top shaker to completely dissolve the DNA and determine the concentration by measuring the A_{260} OD.

[a] The quality of the DNA prep at this stage is appropriate for slot-blot hybridization analysis (see *Protocol 4*). It can also be stored at −20°C for later further purification. However, if the DNA is to be used for Southern hybridization analysis, the protocol should be carefully followed to the end.

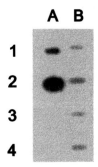

Figure 2. Slot-blot analysis of mouse tail DNA.

Each slot on the membrane is cut in two parts and each piece is hybridized with a different probe. In this example, the left half (A) is hybridized with a radiolabelled probe specific for the detection of human p55 TNF-receptor transgenes and the right half (B) with a radiolabelled probe specific for the single copy endogenous mouse p55 TNF-receptor gene. The latter is used as a quantitative control.

Slot 1: Tail DNA from a low-copy number transgenic mouse
Slot 2: Tail DNA from a high-copy number transgenic mouse
Slot 3: Tail DNA from a non-transgenic littermate
Slot 4: Normal mouse DNA

Protocol 4. Slot-blot analysis of tail DNA

Equipment and reagents

- Slot-blot apparatus (e.g. Schleicher and Schuel)
- Tail DNA; 5 μg of DNA or alternatively 20 μl of the supernatant from the first phenol extraction of tail DNA (see *Protocol 3*, step 5)
- X-ray film
- Hybridization equipment and reagents

- Positively charged nylon membrane (e.g. Hybond N⁺ from Amersham)
- Alkaline solution (0.5 M NaOH, 1.5 M NaCl)
- Neutralizing solution (0.5 M Tris–HCl pH:7.4, 1.5 M NaCl)
- Radiolabelled probe for endogenous single-copy gene
- Radiolabelled probe for transgene detection

Method

1. For each sample, add 5 μg of DNA (or 20 μl of phenol supernatant) in water to a final volume of 180 μl. Then add 20 μl of 4 M NaOH, mix and allow to stand at room temp. for 15 min.

2. Cut a piece of nylon membrane of the correct size, rinse in distilled water and soak in alkaline solution for at least 10 min.

3. Assemble the slot-blot apparatus according to the manufacturer's instructions and wash each well twice with alkaline solution. Apply the samples.

4. Disassemble the apparatus, immerse the membrane in 200 ml of neutralizing solution and shake gently for 10 min at room temp.

5. Dry the membrane and cut it so that each well is divided in two parts of equal size (see *Figure 2*).

6. Hybridize half of the membrane with a radiolabelled probe for an endogenous single-copy gene and the other half with the appropriate radiolabelled probe for the detection of the transgene. This approach controls efficiently for false negatives or false positives and minimizes general quantitative artefacts. It is recommended especially when low-copy transgenic lines are analysed.

7. After hybridization assemble the pieces of the membrane and expose on X-ray film.

Protocol 5. Screening of transgenic animals using PCR

Equipment and reagents

- PCR lysis buffer (50 mM KCl, 10 mM Tris–HCl pH 8.3, 1.5 mM $MgCl_2$, 0.1 mg/ml gelatin, 0.45% NP40, 0.45% Tween 20)
- 10 × PCR buffer (670 mM Tris–HCl pH 8.8, 166 mM $(NH_4)_2SO_4$, 1 mg/ml BSA)
- Waterbaths at 56°C and 95°C
- Microcentrifuge and tubes
- 2000 U/ml Taq polymerase
- Sterile distilled water
- Paraffin oil (depending on the PCR apparatus used)
- 10 mg/ml proteinase K
- 10 mM $MgCl_2$
- dNTP mixture (2.5 mM each)
- Specific oligonucleotides (sense and anti-sense) for the transgene
- 10 × TBE buffer (0.89 M Tris-base, 0.89 M Boric Acid, 25 mM EDTA); adjust pH to 8.3
- Loading buffer
- 1.5% agarose gel in 1 × TBE buffer
- UV transilluminator

Method

1. Cut a small piece (not more than 2 mm) from the tip of the tail and digest overnight at 56°C in 250 μl of PCR-lysis buffer containing 3 μl of 10 mg/ml proteinase K solution.

2. The next day incubate the tail digests at 95°C for 30 min to inactivate the proteinase.

3. Cool on ice for 10 min, spin at full speed for 10 min in a microcentrifuge, and remove 2 μl from each sample into the tubes for PCR.

4. Prepare enough reaction mixture to use for all samples by calculating per reaction:
   - 2 μl 10 × PCR buffer
   - 2 μl 10 mM $MgCl_2$
   - 1.5 μl dNTPs (mixture 2.5 mM each)
   - 1 μl oligonucleotide mix[a]
   - 1 μl Taq polymerase (2000 U/ml)
   - 10.5 μl sterile distilled water

Protocol 5. *Continued*

5. Add 18 µl of the reaction mixture in each tube, overlay each reaction with paraffin oil (depending on the model of the PCR apparatus used) and cycle the tubes using standard appropriate conditions.

6. When cycling is completed, add 5 µl loading buffer to each reaction and run the samples on a 1.5% agarose gel in 1 × TBE buffer.

7. Visualize the bands on a UV transilluminator and take a photograph of the gel.

[a] The optimal concentration for each pair of oligonucleotides should be established by carrying out reactions with concentrations ranging from 0.5 to 10 pmol of each oligonucleotide per reaction.

3. Production of knockout mice

The embryonic stem (ES) cell technology, combined with the methodology of gene targeting by homologous recombination, has made possible the introduction of designed mutations into the mouse germline. With the use of this technology, any desired genetic alteration can be introduced into the genome of ES cells in culture and then transferred into the mouse germ-line, making possible the study of the function of the mutated gene in the animal.

3.1 Embryonic stem cells

Embryonic stem cells are derived directly from the inner cell mass of mouse blastocysts and when cultured *in vitro* retain their undifferentiated pluripotential character even after a significant number of passages (14). The most useful property of ES cells has been their ability to contribute to the development of all cell lineages, including the germ cells, when transferred back into the mouse embryo (15). This feature makes it possible to generate mice derived exclusively from a single cell which has been cultured and manipulated *in vitro*. ES cell-lines have been derived from both inbred and outbred strains of mice. However, most existing ES cell lines are derived mainly from the 129/J and 129/Sv strains but also from the C57Bl/6 mouse strain.

3.2 Culturing ES cells

An extensive description of the methods for the isolation and culturing of ES cells is beyond the aim of this chapter and those who require a detailed laboratory manual for these methods are referred to ref. 16.

Optimal culture conditions are critical to the *in vitro* maintenance of ES cells. Good care should be given to the quality of the culture medium and serum. Since only a limited number of sera are found to be of sufficient quality for ES cells, fetal calf serum should be routinely batch tested for its

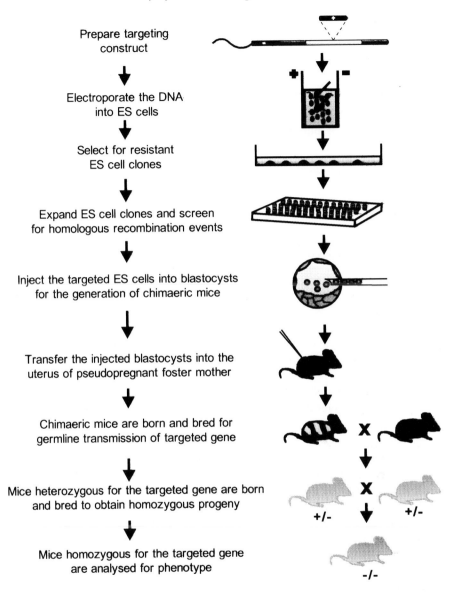

Prepare targeting
construct

Electroporate the DNA
into ES cells

Select for resistant
ES cell clones

Expand ES cell clones and screen
for homologous recombination events

Inject the targeted ES cells into blastocysts
for the generation of chimaeric mice

Transfer the injected blastocysts into the
uterus of pseudopregnant foster mother

Chimaeric mice are born and bred for
germline transmission of targeted gene

Mice heterozygous for the targeted gene are born
and bred to obtain homozygous progeny

Mice homozygous for the targeted gene
are analysed for phenotype

Figure 3. Schematic representation of a knockout protocol.

ability to support their growth. The ability of ES cells to remain pluripotential
in culture is shown to be dependent mainly on the presence in the culture
medium of the cytokine leukaemia inhibitory factor (LIF) (17, 18). Even
though recombinant LIF can be supplemented in the ES cell medium, the
most effective and economical approach for maintaining pluripotential ES

cells is to culture them on layers of mitotically inactivated feeder cells that produce and secrete LIF and other factors in the medium.

3.2.1 Preparation of feeder cell layers

Both permanently growing cell lines (i.e. STO fibroblasts) and primary mouse embryo fibroblasts (PMEFs) have been successfully used as feeders for the maintenance of pluripotential ES cells. We use STO (neo^r) fibroblasts grown routinely in DMEM plus 7% FCS, for the preparation of feeder cells.

Protocol 6. Preparation of feeder cell layers

Equipment and reagents

- DMEM plus 7% FCS
- PBS w/o Ca^{2+} and Mg^{2+}
- Trypsin-EDTA solution (1:250 Trypsin, 0.2 g/litre EDTA, prepared in Modified Puck's saline, GIBCO-BRL)
- 10 cm tissue culture dish
- 37°C, 5% CO$_2$ incubation
- Tissue culture centrifuge

- Mitomycin C (Sigma). Dissolve in PBS at 1 mg/ml, aliquot and store at −70°C. Thaw an aliquot just prior to use
- Gelatin solution 0.1% is prepared by diluting a 2% gelatin solution (Sigma cell culture reagents, type B, from Bovine skin) in sterile ddH$_2$O

Method

1. To a confluent 10 cm dish of STO fibroblasts add mitomycin C at a final concentration of 10 µg/ml and incubate at 37°C, 5% CO$_2$ for 2–3 h.

2. To prepare gelatinized tissue culture dishes add enough 0.1% gelatin solution to cover the surface of the dishes. Incubate at room temp. for at least one hour. Aspirate the gelatin solution before plating the feeder cells.

3. Aspirate the mitomycin C containing medium from the cells and wash them 3 times with 5 ml PBS. Trypsinize cells with 1 ml Trypsin-EDTA solution and collect them in DMEM plus 7% FCS.

4. Centrifuge at 250 *g* for 5 min and resuspend the cell pellet in DMEM plus 7% FCS. Count cells and dispense cell suspension in gelatin coated dishes at a concentration of 6–7 × 10^4 cells/cm^2.

5. Feeder cells should spread to give a monolayer in a few hours and if fed with fresh medium weekly, they can be used for as long as 15–20 days after their preparation. Change the medium to ES medium before adding ES cells.

3.2.2 The ES cell culture medium

In our laboratory we use mitomycin-C treated STO fibroblasts as feeder cells. The medium used is Dulbecco's modified Eagle's medium (DMEM) with high glucose (4.5 g/litre), L-glutamine, non-essential amino acids, and no sodium pyruvate. DMEM and phosphate-buffered saline (PBS, Ca^{2+} and Mg^{2+} free) is purchased in 1 × liquid form, to avoid any variations in the quality of water used for its preparation. L-glutamine is added to a final concentration of 2 mM, penicillin (100 U/ml) and streptomycin (100 µg/ml) are included, and 2-mercaptoethanol is added to a final concentration of 0.1 mM. DMEM and PBS in 1 × liquid form and L-glutamine, MEM non-essential amino acids, and penicillin/streptomycin in 100 × solutions are obtained from GIBCO-BRL. The 100 × solution of 2-mercaptoethanol is prepared by diluting 7 µl of a standard 14 M solution (Sigma) into 10 ml of PBS. The medium is supplemented with 15% fetal calf serum and this formulation is referred to as ES-medium.

3.3 Gene targeting in ES cells

Gene targeting by homologous recombination in mammalian cells is a very infrequent process. The ratio of homologous recombination is found to vary significantly between different experiments. In recent reports, increased homologous integration frequencies have been achieved by using DNA sequences isogenic to the target ES cell DNA sequences (19).

3.3.1 Vectors for gene targeting

A targeting vector is designed to recombine with and mutate a specific chromosomal locus. For this, a vector should contain DNA sequences homologous to the target site and selection markers to enrich for the rare homologous recombination event.

Since homologous recombination events are highly infrequent, it is desirable to apply efficient selection schemes to enrich for the targeted clones. A positive–negative selection approach, devised by Mansour *et al.* (20), has been successfully used for many different targeting experiments. Positive selection, usually conferred by a neomycin resistance gene cassette (*neo*), serves as a transfection marker and in many cases as a means to disrupt or replace coding exons and therefore inactivate genes. Negative selection usually imposed by the HSV thymidine kinase (*tk*) gene product, is useful for the elimination of cells in which random integration has occurred (see *Figure 4*).

3.3.2 Strategies for the introduction of subtle mutations into the ES cell genome

Several methods have been developed to introduce subtle mutations into the ES cell genome. These include microinjection of the targeting construct into ES cells followed by PCR screening to identify the positive clones (21), coelectroporation of a targeting construct with an unlinked selectable marker into ES cells (22) and the two-step 'in-out' or 'hit-and-run' method developed by Valancious and Smithies 1991 (23) and Hasty *et al.* (24).

Recently, a 'double replacement' approach has been developed for the introduction of subtle mutations in ES cells (25). In this approach endogenous

(A)

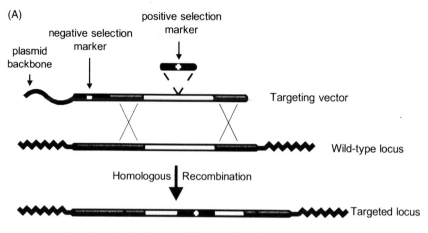

Resistant to Positive and Negative Selection

(B)

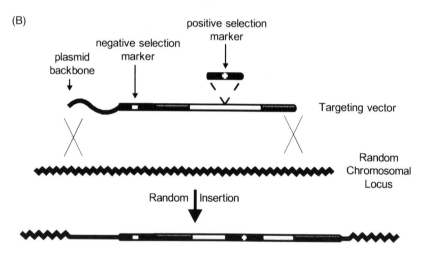

Resistant to Positive, Sensitive to Negative Selection

Figure 4. Schematic representation of vector/target recombination events using the positive–negative selection approach. (A) Gene targeting via homologous recombination, (B) random vector insertion into the ES cell genome.

sequences from the gene of interest are first exchanged by replacement with a positive and a negative selection marker (*neo* and *tk* respectively). In a second step, the introduced *neo/tk* cassette is replaced with sequences containing the desired mutation using a new targeting construct. Targeted clones are then selected for the absence of the negative selection marker. The advantage of this method is that a number of different mutations can be efficiently introduced into the same chromosomal locus by homologous targeting in the engineered *neo/tk* containing ES cell clone.

Using a different approach, conditional or cell-type specific activation or inactivation of gene expression can be achieved (26). In this approach, the Cre/loxP recombination system of bacteriophage P_1 is used to confer site-specific excision of DNA sequences flanked by directly repeated loxP sites. For this, two mouse strains are required. One is constructed by homologous recombination in ES cells to contain loxP sequences in chosen sites of any gene of interest. The second, is a transgenic mouse line engineered to express Cre recombinase in a cell type- or developmental stage-specific manner. Inter-crosses between these two mouse lines should produce mice in which loxP site-directed deletions will occur exclusively in cells where Cre recombinase is expressed. Using variations of this approach, a multitude of genetic manipulations in the mouse may be envisaged, including cell-fate determination and the *in vivo* ablation of specific cell lineages at specific developmental stages.

3.4 Transfection of ES cells

Several methods are available for the introduction of DNA into mammalian cells. Amongst them electroporation is the most widely used for gene targeting in ES cells.

Protocol 7. Electroporation of DNA into ES cells

Equipment and reagents

- Targeting DNA construct purified on a CsCl gradient and linearized with appropriate restriction enzyme
- ES-medium (see Section 3.2.2)
- Trypsin-EDTA solution (1:250 Trypsin, 0.2 g/litre EDTA, prepared in Modified Puck's saline, Gibco-BRL)
- PBS w/o Ca^{2+} and Mg^{2+}
- Plastic tissue culture dishes (100 × 20 mm)
- Gel electrophoresis equipment and reagents
- Electroporation apparatus: Bio-Rad gene

- pulser with a capacitance extender (Bio-Rad)
- Bio-Rad electroporation cuvettes (0.4 cm gap, Bio-Rad)
- G418 (Geneticin, Gibco)
- Ganciclovir (Symevene)
- Phenol/chloroform
- Chloroform
- Ethanol
- Sterile double-distilled water
- Tissue culture centrifuge

Method

1. Prepare the targeting vector DNA on a CsCl gradient and linearize it by digestion with the appropriate restriction enzyme. Check by gel electrophoresis that digestion is complete and extract the DNA by treatment with phenol/chloroform and chloroform. Precipitate with ethanol, wash with 70% ethanol and resuspend in sterile ddH$_2$O to 0.5 μg/ml.

Protocol 7. *Continued*

2. One day before electroporation, passage subconfluent cultures of ES cells 1:3. Change the medium 3–4 h before harvesting them on the day of electroporation to ensure that cells are actively growing.

3. Wash plates twice with 10 ml PBS, add 2 ml Trypsin-EDTA solution and place back in the incubator for 4–5 min. Add 4 ml of medium in each plate and pipette vigorously to achieve a near single-cell suspension.

4. Centrifuge the cells at 250 *g* for 5 min and resuspend them in PBS (5 ml for every 10 cm plate). Count cell numbers.

5. Recentrifuge the cells and resuspend them in PBS at a final concentration of 1.25×10^7 cells/ml.

6. Mix 0.8 ml of the cell suspension with 25 μg of the targeting vector in a sterile 1.5 ml tube and let stand for 5 min at room temp.

7. Transfer the suspension into an electroporation cuvette, place the cuvette in the electroporation chamber and apply a single pulse at 230 V, 500 μF. Tap the cuvette on the bench to suspend the cells and incubate for 5 min at room temp.

8. Plate the contents of each cuvette in two 10 cm plates containing feeder cells freshly fed with ES medium.

9. 24 h after electroporation apply G418 selection (200 μg/ml active concentration). Negative selection (10^{-5} M Ganciclovir) may be applied 72 h after electroporation.

10. Refeed the cells when the medium becomes acidic (indicated by a yellow colour), usually daily for the first 3–5 days. Resistant colonies should be visible 7–8 days after electroporation and ready to be picked 2–3 days later.

3.5 Picking and expansion of ES cell colonies

ES cell colonies are ready to be picked approximately 10 days after electroporation. Single colonies are picked and transferred into single wells with feeder cells in 96-well plates. When the clones have grown to sub-confluence they are split into replica plates for freezing and for DNA isolation.

Protocol 8. Picking and expansion of ES cell colonies

Equipment and reagents

- ES-medium (see Section 3.2.2)
- Trypsin-EDTA solution (1:250 Trypsin, 0.2 g/litre EDTA, prepared in Modified Puck's saline, Gibco-BRL)
- PBS w/o Ca^{2+} and Mg^{2+}
- Tissue culture 96-well plates (with flat-bottom and U-shaped wells)
- 8-channel multichannel pipette (Capp, Denmark)
- Multipipette with 8-channel adapter for dispensing liquid media into wells (Eppendorf)
- 8-place manifold aspirator for 96-well plates (Drummont Scientific Co.)

Method

1. Add 25 μl of trypsin-EDTA solution in each well of a U-shaped 96-well plate.

2. Wash the plate containing the ES cell colonies with 10 ml PBS and fill it with 10 ml PBS.

3. Use a stereo-microscope, a 20 μl micropipettor and sterile disposable tips to pick individual colonies in a small volume of PBS (5–10 μl). Transfer individual colonies to the plate prepared in step 1.

4. After having finished with one 96-well plate, place it in the incubator for 5–10 min.

5. Take a flat-bottomed 96-well plate containing feeder cells, aspirate the medium and add 100 μl of ES cell medium to each well,

6. Remove the plate with the trypsinized colonies from the incubator and add 100 μl of ES cell medium in each well. Use a multichannel pipettor (8–12 channel) and sterile disposable tips (change tips for each set of wells) to dissociate the ES cell colonies by vigorously pipetting them up and down.

7. Transfer the contents of each well to the respective wells of the plate prepared in step 5.

8. Let the cells grow for the next 3–4 days, changing the medium when it becomes acidic.

9. When cells are approaching confluence, split them 1:2. For this, wash each well twice with 100 μl of PBS, add 50 μl of trypsin-EDTA solution and let them stand in the incubator for 5–10 min. Add 100 μl of ES cell medium in each well, dissociate colonies by vigorous pipetting and split the contents in two feeder cell containing 96-well plates.

10. When ES cells in the two replica plates approach confluence, freeze one at −70°C (see *Protocol 9*) and split the other 1:3. From these three new plates, one should contain feeder cells and will be frozen as a second safety stock. Two more will be gelatinized but without feeders and will be used to prepare DNA (see *Protocol 11*).

3.6 Storage and recovery of ES cell clones

Screening of large numbers of ES cell clones may be a time-consuming process. To minimize the time that ES cells are kept in culture, it is preferable to freeze them down until screening is completed. ES cells may be frozen directly in the 96-well plate making it possible to simultaneously freeze a large number of clones.

Protocol 9. Freezing ES cell clones in 96-well plates

Equipment and reagents

- Trypsin-EDTA solution (1:250 Trypsin, 0.2 g/litre EDTA, prepared in Modified Puck's saline, Gibco-BRL)
- PBS w/o Ca^{2+} and Mg^{2+}
- 2 × freezing medium (20% DMSO, 80% FCS)
- Sterile light paraffin oil (Sigma)
- 96-well plates containing feeder cells
- ES clones and ES medium

- 8-channel multichannel pipette (Capp, Denmark)
- Multipipette with 8-channel adapter for dispensing liquid media into wells (Eppendorf)
- 8-place manifold aspirator for 96-well plates (Drummont Scientific Co.)
- Parafilm
- Styrofoam box

Method

1. Grow ES cell clones to subconfluence in feeder cell containing 96-well plates.

2. Feed cells with fresh medium 4 h before freezing.

3. Aspirate the medium and wash wells twice with 100 μl of PBS.

4. Add 50 μl of trypsin solution per well and place back in the incubator for 5–10 min.

5. Remove the plate from the incubator and add 50 μl of 2 × freezing medium in each well. Pipette up and down using the multichannel pipette (avoiding bubbling) until the ES cell clumps are dispersed into a near single-cell suspension.

6. Add 100 μl of sterile light paraffin-oil in each well to prevent evaporation during storage at −70°C. Seal the 96-well plate with parafilm, place in styrofoam box and store at −70°C until analysis of DNA is completed.

Protocol 10. Thawing ES clones from 96-well plates

Equipment and reagents

- ES medium (see Section 3.2.2)
- Sterile distilled water

- Sterilized Pyrex dish
- Tissue culture 24-well plates

Method

1. Prepare in advance 24-well plates containing feeder cells. Before use, feed with 2 ml fresh ES medium.

2. Warm up sterile distilled water to 38–40°C and pour in a sterile Pyrex dish in the laminar flow hood.

3. Remove one frozen 96-well plate from the −70°C freezer and place it directly on the surface of the water taking care not to allow water to enter into the wells.

4. Hold the plate until its contents are thawed. Transfer the selected clones, avoiding most of the paraffin oil, into the individual wells of the 24-well plate prepared in step 1.

5. The next day change the medium to remove the DMSO and the traces of the paraffin oil. Passage the cells when they reach subconfluence.

3.7 Identification of targeted ES cell clones

Screening for targeted ES cell clones can be performed either by PCR or by Southern blot hybridization. PCR is often used for screening of pooled clones and the individual clones of the positive pools are further analysed with Southern blot. To minimize the risk for false negative clones we prefer to directly analyse clones by Southern blotting and hybridization.

Protocol 11. Extraction and restriction enzyme digestion of DNA in 96-well plates[a]

Equipment and reagents

- PBS w/o Ca^{2+} and Mg^{2+}
- Lysis buffer (10 mM NaCl, 10 mM Tris–HCl pH 7.5, 10 mM EDTA, 0.5% Sarcosyl, 0.4 mg/ml freshly added proteinase K)
- Gelatinized 96-well plates (see *Protocol 6*, Step 2)
- ES clones
- Box containing wet paper towels, pre-warmed to 56°C

- Ethanol 70%
- Ethanol 100%
- Restriction digest mix (1 × restriction buffer, 1 mM spermidine, 1 mM DTT, 100 μg/ml BSA, 50 μg/ml RNase A, 20 U of restriction enzyme per reaction)
- 56°C incubator
- Low-power microscope

Method

1. Grow ES cell clones in gelatinized 96-well plates until fully confluent.

2. Wash each well twice with 100 μl of PBS and add 50 μl of lysis buffer.

3. Transfer the plate into a 56°C pre-warmed box containing wet paper towels to create a humidified atmosphere and incubate in a 56°C-oven overnight.

4. Allow the box to cool at room temp. for 1 h.

5. Add 100 μl of 100% ethanol into each well and let the plate stand on the bench for 1 h. Check for a filamentous DNA precipitate which should be visible under low-power magnification.

Protocol 11. *Continued*

6. Invert the plate carefully to discard its contents and drain on paper towel. Most of the DNA should remain attached to the bottom of the wells.

7. Wash the wells 3 × with 100 μl of 70% ethanol, discarding each wash by carefully inverting the plate.

8. After the last wash, air dry the DNA containing plate on the bench. Do not let DNA dry completely as it will be then difficult to dissolve.

9. Prepare the restriction digestion mix.

10. Add 35 μl of the digestion mix into each well and incubate overnight at the appropriate temperature in a humidified atmosphere.

11. The next day load the digested DNA samples on agarose gels and prepare for Southern hybridization analysis.

[a] Adapted from ref. 27

3.8 Generation of chimaeric mice

Once targeted ES cell-clones have been identified, chimaeric mice can be generated by using one of several methods including injection of ES cells into blastocysts; aggregation or co-culture of 8-cell stage embryos with ES cells, and injection of ES cells into 8-cell stage embryos. Methods for aggregation (28) and co-culture (29) of 8-cell stage embryos with ES cells are relatively simple and they do not require the sophisticated equipment used in the micro-injection procedures. However, as yet, in many laboratories blastocyst injection is preferred over these methods as they seem to require tricky setting up procedures. Morula injection has been developed recently (28) as an alternative to the blastocyst injection method. In this method 3–6 ES cells are injected under the zona pellucida of 8-cell stage embryos and placed adjacent to the blastomeres. Initial results obtained with this method have shown lower embryo implantation frequencies in comparison to the blastocyst injection method. However, the extremely high degree of chimaerism obtained in those embryos that finally implant and develop makes the morula injection method a good alternative to the blastocyst injection.

Blastocyst injection is currently the most widely used method for the production of germ-line chimaeras using targeted ES cell clones. For ES cell-lines which are derived from the 129 strain of mice, C57Bl/6 blastocysts are commonly used as hosts since they have been shown to produce the best yields of germ-line chimaeras (30).

Protocol 12. Production of germ-line chimeras by injection of ES cells into mouse blastocysts

Equipment and reagents

- ES medium (see Section 3.2.2)
- Hepes buffered ES medium (ES medium containing 20 mM Hepes)
- Trypsin-EDTA solution (1:250 Trypsin, 0.2 g/litre EDTA, prepared in Modified Puck's saline, Gibco-BRL)
- PBS w/o Ca^{2+} and Mg^{2+}
- 1 ml syringes with 25-gauge needles
- Glass Pasteur pipettes
- Tissue culture dishes (35- and 60-mm)
- Borosilicate glass capillaries, thin-wall w/o inner filament (Intracel Ltd)
- Micrometer head, 0–25 mm, 0.0005 mm (Mitutoyo Co., available from Pillar Engineering Supplies Ltd)
- Microforge (Micro Instruments Ltd)
- Anaesthetic (Hypnorm/Hypnovel/ddH_2O mix 1:1:6)
- C57Bl/6 adult males and females
- F_1 (C57Bl/6 × CBA) adult females and vasectomized males

- Sterile dissecting instruments, incl. blunt forceps
- Wound clips
- Stereo microscope
- Phase-contrast optics, 200 × magnification
- Microdrop cultures overlaid with paraffin oil (*Protocol 1*)
- 37°C, 5% CO_2 incubator
- Holding pipette (Section 2.4.2)
- Injection needle (See *Protocol 12C* steps 2–4)
- Pipette puller
- Silicon rubber sheet
- Sharp scalpel blades
- Tygon tubing
- Glass syringe
- Gelatinized Petri dishes
- Centrifuge
- Ethanol

A. Setting up matings

Day 0: Set up matings between C57Bl/6 males and females. For better mating efficiencies females in oestrus may be selected by examining for the appearance of a pink and swollen vagina.

Day 1: Identify females that have mated by checking for copulation plugs and place them in a separate cage for later collection of 3.5 d blastocysts (day 4). Set up additional matings between F1 (C57Bl/6 × CBA) females and vasectomized males for the production of pseudopregnant females.

Day 2: Check the F1 females for plugs and place those that have mated in a separate cage. They will be used on day 4 (at 2.5 days of pseudopregnancy) as recipients for the injected blastocysts.

B. Recovery of blastocysts

1. On day 4, kill by cervical dislocation the C57Bl/6 females (collected on day 1) and dissect out both uterine horns by two incisions; one next to the oviduct and a second at the distal-end junction of the two uterine horns.

2. Use a 1 ml syringe (25 G needle) filled with Hepes buffered ES medium to flush the blastocysts out of the uteri in a 6 cm tissue culture dish.

3. Collect the blastocysts under a stereo microscope using a mouth controlled heat-drawn Pasteur pipette, wash in ES medium and transfer in

Protocol 12. *Continued*

micro-drop cultures overlaid with light paraffin oil. Store in a 5% CO_2, 37°C incubator. At this stage, some blastocysts may not be fully expanded; they will do so later during the day.

C. *Preparation of the holding and injection pipettes*

1. The holding pipette is essentially the same as the one used for pronuclear injection (see Section 2.4.2) except that a bend of approximately 30° is introduced 2–3 mm from its end, using a microforge.

2. The injection needle is used to collect individual ES cells and introduce them into the blastocoel cavity. Thin-walled borosilicate capillaries and a pipette puller are used to produce needles that have a relatively long section at the appropriate internal diameter, which should be slightly larger than the ES cells (18–20 μm).

3. Place the pulled needle on a transparent silicon rubber sheet under a stereo-microscope and using a sharp scalpel blade, snap it at the region of the appropriate diameter to create a sharp bevelled point.

4. Use the microforge to introduce a bend of approximately 30° close to the end of the injection needle, taking care to keep the bevel facing to the side.

D. *Setting up the injection chamber*

1. The microscopes and micromanipulators described for pro-nuclear injections (see Section 2.1) are also suitable for blastocyst injections. Blastocyst injections are performed on a lid of a 3.5 cm tissue culture dish semi-filled with Hepes-buffered ES cell-medium and overlaid with light paraffin oil. Lids of Petri dishes are conveniently shallow to allow free movements of the holding and injection pipettes.

2. The holding pipette is set as for pro-nuclear injections (see Sections 2.4.2 and 2.6). The injection needle is filled with light paraffin oil and connected through a tygon tubing with a glass syringe controlled by a sensitive micrometer head.

3. The holding and injection pipettes are lowered in the injection dish and appropriate adjustments are made to position their ends parallel to the bottom of the dish.

E. *Preparation of ES cells for blastocyst injections*

1. 3–4 h before harvesting, feed a subconfluent ES cell-containing dish (35- or 60-mm) with fresh ES medium.

2. Wash the plate twice with PBS, add Trypsin-EDTA solution and place it back in the incubator for 4–5 min.

3. Add 3–4 ml of ES medium and pipette vigorously to dissociate colonies into a single cell suspension. Plate the suspension on a gelatinized dish and transfer into the incubator.

4. 1 h later feeder cells attach strongly on the surface of the gelatinized dish while most of the viable ES cells only begin to adhere. Carefully, aspirate and discard the medium containing the non-adherent (non-viable) cells. Add 5 ml of ES medium to the dish and suspend the loosely adhering ES cells by pipetting.

5. Centrifuge the suspension for 5 min at 200 g and resuspend the cell pellet in 1–2 ml ES medium which has been pre-cooled at 8 °C. Transfer a few μl into the injection dish and store the rest of the suspension at 8 °C.

F. *Blastocyst injection*

1. Transfer 10–20 blastocysts into the injection dish.

2. Collect 10–15 ES cells in the injection needle using phase contrast optics at 200 x magnification. Select the round cells that have a light yellow colour. Cells that appear dark are dead or dying.

3. Use the holding pipette to pick up a blastocyst and focus on an appropriate point for injection. These are usually points of intercellular junctions between adjacent trophectoderm cells.

4. Bring the point of entry and the injection needle into the same focal plane. Push the needle into the blastocoel cavity with a steady smooth movement. Too slow movement may result in blastocyst collapse making impossible further penetration of the needle and injection.

5. Release 10–15 ES cells by carefully applying positive pressure. After injection the blastocyst is seen to collapse.

6. After having injected all the blastocysts, place them back in the incubator. One hour later they will start to re-expand and injected cells will be observed in the blastocoel, some of them being attached to the inner cell mass.

G. *Transferring blastocysts into the uteri of pseudopregnant females*

1. Anaesthetize the recipient female by intraperitoneal injection of 0.3 ml of Hypnorm/Hypnovel/ddH$_2$O mix (10).

2. Swab the back of the mouse with ethanol and make a small mid-line incision in the skin at approximately the level of the last rib. Locate the position of the ovary indicated by a pink structure seen through the body wall.

3. Make a small incision through the body wall and use blunt forceps to grasp and gently pull out the fat pad which is attached to the ovary exposing the uterus.

Protocol 12. *Continued*

4. Using a 25-gauge needle, make a hole in the uterus close to the oviduct end.

5. Using a finely drawn Pasteur pipette transfer 6–7 blastocysts into the uterus through the hole created by the needle.

6. Gently push back the uterus into the peritoneal cavity and continue with the opposite uterine horn.

7. When the operation is completed, staple together the edges of the skin where the incision was made using one or two wound clips.

8. Keep the operated females warm until they have recovered from the anaesthesia. They should give birth 17–18 days later.

9. Identify chimaeric newborns at around 7 days after birth by the presence of the agouti coat colour (derived from the ES cells) on the black (C57Bl/6) background. Since ES cell lines with a male karyotype are more often used, a distortion of sex rate (towards males) should be expected in chimaeric mice.

10. Cross male chimaeras with C57Bl/6 females to obtain ES cell-derived progeny (Agouti coat colour). Germ-line transmission of the targeted allele which is expected in 50% of the Agouti progeny, is confirmed by DNA analysis of mouse tail fragments.

11. Cross-mice heterozygous for the targeted allele to obtain homozygous knockout mice.

4. Transgenic and knockout mice in cytokine research

Transgenic and knockout systems offer clear advantages over cellular systems in the analysis of the functional potency of factors participating in complex multicellular processes. This is especially true when pleiotropic and redundant activities of factors are studied. For example, fine analysis of cytokine functioning in the immune system necessitates the use of experimental settings where faithful measurement of *in vivo* reactivities, due to the presence or absence of a certain cytokine, may be easily performed.

Over-expressing or knocking out cytokine and cytokine receptor genes in transgenic mice is currently providing much insight into the contribution of these factors to the maintenance of homeostasis or the triggering of disease in the course of immune responses. Further understanding of cytokine functioning should come mainly through studies addressing susceptibility or resistance of such 'mutant' mice to infectious or genetic disease. Current advances in the 'genetic engineering of the mouse' including the tissue specific activation

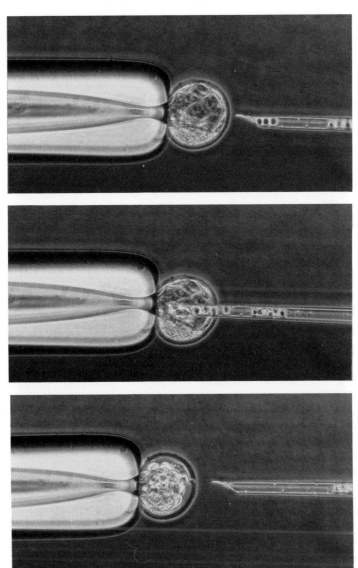

Figure 5. Injection of ES cells into the blastocoel cavity of a mouse blastocyst. (A) The blastocyst is immobilized on the holding pipette and the needle containing the ES cells is brought into focus, (B) ES cells are expelled in the blastocoel cavity, (C) the Blastocyst is starting to collapse after the ES cells have been injected.

or inactivation of gene expression combined with developing technologies for switching gene expression on and off at will, should provide experimental settings unprecedented in their potential to offer answers to long-standing questions or to even inspire questions currently unthought of.

Acknowledgements

We are indebted to our friend Dr Vasso Episkopou for teaching us the knockout technology. Work in the authors' laboratory is supported by EEC grants SC1*-CT91–0653, BIO2-CT92–0002 and BIO2-CT94–2092.

References

1. Hogan, B., Beddington, R., Costantini, F., and Lacy, E. (1994). *Manipulating the mouse embryo: a laboratory manual* (2nd edn). Cold Spring Harbor Press, New York.
2. *Guide to techniques in mouse development. Methods in enzymology* (1993). (ed. P. M. Wassarman and M. L. DePamphilis), **Vol. 225**. Academic Press, London.
3. Taketo, M., Schroeder, A. C., Mobraaten, L. E., Gunning, K. B., Hanten, G., Fox, R. R., Roderick, T. H., Stewart, C. L., Lilly, F., Hansen, C. T., and Overbeek. P. A. (1991). *Proc. Natl. Acad. Sci. USA*, **88**, 2065.
4. Brinster, R. L., Chen, H. Y., Trumbauer, M. E., Yagle, M. K., and Palmiter, R. D. (1985). *Proc. Natl. Acad. Sci. USA*, **82**, 4438.
5. Towens, T. M., Lingrel, J. B., Chen, H. Y., Brinster, R. L., and Palmiter, R. D. (1985). *EMBO J*, **4**, 1715.
6. Brinster, R. L., Allen, J. M., Behringer, R. R., Gelinas, R. E., and Palmiter, R. D. (1988). *Proc. Natl. Acad. Sci. USA*, **85**, 836.
7. Sambrook, J., Fritch, E. F., and Maniatis, T. (ed.) (1989). *Molecular cloning, a laboratory manual* (2nd edn). Cold Spring Harbor Laboratory Press, NY.
8. Schedl, A., Larin, Z., Montoliu, L., Thies, E., Kelsey, G., Lehrach, H., and Schutz, G. (1993). *Nucleic Acids Res.*, **21**, 4783.
9. Strouboulis, J., Dillon, N., and Grosveld, F. (1992). *Genes Dev.*, **6**, 1857.
10. Mann J. R. (1993). In *Methods in enzymology* (ed. P. M. Wassarman and M. L. DePamphilis), **Vol. 225**, pp. 782. Academic Press, London.
11. Probert, L., Keffer, J., Corbella, P., Cazlaris, H., Patsavoudi, E., Stephens, S., Kaslaris, E., Kioussis, D., and Kollias, G. (1993). *J. Immunol.*, **151**, 1894.
12. Grosveld, F., Blom van Assendelft, G. B., Greaves, D. R., and Kollias, G. (1987). *Cell*, **51**, 975.
13. Kollias, G. and Grosveld, F. (1992). In *Transgenic animals* (ed. F. Grosveld and G. Kollias), pp. 79–98. Academic Press, London.
14. Evans, M. J. and Kaufman, M. H. (1981). *Nature*, **292**, 154.
15. Bradley, A., Evans, M. J., Kaufman, M. H., and Robertson, E. J. (1984). *Nature*, **309**, 255.
16. Robertson, E. J. (1987). In *Teratocarcinomas and embryonic stem cells: a practical approach* (ed. E. J. Robertson), pp. 71–112. IRL Press, Oxford.
17. Williams, R. L., Hilton, D. J., Pease, S., Wilson, T. A., Stewart, C. L., Gearing, D. P., Wagner, E. F., Metcalf, D., Nicola, N. A., and Gough, N. M. (1988). *Nature*, **336**, 684.
18. Smith, A. G., Heath, J. K., Donaldson, D. D., Wong, G. G., Moreau, J., Stahl, M., and Rogers, D. (1988). *Nature*, **336**, 688.
19. te-Riele, H., Maandag, E. R., and Berns, A. (1992). *Proc. Natl. Acad. Sci. USA*, **89**, 5128.

20. Mansour, S. L., Thomas, K. R., and Capecchi, M. R. (1988). *Nature*, **336**, 348.
21. Zimmer, A. and Gruss, P. (1989). *Nature*, **338**, 150.
22. Reid, L. H., Shesely, E. G., Kim, H. S., and Smithies, O. (1991). *Mol. Cell. Biol.*, **11**, 2769.
23. Valancius, V. and Smithies, O. (1991). *Mol. Cell. Biol.*, **11**, 1402.
24. Hasty., P., Ramirez-Solis, R., Krumlauf, R., and Bradley, A. (1991). *Nature*, **350**, 243.
25. Wu, H., Liu, X., and Jaenisch, R. (1994). *Proc. Natl. Acad. Sci. USA*, **91**, 2819.
26. Gu, H., Marth, J. D., Orban, P. C., Mossmann, H., and Rajewsky, K. (1994). *Science*, **265**, 103.
27. Ramirez-Solis, R., Rivera-Perez, J., Wallace, J. D., Wims, M., Zheng, H., and Bradley, A. (1992). *Anal. Biochem.*, **201**, 331.
28. Stewart, C. L. (1993). In *Methods in enzymology* (ed. P. M. Wassarman and M. L. DePamphilis), **Vol. 225**, pp. 843–8. Academic Press, London.
29. Wood, S. A., Pascoe, W. S., Schmidt, C., Kemler, R., Evans, M., and Allen, N. (1993). *Proc. Natl. Acad. Sci. USA*, **90**, 4582.
30. Schwarzberg, P. L., Goff, S. P., and Robertson E. J. (1989). *Science*, **246**, 799.

19

Development of antibodies to cytokines

S. POOLE

1. Introduction

Antibodies to cytokines can greatly facilitate research on many aspects of these mediators: including the development of specific and sensitive immuno-assays (see Chapter 22); the detection of cytokines in biological fluids, tissues, and cells using immunochemical and immunocytochemical techniques; immunoneutralization of biological activities, and delineation of external regions (epitopes) of the cytokines.

2. Choice of immunogen

Recombinant cytokines are generally used although purified 'natural' cyto-kines, conjugates of cytokines to carrier proteins, synthesized peptides with amino acid sequences corresponding to selected portions of cytokine sequences, and enzymatic digests of cytokines, may be used.

2.1 Recombinant cytokines

These offer advantages in that it is possible to prepare them in large quantities and with low levels of contaminating proteins. A disadvantage for glyco-sylated cytokines is that non-glycosylated analogues are synthesized in *Escherichia coli* and glycosylation patterns in yeast and in Chinese hamster ovary (CHO) cells may be different from those of the 'natural' cytokine. Also, the extraction procedures may result in inappropriate folding of these molecules.

2.2 Purified 'natural' cytokines

'Natural' cytokines, e.g. ones prepared from cell-conditioned media, require similar purification procedures to recombinant proteins but they are more likely to be contaminated with other polypeptides/proteins.

2.3 Conjugates of cytokines

Where there is a high degree of homology between a human cytokine and the equivalent cytokine in the species to be immunized, immunogenicity may be enhanced by conjugating the cytokine to a protein carrier. The choice of proteins for conjugation and coupling methods will be discussed below in relation to small (synthesized) peptide homologues of cytokines. As far as conjugates of the cytokines themselves are concerned, the author has found that conjugates of recombinant human interleukin-1α (IL-1α), IL-1β, rat IL-1β and murine IL-1β, when injected into rabbits and guinea-pigs, were less effective than the unconjugated cytokines, at generating high titre, high affinity antisera.

2.4 Small peptide homologues

These are often used when the amino acid sequence of a cytokine is known but where only limited amounts of pure antigen are available. Also, use of peptide homologues avoids the circular logic that sometimes arises when characterizing antisera with the same antigens to which they were raised, e.g. a partially purified preparation of a cytokine. A major disadvantage of anti-peptide antibodies is that their affinities for the relevant cytokine (i.e. native molecule) is generally much lower than their affinities for the peptides. This renders the anti-peptide antibodies generated of little use in developing sensitive immunoassays for the cytokine although useful for immunochemical/immunohistochemical procedures and for epitope mapping studies. Indeed, the capacity of an anti-peptide antibody to immuno-precipitate a cytokine in a liquid-phase assay is the only test that unambiguously defines external sites on the native molecule.

Peptides with molecular weights of less than 2000–3000 are not usually immunogenic but specific antibodies against these low molecular-weight immunogens can be produced by conjugation to high molecular-weight proteins or other carriers. A chemical group rendered immunogenic in this way is known as a hapten and the process of coupling it to a larger protein is known as haptenization. Small peptide homologues of cytokines used as immunogens usually comprise 15–20 amino acid residues and therefore need to be haptenized to generate useful antibodies.

A wide variety of high molecular-weight substances have been used successfully for haptenization of peptides. These include bovine thyroglobulin (BTG), bovine serum albumin, ovalbumin (OVA), keyhole limpet haemocyanin (KLH), and purified protein derivative of tuberculin (PPD). PPD offers several advantages—at least in theory (1). PPD elicits delayed-type hypersensitivity reactions in animals (and in humans) previously exposed to the tubercle bacillus. Thus, animals previously immunized with Bacillus Calmette-Guérin (BCG) are injected subsequently with the peptide-PPD conjugate. PPD provides T cell help and gives rise to virtually no antibody response in itself: this is particularly useful if it is intended to go on to make monoclonal antibodies, where the presence of significant anticonjugate antibodies are undesirable. The author has found little to choose between the proteins given above although most workers have their favourites. The choice of coupling reagent is more significant.

A number of chemical procedures have been developed for conjugating peptides to proteins. All depend upon the use of cross-linking reagents which are of two basic types: (a) homobifunctional cross-linkers, which have the same functional group at either end and couple via the same amino acid side-chains in peptide and protein; and (b) heterobifunctional cross-linkers, which couple via two different side-chains.

2.4.1 Homobifunctional cross-linkers

The most widely used (and easy to use) cross-linker is glutaraldehyde (Pentane-1,5 dial). The reaction is complex, involving many different side chains and produces a heterogeneous, uncharacterized conjugate. With shorter peptides glutaraldehyde will almost invariably couple via the n-terminal amino group which may be a disadvantage, e.g. if the n-terminal amino acids comprise part of an epitope.

Protocol 1. Coupling peptides to KLH with glutaraldehyde

Equipment and reagents
- Keyhole limpet haemocyanin (KLH)
- 0.01 M NaOH
- 0.5 M sodium bicarbonate
- Peptide
- DMSO
- PBS
- 1 M Tris–HCl, pH 8.5

Method

1. Dissolve 5 mg KLH in 1.0 ml 0.01 M NaOH and then add 1.0 ml 0.5 M sodium bicarbonate.

2. Dissolve 250 µg peptide in 25 µl DMSO.

3. Mix 100 µl KLH solution with 10 µl peptide solution and, while stirring, add 2.5% glutaraldehyde dropwise over 1 h to a final concentration of 0.1%.

4. After stirring for further 4 h in the dark, add 0.25 ml PBS + 25 µl 1.0 M Tris/HCl, pH 8.5. Store the conjugate at −40°C.

When a peptide contains a tyrosine group, either as part of the naturally occurring sequence or added during synthesis, a more specific reagent, bis-diazotolidine (BDT), may be used to couple together tyrosine side chains of the peptide and protein.

Protocol 2. Coupling peptides to OVA with BDT

Equipment and reagents

- *o*-toluidine
- 0.2 M HCl
- Sodium nitrite (NaNO₂)
- Ovalbumin 5 mg/ml in borate saline, pH 9.0
- Peptide

- Bis-diazo-toluidine
- Dialysis membrane
- 0.9% NaCl
- Cold acetone
- Centrifuge

Method

1. Dissolve 0.23 g *o*-toluidine in 45 ml 0.2 M HCl and place on ice. Dissolve 0.175 g sodium nitrite ($NaNO_2$) in 5 ml water.

2. Add the $NaNO_2$ solution to the O-tolidine drop-wise, over 5–10 min with constant stirring, ensuring that the mixture remains cold. The colour of the solution changes from clear to yellow, to orange-red and back to yellow.

3. Dissolve OVA at 5 mg/ml in borate saline at pH 9.0.

4. Dissolve 5.0 mg peptide in 2.5 ml of the solution of OVA.

5. Briskly add drop-wise, 1.0 ml freshly prepared bis-diazotolidine (which is unstable).

6. Incubate for 4 h in the dark, and then remove any excess reagents by dialysis or de-salting. (Uncoupled peptide is lost through the dialysis membrane.) Before immunization dialyse conjugate into 0.9% NaCl or precipitate with 4 vol. of cold acetone ($-70°C$), pellet at 10 000 *g*, air-dry, and redisperse in 0.9% NaCl. Store the conjugate at $-40°C$.

2.4.2 Heterobifunctional cross-linkers

The water soluble carbodiimides (e.g. 1-ethyl-3-(3-dimethyl aminopropyl-carbodiimide) couple amino groups to carboxyl groups, and thus form heterogeneous conjugates.

Protocol 3. Conjugation of peptides to BTG with carbodiimide

Equipment and reagents

- Peptide
- 0.1 M sodium bicarbonate
- Bovine thyroglobulin (BTG)
- 0.5 M glycine

- Sephadex G-50, 6 × 100 mm, column (Pharmacia)
- Column buffer: 0.01 M sodium bicarbonate containing 0.9% NaCl

Method

1. Dissolve 0.5 mg peptide in 0.5 ml 0.1 M sodium bicarbonate and add to BTG, 1.0 mg in 0.5 ml 0.1 M sodium bicarbonate.

---

2. Add 1 mg carbodiimide and incubate the mixture for 1 h in the dark at 20°C.

3. Add glycine (0.5 M, 20 μl) to quench unreacted groups and de-salt the conjugate on a Sephadex G-50 column (6 × 100 mm) using 0.01 M sodium bicarbonate containing 0.9% NaCl as column buffer. Store the conjugate at −40°C.

When a peptide contains a cysteine residue, either as part of the naturally occurring sequence or added during synthesis, the hetero-bifunctional reagent *m*-maleimidobenzoic acid *N*-hydroxy succinimide (MBS) ester may be used to couple sulphydryl to amino groups.

Protocol 4. Coupling peptides to KLH with MBS

Equipment and reagents

- Dialysis membrane
- Keyhole limpet haemocyanin (KLH)
- 10 mM potassium phosphate buffer, pH 7.2
- Peptide
- 3 mg/ml MBS stock in dimethyl formamide
- 20 ml P30 column (Bio-Rad)
- 50 mM phosphate buffer, pH 6.0

Method

1. Dialyse KLH against 10 mM potassium phosphate buffer pH 7.2. Adjust KLH concentration to about 20 mg/ml.

2. Ideally, freshly weighed-out peptide should be used. If using stock solution of peptide (e.g in PBS), it may be necessary to reduce the peptide before use. Otherwise, dissolve peptide at 5 mg/ml in 10 mM phosphate buffer. 4 mg of KLH is required for every 5 mg of peptide.

3. Add 55 μl of 10 mM phosphate buffer to each 4 mg KLH solution.

4. Slowly add 85 μl MBS stock at 3 mg/ml in dimethyl formamide to each 4 mg KLH. Stir for 30 min at room temperature.

5. De-salt activated KLH on a 20 ml P30 column. Pre-equilibrate and run column in 50 mM phosphate buffer, pH 6.0. Collect 1 ml fractions (typically about 15). Protein elutes as a visible grey peak at the exclusion volume (fractions 6–8). About 95% of KLH is recovered.

6. Add activated KLH to peptide solution while stirring at room temperature. Adjust pH to 7.4 and stir at room temperature for 3 h. Store the conjugate at −40°C.

2.5 Enzymatic digests of cytokines

Purified fragments of cytokines offer advantages and disadvantages similar to those of small peptide homologues. Small fragments, i.e. <3000 kDa, will require haptenization.

3. Choice of animal

Where large quantities of antibodies are required, e.g. for commercial purposes or to provide reference reagents, hybridoma technology has made possible the production of monoclonal antibodies with predefined binding characteristics, which can be produced in large amounts from immortal cell lines. Another advantage of monoclonal antibodies is that they can be raised against impure immunogens provided that pure material is available for screening. However, monoclonal antibodies to cytokines often are not neutralizing and (like monoclonal antibodies to most antigens) have lower affinities than polyclonal antibodies to the same antigens. Polyclonal antibodies also may be obtained in quantity from large animals such as sheep, goats, donkeys, and horses.

Where very large volumes of antiserum are not required, rabbits are the first choice provided that sequence homology between the rabbit and human cytokine is not too great. Guinea-pigs usually will be the second choice, although they produce less antiserum and are more difficult to bleed than rabbits.

Whichever of these small species is chosen, it is recommended that several individuals are immunized since individual variation in response can be marked. Groups of at least three, but preferably more rabbits or guinea-pigs should be used. The sex of the animals is not critical, young animals of either sex can be used.

4. The adjuvant

A wide variety of substances potentiate the humoral antibody response to injected immunogen. These include inorganic adsorbents, e.g. aluminium hydroxide, mineral oils, such as liquid paraffin, and bacterial cell-wall components. This diversity of materials having adjuvant properties makes it difficult to identify a simple mechanism of action. The most important advance in adjuvant technology was the development by Freund and coworkers of adjuvants containing mycobacteria, mineral oil, and emulsifier (2). The simple oil-detergent mixtures are termed Freund's 'incomplete' adjuvant (FIA); incorporation of heat-killed *Mycobacterium tuberculosis* or *M. butyricum* (0.5 mg/ml) into the oily mixture yields Freund's 'complete' adjuvant (FCA). The latter is more effective, probably as a result of greater stimulation of the local cellular response. For the immunization schedules described below FCA is used for the primary immunization and FIA for all boosts.

For maximum efficiency, it is necessary to prepare a stable, water-in-oil emulsion. This can be achieved in a number of ways, but the simplest is by using the double-hub connector method described in detail by Hurn and Chantler (3). One vol. of aqueous immunogen is emulsified in 2–4 vol. of oily adjuvant.

5. Immunization schedules

The protocols described below have proved successful although scientists in the UK may be required to modify certain of these to comply with 'Home Office guidelines on antibody production: advice on protocols for minimum severity'. This will depend upon the procedures permitted under Project Licences and Personal Licences held under the Animals (Scientific Procedures) Act 1986. Consequently scientists in the UK are advised to discuss immunization protocols with their Home Office Inspectors before commencing immunizations. Similarly, scientists in other countries should ensure that immunization protocols are consistent with local legislation before commencing work of this nature.

5.1 Mice

A variety of protocols have proved successful. A protocol which yielded monoclonal antibodies to IL-1α using Balb/C mice (4) is given below.

Protocol 5. Generation of monoclonal antibodies to IL-1α

Equipment and reagents

- Balb/C mice
- IL-α
- Freund's complete adjuvant (FCA)
- Freund's incomplete adjuvant (FIA)
- PBS
- Syringes and needles
- Blood-clotting tubes

Method

1. Emulsify IL-1α (25 μg in 0.25 ml PBS) in FCA (0.75 ml) and inject subcutaneously (s.c.) into three sites on day 1.

2. On days 30 and 60 after the primary immunization, emulsify 10 μg IL-1α (in 0.1 ml PBS) in FIA (0.2 ml) and inject intraperitoneally (IP).

3. On day 150 inject the mice IV with 10 μg IL-1α in PBS (0.3 ml). On day 153 remove spleens for fusion. A detailed protocol for production of monoclonal antibodies is given elsewhere (5).

S. Poole

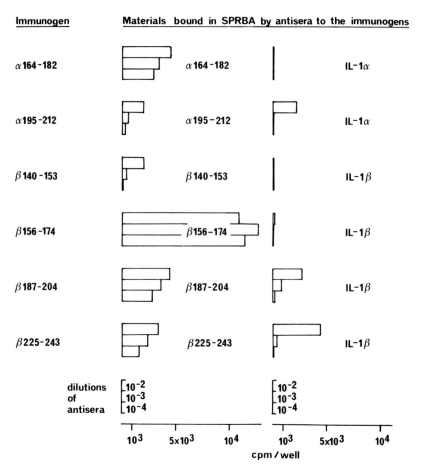

Figure 1. Materials bound by rabbit antisera to IL-1α and IL-1β related peptide immunogens in a solid phase radiobinding assay (SPBA). There was no detectable binding of any
of the antisera to irrelevant peptides/proteins. Sera from rabbits immunized with IL-1β
121–134/KLH/BDT did not bind IL-1β 121–134 or IL-1β.

5.2 Rabbits

The multiple intradermal method using Dutch rabbits has worked well for rh
IL-1α, rh IL-1β, IL-6, and a variety of IL-1β related peptides (see *Figure 1*).

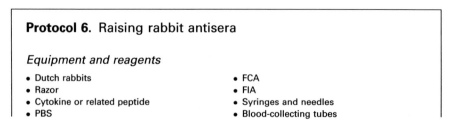

Protocol 6. Raising rabbit antisera

Equipment and reagents

- Dutch rabbits
- Razor
- Cytokine or related peptide
- PBS

- FCA
- FIA
- Syringes and needles
- Blood-collecting tubes

334

Method

1. Shave the hair on the back and on the proximal parts of all four limbs. Emulsify cytokine or related peptide, about 70 μg (in 0.5 ml PBS), in 1.5 ml FCA for the primary immunization.

2. Make intradermal injections, spaced evenly over the back (24 × 0.05 ml) and the rest of the shaved area (16 × 0.05 ml).

3. Boost animals by the intramuscular (IM) route, 10 weeks after the primary immunization and at intervals of 4–8 weeks thereafter. For boosts, emulsify about 50 μg cytokine in 0.25 ml PBS in 0.75 ml FIA and inject 0.5 ml of emulsion IM into each hind limb or into each fore limb, alternately. Bleeds (20–30 ml) may be taken 8–10 weeks after the primary immunization and 8–11 days after each boost. Variations of this protocol have also proved successful (6, 7).

5.3 Guinea-pigs

Although guinea-pigs have responded well to a number of proteins poorly immunogenic in other species, e.g. insulin and parathyroid hormone, the author has not found guinea-pigs to be good responders to human IL-1β related peptides nor to rat and mouse IL-1β (possibly because of good sequence homology between rat/mouse IL-1β and guinea-pig IL-1β).

Protocol 7. Raising guinea-pig antisera

Equipment and reagents

- Guinea-pigs
- Cytokine
- PBS
- FCA
- FIA
- Syringes and needles
- Blood-collecting tubes

Method

1. For the primary immunization emulsify about 50 μg of cytokine (in 0.25 ml PBS) in 0.75 ml FCA and inject sc (4 sites) into the abdominal wall just on either side of the mid-line.

2. Boost animals with about 50 μg cytokine in FIA by the intramuscular route, 4 weeks after the primary immunization and at intervals of 4–8 weeks thereafter.

3. Animals may be bled by cardiac puncture, which yields only 3–5 ml of serum, about 8–10 days after each boost. However, because of the low yield of serum and the risk of killing the animals when bleeding by cardiac puncture, it is less practicable to bleed guinea-pigs repeatedly than it is to bleed rabbits. Therefore it is best to immunize a relatively large number of animals for a comparatively long period of time, then bleed them out and select the best antisera.

An intramuscular schedule has worked well for a wide range of cytokines, including interleukins (8), interferons, and colony-stimulating factors.

Protocol 8. Raising sheep and goat antisera

Equipment and reagents

- Sheep or goat
- Cytokine
- FCA
- FIA

- PBS
- Syringes and needles
- Blood-collecting tubes

Method

1. For the primary immunization emulsify about 500 μg of cytokine (in 0.5 ml PBS) in 1.5 ml FCA. Inject 0.5 ml of emulsion intramuscularly, deeply into the haunches and shoulders.

2. Boosts of about 200 μg cytokine emulsified in FIA are also by the intramuscular route, either into the haunches or shoulders at intervals after the primary immunization of 1 month, 1 month, 2 months, and 4–6 months.

3. Collect bleeds on a regular schedule throughout the immunization. The highest titres occur usually around 10 days after a boost. Depending on the size of the animal, two or three bleeds of 300–600 ml may be taken in the period between one and two weeks after a boost.

---

6. Screening

Antibodies may be evaluated in a variety of ways depending upon their intended use. Neutralizing activities usually are assessed in specific *in vitro* bioassays (see Chapter 21). Suitability for use in immunoassays or for delineation of external regions of molecules usually is assessed in solid-phase or liquid-phase binding assays (SPBA and LPBA). SPBAs, which are the basis of the IRMA and ELISA procedures described in Chapter 21 have been described in detail elsewhere (6, 9). Cytokines and related peptides are used at about 10 μg/ml. Incubation steps comprise 1 h at $+37°C$ and $\geqslant 1$ h at $+4°C$ using [^{125}I]labelled or biotinylated 'second' antibody, i.e. one raised against the immunoglobulin of the species in which the anti-cytokine antibodies were raised.

LPBAs are the basis of the RIAs described in Chapter 21. A procedure used to quantify the antigen-binding activity of polyclonal antiserum or ascitic fluid is described in *Protocol 9*.

Protocol 9. Liquid phase binding assay

Equipment and reagents

- LP3 test tubes
- PBS, pH 7.4
- BSA
- Antiserum
- [^{125}I]labelled cytokine
- Polyethylene glycol 6000 (PEG)

- Bovine gammaglobulin
- Tris–HCl powder
- Deionized distilled water
- Vortex mixer
- Centrifuge
- Gamma counter

Method

1. Using LP3 test-tubes, to 300 μl PBS, pH 7.4, containing 0.5% BSA add 100 μl of antiserum at 1/100–1/10 000 final dilution.

2. Add 100 μl (10 000 c.p.m.) [^{125}I]labelled cytokine and incubate the mixture at room temp. for 20 h (labelling methodologies are described in Chapter 21).

3. Separate bound from free [^{125}I]cytokine using a polyethyleneglycol (PEG)-assisted second antibody procedure. Dissolve 25 g PEG 6000, 150 mg bovine gamma-globulin and 0.788 g Tris–HCl in 100 ml deionized distilled water and add 500 μl of this solution to each tube, vortex the mixture and incubate for 1 h at room temp.

4. Centrifuge the tubes at 1000 *g* for 30 min and aspirate and discard the supernatant. Quantify the radioactivity in the precipitate in a gamma-counter.

---

7. Purification of antibodies

Depending upon their intended use, anticytokine sera and monoclonal antibodies in ascitic fluid may be used without purification, e.g. for RIA procedures. However, for many immunochemical procedures, e.g. radiolabelling or enzyme conjugation for use in immunoassays, it is usually necessary to carry out some type of chromatographic technique, e.g. ion-exchange, gel filtration, or affinity chromatography. Mouse and rat immunoglobulins are generally less stable than those of higher mammals and can prove more difficult to purify and successfully fragment. Detailed methods for purification of immunoglobulin and murine monoclonal antibodies are described elsewhere (10, 11).

Acknowledgements

The author is grateful to Dr A F Bristow (NIBSC) for his help with the preparation of *Protocols 1* and *2*.

References

1. Lachman, P. J., Strangeways, L., Vyakarnam, A., and Evans, G. (1986). In *Synthetic peptides as antigens*, p. 25. Ciba Foundation Symposium 119. Wiley, Chichester.
2. Freund, J. and McDermott, K. (1942). *Proceedings of the Society for Experimental Biology and Medicine*, **49**, 548.
3. Hurn, B. A. L. and Chantler, S. M. (1980). In *Methods in enzymology*, Vol. 70, *Immunochemical techniques* (ed. H. Van Vunakis and J. L. Langone), p. 104. Academic Press, New York and London.
4. Thorpe, R., Wadhwa, M., Gearing, A. J. H., Mahon, B., and Poole, S. (1988). *Lymphokine Research*, **7**, 119.
5. Bastin, J. M., Kirkley, J., and McMichael, A. M. (1982). In *Monoclonal antibodies in clinical medicine* (ed. A. J. McMichael and J. W. Fabre), p. 503 Academic Press, New York and London.
6. Limjuco, G., Galuska, S., Chin, J., Cameron, P., Boger, J., and Schmidt, J. (1987). *Proceedings of the National Academy of Sciences USA*, **83**, 3972.
7. Bomford, R., Abdulla, E., Hughes-Jenkins, C., Simkin, D., and Schmidt, J. (1987). *Immunology*, **62**, 543.
8. Poole, S., Bristow, A. F., Selkirk, S., and Rafferty, B. (1989). *Journal of Immunological Methods*, **116**, 259.
9. Johnstone, A. and Thorpe, R. (1987). In *Immunochemistry in practice*, p. 241. Blackwell Scientific, Oxford.
10. Baines, M. G. and Thorpe, R. (1990). In *Methods in molecular biology*, Vol. 10: *Immunochemical protocols* (ed. M. Manson), p. 79. Humana Press, Clifton, New Jersey.
11. Baines, M. G., Gearing, A. J. H., and Thorpe, R. (1990). In *Methods in molecular biology*, Vol. 5, Animal cell culture (ed. J. W. Pollard and J. W. Walker), p. 647. Humana Press, Clifton, New Jersey.

20

Visualizing the production of cytokines and their receptors by single human cells

C. E. LEWIS and A. CAMPBELL

1. Introduction

In recent years, fundamental problems have emerged in the analysis of cytokine production. Aspects of cytokine biology such as their overlapping biological effects, association with carrier proteins and soluble receptors, competition with inhibitors, rapid induction, transient storage, and ability to bind to surface sites on producer cells, have meant that considerable care must be taken in the selection of appropriate and meaningful techniques for their accurate detection.

Cytokines are not produced in isolation but usually together with enzymes, hormones, inhibitors and other cytokines/cytokine receptors. This has lead to problems of specificity in the use of bioassays to quantify their release, and subsequently to the development of other strategies for the study of cytokine production. One such approach has been to employ various immunostaining techniques in order both to highlight intracellular production of cytokines and to identify the immunophenotype of producer cells. However, definitive studies correlating the synthesis and intracellular storage of cytokines with their subsequent release have yet to be performed. With this in mind, attention has focused mainly on the measurement of the secreted product by radio- and enzyme-linked immunoassays which quantify the accumulated level of cytokine released into culture supernatant by entire cell populations (see Chapter 21).

More recently, this technology has been extended to visualize cytokine release by individual cells. This has yielded important information about the frequency of cytokine-producing cells in a population at any one time. In some instances, these sensitive immunoassays have also been adapted to: (i) measure the amount of cytokine released per cell; and/or (ii) identify producer cells. A further advantage of their application has been to demonstrate marked differences in cytokine secretion by cells of the same phenotype.

C. E. Lewis and A. Campbell

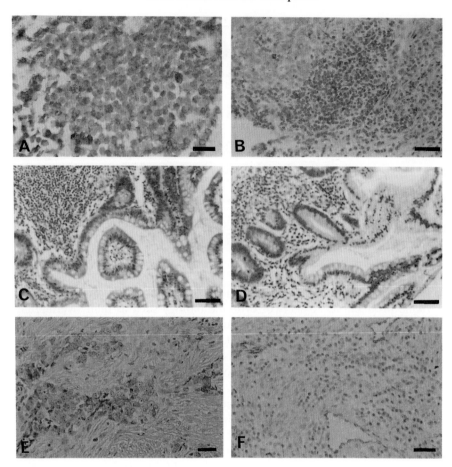

Figure 1. Photomicrographs to illustrate three immunohistochemical procedures for labelling cytokines/cytokine receptors in human tissue sections. The APPAP method was used to label TNF-α receptors, p55 (A) and p75 (B) in cryostat sections of breast carcinomas (immunoreactive cells are stained red)*. The streptavidin–biotin-peroxidase method was used to label TGF-β in colonic crypt epithelial cells in paraffin-embedded inflamed bowel tissue (C) (immunoreactive cells are stained brown). TGF-β staining was completely abolished when the antibody was pre-absorbed overnight with excess TGF-β (D). The streptavidin–biotin-alkaline phosphatase method was used to label VEGF$_{165}$ in paraffin-embedded breast carcinomas (following protease digestion). Red immunoreactivity can be seen in tumour cell islands (E) and on endothelial cells lining blood vessels (F). Nuclei were counterstained blue with haematoxylin. Magnification bars = 50 μm. * Reproduced from Pusztai *et al.* (3) with the kind permission of the *British Journal of Cancer*.

This review outlines various methods currently available for immunolabelling cytokines (and their receptors) in human tissue sections. In addition, two sensitive assays for the measurement of their release at the single-cell level are described; the reverse haemolytic plaque assay and a modification of the ELISA assay, now commonly known as Elispots.

2. Immunolocalization of cytokines

We recently outlined a variety of methods for the immunolocalization of cytokines in cytospins of human cells (1). In the present chapter, we describe two widely used methods for labelling cytokines and their receptors in frozen and paraffin-embedded human tissue sections.

2.1 Alkaline phosphatase anti-alkaline phosphatase (APAAP) method

This technique involves the immunoenzymatic labelling of primary monoclonal or polyclonal antibodies using an immune complex of alkaline phosphatase and monoclonal anti-alkaline phosphatase (2) (see *Figure 1a* and *b*). This technique has been used to immunolocalize TNF-α and its receptors, p55 and p75 (3), PDECGF (4), and EGF receptors (5).

The protocol outlined below for the APAAP method is for immunolocalization in cryostat sections. For routinely processed, paraffin-embedded tissue sections, steps 1–3 should be replaced by steps 1 and 2 of *Protocol 2*.

Protocol 1. APAAP method

Equipment and reagents

- Liquid nitrogen
- Cryotome
- Silane- or polylysine-coated slides
- Acetone
- Tris-buffered saline: 0.05 M Tris, pH 7.6, 0.15 M NaCl
- Human Fc fragments, 20 μg/ml in TBS
- 10% normal human AB serum (NHS), filtered and heat-inactivated for 1 h at 56°C
- Antibody raised against the purified, native, or recombinant cytokine
- See step 10: 1/50 mouse anti-rabbit IgG in TBS/10% NHS *and/or* 1/50 rabbit anti-mouse IgG in TBS/10% NHS

- 1/50 APAAP complex in TBS
- Fast Red substrate solution
- Fast Red TR salt
- Naphthol AS-MX phosphate
- *N,N*-dimethyl formamide
- Glass tubes
- 0.1 M Tris buffer, pH 8.2
- 1 M levamisole
- Light microscope
- Aqueous-based mounting medium (e.g. Kaiser's mounting medium)
- Double-distilled water
- Haematoxylin

Methods

1. Snap freeze tissue using liquid nitrogen.
2. Cut 8 μm sections and mount on silane or polylysine coated slides.
3. Fix sections in acetone at −20°C for 10 min.

C. E. Lewis and A. Campbell

Protocol 1. *Continued*

4. Directly immerse slides in Tris-buffered saline (TBS; 0.05 M Tris, pH 7.6, 0.15 M NaCl) for 5–10 min.

5. Incubate slides in 20 μg/ml human Fc fragments (if these are not available go straight to step 7) in TBS for 30 min to block endogenous Fc receptors.

6. Rinse slides twice in TBS for 2 min each.

7. Incubate in 10% normal human AB serum in TBS for 45 min.

8. Tap off serum and incubate slides for 1–4 h at room temp., and then 16–24 h at 4°C in the optimal concentration of antibody raised against the purified, native, or recombinant cytokine. (This is usually in the range, 1/50 to 1/5000, although this will vary with different antibodies.
Check this with preliminary titrations for each antibody.)

9. Rinse slides twice in TBS for 2 min each. (Wash between all remaining steps in this way.)

10. (a) If primary antibody was raised in rabbit, apply 1/50 mouse anti-rabbit IgG in TBS/10% NHS for 60 min, followed by 1/50 rabbit anti-mouse in TBS/10% NHS for 30 min.
    (b) If primary antibody was raised in mouse, apply 1/50 rabbit anti-mouse alone in TBS/10% NHS for 30 min.

11. Incubate slides in 1/50 APAAP complex in TBS for 30 min.

12. Repeat steps 8 and 9, but incubating for a maximum of 10 min each.

13. Incubate slides with Fast Red substrate solution for 10–30 min at room temp., after which immunoreactive cells will be red in colour. This is done as follows:
    (a) Dissolve 2 mg naphthol AS-MX phosphate in 0.2 ml *N,N*-dimethyl formamide in a glass tube.
    (b) Add 9.8 ml 0.1 M Tris-buffer (pH 8.2).
    (c) Add 0.01 ml 1 M levamisole to block endogenous alkaline phosphatase. (NB. If a given tissue expresses high levels of endogenous phosphatase this step may not be fully effective. In this case the method outlined in *Protocol 2* may be more suitable.)
    (d) Immediately before staining, dissolve 10 mg Fast Red TR salt in the above solution and filter.
    (e) Incubate for 5–60 min (monitor colour development using a light microscope).
    (f) Rinse with double-distilled water.

14. Counterstain nuclei (blue) with haematoxylin for 10–30 sec and rinse in tap water for 1 min. Mount preparations in an aqueous-based medium (such as Kaiser's mounting medium).

2.2 Streptavidin–biotin (AB) method

This technique employs a 'high sensitivity' detection system based on the high affinity interaction between streptavidin and biotin. This has now become widely incorporated into various forms of immunocytochemistry, ELISA, immunoblotting and *in situ* hybridization. Briefly, it involves the use of a biotinylated second antibody to detect the primary antibody (which can be monoclonal or polyclonal) bound to specific sites on the tissue section, and a streptavidin–peroxidase complex. Alternatively, a complex of streptavidin–alkaline phosphatase can be used as the third detection step.

The AB method has been used to detect numerous cytokines and their receptors in tissue sections including VEGF (4), TGF's-α and -β (6), IL-8 and MCP-1 (7), EGF (8), and bFGF (9).

The following two protocols outline the peroxidase and alkaline phosphatase versions of the AB method, as used for the immuno localization of TGF-β (*Figure 1c* and *d*) and $VEGF_{165}$ (*Figure 1e* and *f*) respectively in routinely processed, paraffin-embedded tissue sections. When using cryostat tissue sections, replace steps 1 and 2 below with steps 1–3 of *Protocol 1* and reduce the incubation time for the primary antibody to 60–120 min.

Protocol 2. AB-peroxidase method

Equipment and reagents

- Paraffin-embedded tissue sections
- Microtome
- Glass slides
- De-wax with xylene
- Graded alcohol series
- 3% hydrogen peroxidase in methanol
- 0.03% hydrogen peroxide
- TBS buffer
- 10% heat-inactivated normal human AB serum
- polyclonal anti-human cytokine (e.g. chicken–anti-TGF-β; R & D Systems)
- 37°C humidity chamber
- TBT: TBS containing 3% BSA and 0.05% Triton X-100
- Biotinylated swine or goat anti-rabbit, or rabbit anti-chicken or goat IgG
- Peroxidase-conjugated streptavidin
- Dried skimmed milk (e.g. Marvel)
- Diaminobenzodine (0.5 mg/ml TBS)
- Haematoxylin
- DPX mountant

Method

1. Cut 5 μm sections, mount on glass slides and dewax.

2. Rehydrate sections in a series of graded alcohols.

3. Immerse the slides in a solution of 3% hydrogen peroxidase in methanol for 10 min (this blocks endogenous peroxidase activity in the tissue sections).

4. Wash slides in TBS for 10 min.

5. Incubate sections in 10% heat-inactivated normal human AB serum for 30 min (this blocks non-specific protein binding sites).

343

Protocol 2. *Continued*

6. Incubate sections in polyclonal (rabbit, chicken or goat usually) anti-human cytokine (e.g. chicken anti-TGF- β; R and D Systems), diluted in TBS to 1/20 to 1/1000 (i.e. the optimal 'working dilution' varies between diferent antibodies and for different tissues). This can be done overnight at 4°C or for 1–3 h at room temp. or 37°C (in a humidity chamber).

7. Wash slides twice in TBS (5 min per wash) and once in TBT for 10 min. (TBT is TBS containing 3% bovine serum albumin and 0.05% Triton X-100.)

8. Incubate sections with biotinylated swine or goat anti-rabbit, or rabbit anti-chicken or goat IgG diluted to 1/200 in TBT for 30 min.

9. Repeat step 7.

10. Incubate sections in peroxidase conjugated streptavidin diluted to 1/50 in TBS plus 0.5% (w/v) dried skimmed milk (e.g. 'Marvel') for 30 min.

11. Wash sections three times in TBS.

12. Incubate sections with diaminobenzidine (0.5 mg/ml TBS) in the presence of 0.03% hydrogen peroxide for 10 min. NB. The end product is visible as a brown colour reaction (see *Figure 1c*).

13. Wash slides in running tap water and lightly counterstain the sections with haematoxylin. Dehydrate in alcohol, clear in xylene and mount in DPX.

Protocol 3. AB-alkaline phosphatase method

Equipment and reagents

- Paraffin-embedded tissue sections
- Microtome
- Glass slides
- Xylene
- Graded series of alcohol
- TBS buffer
- 10% milk proteins (e.g. Marvel)/10% normal rabbit serum in TBS

- Polyclonal goat anti-human cytokine (e.g. goat anti-human VEGF$_{165}$; R & D Systems)
- 37°C humidity chamber
- Biotinylated rabbit anti-goat IgG
- Alkaline phosphatase-conjugated streptavidin
- Fast Red substrate solution (see *Protocol 1*)

Method

1. Cut 5 μm sections, mount on glass slides and dewax.

2. Rehydrate sections in a series of graded alcohols.

3. Incubate sections in 10% milk proteins (e.g. 'Marvel')/10% normal rabbit serum in TBS for 60 min (this blocks non-specific protein binding sites).

4. Incubate sections in polyclonal goat anti-human cytokine (e.g. goat anti-human VEGF$_{165}$; R & D Systems), diluted in TBS to 1/20 to 1/500 (i.e. the optimal 'working dilution' varies between diferent antibodies and for different tissues). Affinity-purified antibodies should always be used where possible as this reduces background staining. This incubation can be done overnight at 4°C or for 1–3 h at room temp. or 37°C (in a humidity chamber).

5. Wash slides twice in TBS (5 min per wash).

6. Incubate sections with biotinylated rabbit anti-goat IgG diluted to 1/600 in TBS for 30 min.

7. Repeat step 5.

8. Incubate sections in alkaline phosphatase-conjugated streptavidin diluted to 1/200 in TBS for 30 min.

9. Repeat step 5.

10. Incubate sections in Fast Red substrate solution for 10–30 min at room temp., as described in step 13 of *Protocol 1*. Immunoreactive cells are red in colour.

General points

Care must be taken to control for non-specific reactions in the above immunostaining procedures. Controls must be included to ascertain whether the primary and secondary antibodies are specific for their target antigens. This can be done by replacing the cytokine antibody with either diluent alone or preimmune serum (or the IgG fraction of pre-immune serum, whichever is more appropriate) of the same species as the donor of the antibody. Alternatively, an inappropriate antibody (i.e. of irrelevant specificity) can be used. Care should be taken to substitute control preparations at the same protein concentrations as that of the antibody applied. It is also recommended that antibodies and their control solutions be of comparable age, and thus contain similar quantities of protein aggregates, since the presence of such contaminants may contribute to non-specific staining. To ensure the removal of protein aggregates, all antibody and control preparations should be spun down at 14 000 r.p.m. for 10 min at their final working concentrations, and the supernatants used in the staining procedure. If possible, abolition of staining by pre-absorption of the primary antibody with either a highly purified preparation of the native form of the cytokine, or its recombinant analogue (50–1000 µg/ml for 24 h at 4°C) prior to use in these procedures is the ideal way of checking the specificity of the staining achieved (see *Figure 1d*). In addition, the specificity of secondary and/or tertiary antibodies in the above techniques can be checked by their omission and substitution by the appropriate buffer.

Some cytokines, such as FGF, VEGF, TGF-β, are known to bind to components of the extracellular matrix. This may lead to non-cellular components in tissue sections being labelled in the immunohistochemistry protocols outlined above. This pattern of staining should not be confused with the more general, non-specific, background staining obtained with some antibodies in these procedures (particularly if the early protein blocking step is missing or inadequate). The former but not the latter staining pattern should be abolished when the antibody is pre-absorbed with the appropriate antigen.

In some instances, a process called 'antigen retrieval' may prove necessary to unmask hitherto concealed antigens (or simply to enhance their immuno-reactivity) in paraffin-embedded tissue sections (10)). This involves micro-waving or protease-digesting tissue sections prior to the first immunodetection step in *Protocols 1* and *2* (usually a protein blocking agent followed by exposure to the primary antibody).

Protocol 4. Antigen retrieval for paraffin-embedded tissue sections

Equipment and reagents

- Silane-coated slides
- Xylene
- Graded series of alcohols
- Glass slide rack and glass trough
- 0.01 M sodium citrate in double-distilled water (DDW)

- Microwave oven
- TBS buffer
- PBS
- Protease type 24
- DDW

A. *Microwaving tissue sections*

1. Mount wax sections on silane coated slides.

2. De-wax and rehydrate sections.

3. Place slides in a glass slide rack in a glass trough containing 0.01 M sodium citrate in doubled-distilled water (DDW). Boil in the microwave for 4 min on full power.

4. Top up with fresh sodium citrate and boil again for 4 min in the microwave.

5. Let slides cool down to room temp., wash in TBS and procede with step 4 onwards of *Protocols 1* or *2*.

B. *Proteinase digestion of tissue sections*

1. De-wax and rehydrate sections as usual.

2. Incubate slides in a trough of DDW at 37°C for 10 min.

3. Incubate slides in 200 ml of PBS containing 25 mg protease type 24 at 37°C for 20 min.

---

4. Incubate slides in DDW at room temp. for 30 min.

5. Procede with *Protocols 1* or *2*.

3. Detection of cytokine release by individual cells

3.1 Reverse haemolytic plaque assay (RHPA)

This is a variant of the plaque assay established by Jerne *et al.* (11) to detect and enumerate immunoglobulin-secreting B cells, which has been adapted to detect antigen secretion (12). Recent studies have indicated the lower level of sensitivity of this immunoassay to be 10^{-18} M of secreted product. Secretory cells in a purified or heterogenous cell population form plaques (zones of haemolysis around secretory cells) when incubated in a monolayer with protein A-coated ovine erythrocytes in the presence of a specific antiserum and complement. Since the size of plaques is directly proportional to the amount of product secreted per cell (13, 14), this technique can be used to measure the amount secreted by single cells, as well as to provide an estimate of the frequency of cytokine-secreting cells in a given population. Producer cells can then be identified by routine immunocytochemistry of cells in the monolayer (1, 15). In cytokine biology, the RHPA has recently been adapted to enumerate murine T cells secreting IFN-γ (16), and to quantify the release of various cytokines by human blood and tumour-infiltrating leukocytes (17, 18). Moreover, we have recently used this technique to visualize the release of soluble TNF-α receptors by macrophages and malignant cells from breast carcinomas (Waterworth, Leek and Lewis, unpublished observations).

Protocol 5. Reverse haemolytic plaque assay

Equipment and reagents

- 8 ml sheep erythrocytes (in Alsevers solution)
- 0.9% NaCl
- Ficoll-Hypaque
- Centrifuge and tubes
- Glass Pasteur pipettes
- Protein A (PrA from *Staphylococcus aureus* diluted to 0.5 mg/ml in 0.9% NaCl)
- 0.2 mg/ml chromium chloride hexadrate
- Aluminium foil
- Dulbecco's modified essential medium (DMEM) supplemented with 0.1% BSA and penicillin/streptomycin
- Acid alcohol
- DDW
- 1% DMSO
- 0.05 mg/ml poly-L-lysine in DDW
- Cunningham chambers
- Double-sided sticky tape
- 22 mm^2 coverslips
- PrA-sRBC from *A*
- Test cells (final density of 1–10 × 10^6 cells/ml)
- Humidity chambers
- 37°C, 5% CO$_2$ incubator
- Absorbent paper
- Light microscope
- Polyclonal anti-cytokine
- RPMI 1640 medium supplemented with 0.1% BSA/antibiotics
- 1/50 Guinea-pig complement
- Trypan blue

Protocol 5. *Continued*

- Fixative (e.g. 2% glutaraldehyde in phosphate buffer or cold 50% methanol/50% acetone)
- TBS buffer
- Monoclonal antibody (e.g. anti-CD3 for human T cells) diluted in TBS/0.05% dried milk solids
- 1/50 rabbit anti-mouse IgG
- 1/50 APAAP in TBS
- Fast Red substrate solution (*Protocol 1*)
- Haematoxylin
- Aqueous-based mountant medium

A. *Conjugation of protein A to red blood cells*

1. Dilute two 4 ml aliquots of sheep erythrocytes (sRBC; supplied in Alsevers solution. Do not store for longer than 2 weeks at 4°C before use) 1:1 with a solution of 0.9% NaCl. Any sheep leukocytes present in this preparation are removed by layering each 8 ml of diluted sheep blood over 4 ml aliquots of Ficoll-Hypaque in a centrifuge tube. Spin these tubes at 444 *g* for 25 min.

2. Discard the supernatant from each gradient and harvest the sRBC pellet at the base of each tube. Wash each pellet several times by repeated suspension with a glass pipette in 0.9% NaCl solution and centrifugation at 444 *g* for 5–10 min.

3. Dilute each 1 ml pellet of sRBC in 5 ml of 0.9% NaCl and add 1 ml of 0.5 mg/ml protein A and 5 ml of 0.2 mg/ml chromium chloride hexahydrate ($CrCl_3$). This solution of $CrCl_3$ must be made up and stored at 4°C for at least 1 week before use in the conjugation procedure. It can be stored at 4°C, wrapped in foil, and used repeatedly for up to 6 months. Gently resuspend the sRBC pellet in this solution using a glass pipette and incubate tubes at 30°C for 1 h.

4. Harvest each 1 ml of PrA-conjugated sRBC (PrA-sRBC) by centrifugation at 300 *g* for 6 min and wash several times by repeated suspension and centrifugation steps in either 0.9% NaCl (first wash) or DMEM supplemented with 0.1% BSA and penicillin-streptomycin (for subsequent washes). If considerable lysis of the cells is visible after the first centrifugation, or if the pellet does not resuspend readily in 0.9% NaCl, discard the preparation and repeat procedure using a different batch of sRBC.

5. Resuspend and store PrA-sRBC in 100 ml of DMEM [i.e. as a 2% solution (v/v)] supplemented with 0.1% BSA and antibiotics, for a maximum of 3 weeks at 4°C. The best monolayers in the RHPA are achieved using cells conjugated 1–5 days before use in the assay.

B. *Reverse haemolytic plaque assay (see Figures 2 and 3)*

1. Clean glass slides by immersion in acid alcohol for 5 min, followed by several rinses in double-distilled water, and then immersion in 1% DMSO for 5 min. Wash slides several times in double-distilled water and then leave to air-dry in a clean environment.

2. Coat slides with poly-L-lysine (0.05 mg/ml in double-distilled water) for 10 min. Rinse several times in double-distilled water and air-dry at room temp. or with a warm air current. (Do not heat the coated slides above 40°C during drying.)

3. Construct Cunningham chambers in the following way. Attach two pieces of double-sided sticky tape, separated by a distance of 15–20 mm, to each dry, polylysine-coated slide. Then lower a 22 mm^2 coverslip on to the slide, so that it forms the roof of the chamber, by attaching to the edge of each strip of tape. The position of the coverslip is secured by pressing gently down on it so that it adheres to the tape on both sides of the chamber.

4. For a 10-slide assay, spin down 10 ml of 2% (v/v) suspension of PrA-1 sRBC at 300 g for 5–10 min, and decant and discard all but 0.5 ml of the supernatant. Resuspend pellet of PrA-sRBC in this supernatant, to yield a 40% solution (v/v).

5. Add test cells (at a final working cell density of 1–10 × 10^{-6} cells/ml) to PrA-sRBC at a 1:1 ratio (v/v). For example, 0.5 ml of test cells are added to 0.5 ml of 40% PrA-sRBC [thereby yielding a final, working dilution of 20% PrA-sRBC (v/v)]. Mix gently but thoroughly and apply 100 μl of this cell suspension to the entrance of each Cunningham chamber. This will fill the chamber by capillary action, leaving a small amount at the entrance. Slides are then placed into humidity chambers and incubated at 37°C in 5% CO_2/air for 35–45 min. During this period, the PrA-sRBC and human cells settle down on to the slides to create a monolayer. (Do not leave the slides for longer than 45 min at this stage as multilayers will form which impede plaque formation.)

6. Remove excess unattached cells from the chambers at the end of this period by four rapid washes with warm (37°C) incubation medium (RPMI 1640, supplemented with 0.1% BSA and antibiotics). This is achieved by placing 50 μl of medium at one entrance to the chamber and drawing it through with absorbent paper applied gently to the other side. Using a light microscope, inspect the appearance of the cells remaining in the chamber to check their confluency.

7. Infuse polyclonal anti-cytokine at 1/50 to 1/100 in RPMI 1640/0.1% BSA/antibiotics into the chambers as 3 × 30 μl washes. After the final infusion of antibody, leave a small amount at one entrance to the chamber to avoid excessive drying out of the cells during the subsequent incubation period (i.e. a final volume of 90–100 μl of antibody is needed per slide). Incubate the slides in clean humidity chambers at 37°C in 5% CO_2/air for up to 12 h.

8. At the end of this period, infuse a solution of 1/50 guinea-pig complement in warm (37°C) incubation medium into the chambers as 3 ×

Protocol 5. *Continued*

30 µl washes and incubate in 5% CO_2/air for 20–30 min. Complement-mediated erythrocyte lysis will occur during this period around cytokine-secreting cells. Cell viability can be checked at this stage (i.e. prior to step 9), using the Trypan blue exclusion test. (NB It is advisable to do this step in the absence of such serum proteins as BSA.)

9. Infuse fixative into the chambers (e.g. 2% glutaraldehyde in phosphate buffer or cold 50% methanol/50% acetone. These fixatives do not cause lysis of the red cells in the monolayer), followed by TBS. The slides can then be stored at 4°C until measurement of individual plaque size and/or number (this can be done using a simple image analysis device).

C. *Immunophenotyping of cells in the RHPA*

1. After infusion of fixative into chambers, immerse slides in TBS and remove coverslips forming the roof of each chamber, and the double-sided sticky tape.

2. Rinse monolayers with TBS for 5 min.

3. Incubate slides in the appropriate monoclonal antibody for a given human or murine cell-marker (e.g. anti-CD3 for human T cells), diluted 1/10 to 1/200 in TBS/0.05% dried milk solids for 2–4 h at room temp.

4. Wash slides three times in TBS (do not wash too vigorously as this can dislodge cells).

5. Incubate in 1/50 rabbit anti-mouse IgG in TBS for 30 min at room temp.

6. Repeat step 4.

7. Incubate slides in 1/50 APAAP in TBS for 30 min at room temp.

8. Repeat step 4.

9. Repeat steps 5 and 7, but for only 10 min each.

10. Add Fast Red substrate (see step 13 of *Protocol 1*) for 5–60 min at room temp., after which immunoreactive cells will appear red in colour. The nuclei of stained and unstained cells are then lightly counterstained with haematoxylin and the monolayers mounted in an aqueous-based medium.

Control experiments to assess the dependency of plaque formation on the secretion of a specific cytokine by cells in the RHPA should include:
(i) substitution of the polyclonal anti-cytokine with an equivalent concentration of either an inappropriate antibody, or whole (or if more appropriate, the purified IgG fraction) non-immune serum from the same species as the donor animal;
(ii) pre-absorption of the polyclonal antibody with the appropriate recombinant or native, purified form of cytokine for 24 h at 4°C.

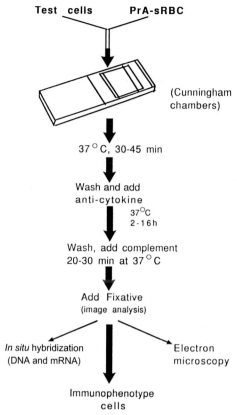

Figure 2. Protocol for the reverse haemolytic plaque assay.

3.2 Elispot assay

This technique is based on the ELISA-spot assay for the detection of individual antibody-secreting B cells (18). In the reverse form of this assay outlined below, single cells are incubated on a solid surface pre-coated with a specific cytokine antibody, which then binds cytokine secreted during the assay. The cells are washed away and the bound cytokine detected by routine immunocytochemical methods using a second monoclonal antibody which recognizes a different epitope on the cytokine molecule (see *Figure 4*).

A number of studies have described the application of the Elispot assay to detect and enumerate:

- human lymphocytes secreting IFN-γ (19)
- human T and monocytic cell-lines producing IFN-γ and TNF-α (20)
- human cells from kidney allografts secreting IFN-γ, IL-6, and IL-10 (21)
- murine T cells secreting IFN-γ and TNF-α (22)
- murine splenic T cells secreting IL-5 and IFN-γ (23).

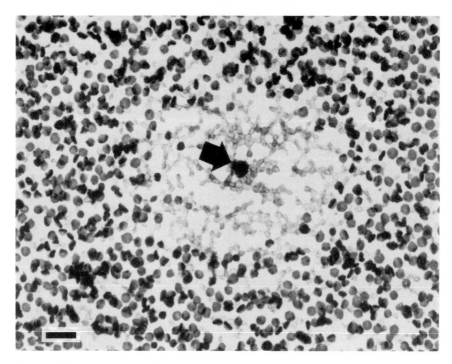

Figure 3. Photomicrograph of a human monocyte (arrowhead) at the centre of an IL-1 plaque (area of sRBC lysis) in the reverse haemolytic plaque assay. Magnification bar = 20 μm.

Protocol 6. Elispot assay

Equipment and reagents

- 24-well tissue culture plates
- Specific cytokine antibody
- Polyclonal rabbit or monoclonal mouse anti-cytokine
- Alkaline phosphatase-conjugated goat anti-rabbit IgG
- PBS
- PBS containing 1% BSA/0.5% Tween, *or* 5% FCS, *or* 5% dried milk
- PBS/0.5% Tween

- Test cells
- RPMI 1640 supplemented with 5 M L-glutamine, 5×10^{-5} M 2-mercaptoethanol, antibiotics, and 10% FCS
- 37°C, 5% CO_2, humidified incubator
- Enzyme substrate: 1 mg/ml S-bromo-4-chloro-3-indolyl phosphate in 2-amino-2-methyl-1-propanol buffer in 6% agarose
- Dissecting microscope

Method

1. Coat 24 well tissue-culture plates with 500 μl of the optimal dilution (1/200 to 1/5000) of monoclonal or polyclonal anti-human cytokine or hamster anti-mouse cytokine in PBS at 4°C for 16–24 h. 96-well tissue

culture plates can also be used with the appropriate reduction in coating solution (50–100 μl).

2. Wash plates twice with PBS for 5 min each.

3. Block non-specific protein-binding sites with PBS containing either 1% BSA/0.5% Tween, or 5% fetal calf serum (FCS), or 5% dried milk for 1 h at room temp. All subsequent antibodies are diluted in one of these protein solutions.

4. Aliquot test cells at a density of 10^4 cells/100 μl/well in the supplemented RPMI 1640 medium. Incubate for 2–24 h at 37°C in a humid atmosphere of 5% CO_2 in air. The optimal length of this incubation period for good spot formation will vary for different cell types and their level of cytokine-secreting activity.

5. Remove cells from plates by repeated washing with cold PBS/0.5% Tween.

6. Add 500 μl of the optimal concentration of the appropriate polyclonal rabbit or monoclonal mouse anti-cytokine (likely to be in the range of 1/100 to 1/5000) and incubate for 1–2 h at 37°C. (NB Sterile conditions must be employed up to and including this step.)

7. Wash plates three times with PBS to remove unbound antibody.

8. Add 500 μl of 1/1000 to 1/5000 alkaline phosphatase-conjugated goat anti-rabbit IgG to wells for 1–2 h at 37°C.

9. Wash plates twice with PBS for 5 min each.

10. Add enzyme substrate (15). After 5–10 min this mixture will solidify (do not move the plates during this time). After 3–5 min at 37°C, blue spots will appear. The number of cytokine secreting cells can then be estimated, using a dissecting microscope with the plates viewed against a white background.

Controls for the Elispot assay are essentially those described for the immunohistochemical methods in *Protocols 1, 2* and *3*.

4. Discussion

Over the past few years increasing use has been made of immunohistochemical methods to investigate the *in situ* production of cytokines and their receptors in human tissues. The two protocols for this outlined here are the most commonly used in this context. It should be noted, however, that such an approach is not without drawbacks. The absence of immunostaining may be due to the absence of cytokine/cytokine expression, but also the sub-threshold expression of cytokine/cytokine receptor and/or the masking of the epitope (for example, if the cytokine/cytokine receptor is bound to other

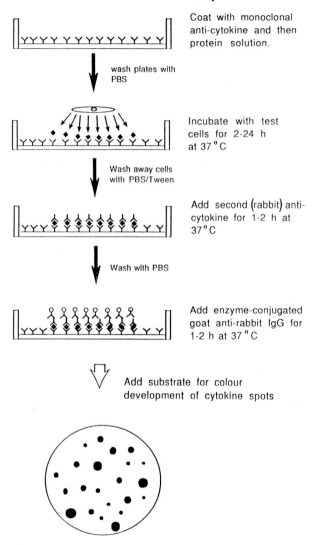

Coat with monoclonal anti-cytokine and then protein solution.

wash plates with PBS

Incubate with test cells for 2-24 h at 37 °C

Wash away cells with PBS/Tween

Add second (rabbit) anti-cytokine for 1-2 h at 37°C

Wash with PBS

Add enzyme-conjugated goat anti-rabbit IgG for 1-2 h at 37 °C

Add substrate for colour development of cytokine spots

Figure 4. Protocol for the Elispot assay.

proteins within the cell). Moreover, positive staining may indicate the uptake rather than the synthesis of cytokine/cytokine receptors by cells. For these reasons, the above methods are now usually combined with *in situ* hybridization or *in situ* PCR to demonstrate and correlate the production of both the mRNA and the corresponding protein for cytokines/cytokine receptors in tissue sections.

Although the two single-cell cytokine release assays outlined in this chapter have the advantage of increased sensitivity as compared to conventional bulk release assays (e.g. ELISA), a number of salient points should be noted concerning their general applications and interpretation. First, by definition, producer cells in such single-cell techniques secrete their products in isolation from their normal intercellular contacts and communication. Thus, the possibility exists that these assays may yield artefactual results when the cells used are from tissue dispersants rather than body fluids (where it could be argued that most cells normally exist as single, unattached cells). Second, there is considerable debate at present concerning the use of such techniques which detect the antigenic, but not necessarily biologically active, form of cytokine. However, bioassays are equally limited in as much as they are not always specific for a given cytokine, and when two or more cytokines are present synergistic or antagonistic interactions can also produce confusing data.

In summary, each of the techniques outlined in this chapter has both its advantages and its limitations. With this in mind, it may be advisable to use several different forms of cytokine detection in tandem, rather than consider them as alternatives.

Acknowledgements

The authors would like to thank Mr R. D. Leek, and Professors A. Harris and J. O'D. McGee for their helpful advice in the preparation of this manuscript. They would also like to express their gratitude to R&D Systems Europe, Ltd. for the generous gift of recombinant human TGF-β and VEGF as well as polyclonal antibodies to these products.

References

1. Lewis, C. E. (1991). Cytokine production by individual cells. In *Cytokines: a practical approach* (ed. F. R. Balkwill). Oxford University Press.
2. Cordell, J. L., Fanili, B., Erber, W. N., Ghosh, A. K., Abdulaziz, Z., Macdonald, S. *et al.*, (1984). *Journal of Histochemistry and Cytochemistry*, **32**, 219.
3. Pusztai, L., Clover, L. M., Cooper, K., Starkey, P. M., Lewis, C. E., and McGee, J. O'D. (1994). *British Journal of Cancer*, **70**, 289.
4. Leek, R. D., Harris, A. L., and Lewis, C. E. (1994). *Journal of Leukocyte Biology*, **56**, 423.
5. Gullick, W. J., Hughes, C. M., Mellon, K., Neal, D. E., and Lemoine, N. R. (1991). *Journal of Pathology*, **164**, 285.
6. Mizukami, Y., Nonomura, A., Yamada, T., Kurumaya, H., Hayashi, M., Koyasaki, N., Taniya, T., Noguchi, M., Nakamura, S., and Matsubara, F. (1990). *Anticancer Research*, **10**, 1115.
7. Koch, A. E., Kunkel, S. L., Pearce, W. H., Shah, M. R., Parikh, D., Evanoff, H. L., Haines, G. K., Burdick, M. D., and Strieter, R. M. (1993). *American Journal of Pathology*, **142**, 1423.
8. Mizukami, Y., Nonomura, A., Noguchi, M., Taniya, T., Koyasaki, N., Saito, Y., Hashimotot, T., Matsubara, F., and Yanaihara, N. (1991). *Anticancer Research*, **11**, 1485.

9. Cordon-Cardo, C., Vlodavsky, I., Haimovitz-Friedman, A., Hicklin, D., and Fuks, Z. (1990). *Laboratory Investigation*, **63**, 832.

10. Cattoretti, G., Pileri, S., Parravicini, C., Becker, M. H., Poggi, S., Bifulco, C., Key, G., D'amato, L., Sabattini, E., and Fuedale, E. (1993). *Journal of Pathology*, **171**, 83.

11. Jerne, N. K., Henry, A. A., Nordin, H., Fuji, L., Koros, A. M. C., and Leftkovits, S. (1974). *Transplant Reviews*, **18**, 130.

12. Molinaro, G. A. and Dray, S. (1974). *Nature*, **248**, 515.

13. Neill, J. D., Smith, P. F., Luque, E. H., Munoz de Toro, M., Nagy, G., and Mulchahey, J. J. (1987). *Recent Progress in Hormone Research*, **43**, 175.

14. Allaerts, W., Wouters, A., Van der Massen, D., Persons, A., and Denef, C. (1988). *Journal of Theoretical Biology*, **131**, 441.

15. Lewis, C. E., Ramshaw, A. L., Lorenzen, J., and McGee, J. O'D. (1991). *Cellular Immunology*, **132**, 158.

16. Palacios, R., Martinez-Mason, O., and De Ley, M. (1983). *European Journal of Immunology*, **13**, 221.

17. Lewis, C. E., Horak, E., Lorenzen, J., McCarthy, S. P., and McGee, J. O'D. (1989). *European Journal of Immunology*, **19**, 2037.

18. Lewis, C. E., McCarthy, S. P., Richards, P. S. M., Lorenzen, J., Horak, E., and McGee, J. O'D. (1991). *Journal of Immunological Methods*, **127**, 51.

19. Czerkinsky, C., Andersson, G., Ekre, H.-P., Nilson, L.-A., Kloreskog, L., and Ouchterlony, O. (1988). *Journal of Immunological Methods*, **111**, 29.

20. Hutchings, P. R., Cambridge, G., Tite, J. P., Meager, T., and Cooke, A. (1989). *Journal of Immunological Methods*, **120**, 1.

21. Merville, P., Pouteil-Noble, C., Wijdeness, J., Potaux, L., Touraine, J. L., and Bancherau, J. (1993). *Transplantation*, **55**, 639.

22. Skidmore, B. J., Stamnes, S. A., Townsend, K., Glasebrook, A. L., Sheehan, K. C. F., Schreiber, R. D., and Chiller, J. M. (1989). *European Journal of Immunology*, **19**, 1591.

23. Taguchi, T., McGhee, J. R., Coffman, R. L., Beagley, K. W., Eldridge, J. H., Takatsu, K., and Kiyono, H. (1990). *Journal of Immunological Methods*, **128**, 65.

Quantitative biological assays for individual cytokines

MEENU WADHWA, CHRISTOPHER BIRD, LISA PAGE,
ANTHONY MIRE-SLUIS, and ROBIN THORPE

1. Introduction

Cytokines have often been discovered in complex mixtures by observing their effect on a biological system. This usually becomes the definitive property (or one of several properties) of the cytokine and can become the basis for its bioassay.

Cytokine bioassays are based on various biological effects such as cell proliferation, cytotoxic/cytostatic activity, kinetic effects, antiviral activity, colony formation, or induction of secretion of other cytokine/non cytokine molecules (1).

Bioassays can use primary cultures of cells obtained from animals or more conveniently, continuous cytokine-dependent or autonomous cell-lines. Biological assays are rarely entirely specific for a particular cytokine and can respond to a number of cytokines and other molecules (e.g. thymocyte assay; assays based on acute myeloid leukemic cell-lines, *Table 1*). In these assays, specificity for a particular cytokine can be established by using a monospecific neutralizing antibody and this approach is crucial for confirming the presence of cytokines in many biological (and particularly clinical) samples. In some cases, bioassays may underestimate cytokine protein content of a biological sample due to the presence of inhibitory molecules, e.g. binding proteins, soluble receptors, receptor antagonists, or inhibitory cytokines such as TGF-β (1, 2–4). It is also possible that bioassays may overestimate cytokine content of a sample due to synergistic interactions between individual cytokines in the sample (1).

1.1 Bioassays compared with immunoassays and receptor binding assays

It seems to be impossible to distinguish between different molecular species of some cytokines by using bioassays. The classic example of this is IL-1α and β, which although very different structurally, seem to possess apparently identical biological properties, at least on homologous cells. In such instances, immunoassays can be used to specifically quantify the different molecular forms on a protein basis (5). Immunoassays can in all cases be used as an alternative to bioassays although they can detect denatured biologically inactive cytokine molecules and fragments as well as 'inactive' receptor–cytokine complexes (1). Immunoassays need to be very carefully validated to eliminate non-specific artefacts, particularly if they are used for assaying clinical samples (1, 5, 6). Receptor binding assays are often regarded as a compromise between bioassays and immunoassays as they are based on the ability of cytokine molecules to bind to natural receptors. However, these assays may produce results which do not correlate with bioassay data if some ligand molecules are unable to induce signal transduction following binding to receptor.

2. Biological assays for interleukins

2.1 Bioassay of interleukin-2 (IL-2)

The original bioassays for IL-2 used short-term lectin-induced T cell blasts but these respond to other cytokines and some non-cytokine molecules. IL-2 can be measured by its proliferative effect on IL-2 dependent cell lines such as HT-2 and CTLL clones. The assay based on the CTLL-2 murine T cell-line provides a reliable and easy method which responds to IL-2 from most mammalian species.*

2.1.1 Maintenance of CTLL cell-line

Culture CTLL cells in RPMI 1640 medium containing 10% fetal calf serum (FCS) supplemented with partially purified rat splenocyte conditioned medium (or recombinant IL-2) in upright 25 cm^2 flasks. Maintain the cultures using a 3-day feeding schedule. Seed the cells at approximately 2×10^4 cells/ml and feed with 1–5% purified conditioned medium (7); this corresponds to a concentration of approximately 15–20 international units (IU/ml) IL-2. After 3 days when the cell density is approximately 2×10^5 cells/ml, cultures are split to 2×10^4 cells/ml and refed with IL-2.

Protocol 1. Bioassay of IL-2 using CTLL cell-line

Equipment and regents

- CTLL cell culture (Section 2.1.1)
- RPMI 1640 medium
- RPMI 1640 medium containing 10% FCS
- Centrifuge (MSE-benchtop)
- Trypan blue
- IL-2 standard (see Section 6 and *Table 3*)

- Test samples
- 96-well microtitre plates
- 37°C, 5% CO_2, humidified incubator
- [^{3}H]thymidine (25 Ci/m Mol, 5 mCi/5 mls)
- Filter mats
- Liquid scintillation counter system

---

* CTLL-2 cells also respond to murine interleukin-4 (mIL-4); a neutralizing antibody to murine IL-4, e.g. 11B11 (available from Biological Response Modifiers Program) can be used to discriminate between mouse IL-2 and IL-4 in these assays.

Method

1. Wash CTLL cells (3 days after feeding) three times with RPMI 1640 by centrifuging the cells at 250 g for 10 min.

2. Determine viability of the cells, e.g. by Trypan blue dye exclusion[a] and resuspend cells to a final concentration of 1 × 10^5 cells/ml in RPMI 1640 medium containing 10% FCS.

3. Titrate the IL-2 standard (see Section 6 and *Table 3*) in triplicate in 96-well microtitre plates. Start the titration at 40 IU/ml IL-2 and then make serial two-fold dilutions down to 0.019 IU/ml IL-2. Prepare dilutions of the samples in triplicate. Include a negative control, i.e. culture medium alone. Each well should contain a volume of 50 μl.

4. Add 50 μl of the cell suspension to each well and incubate the plates for 18 h at 37°C in a humidified CO_2 incubator.

5. Add 0.5 μCi of tritiated thymidine to each well and return the plates to the incubator for approximately 4 h.

6. Harvest the contents of each well on to filter mats and determine the radioactivity by liquid scintillation counting.

7. Plot a standard curve of c.p.m. versus concentration of IL-2. For quantitation of activity in unknown samples, compare test results with standard curve (see Section 5).

[a] Cells should be > 80% viable. The viability test should be routinely conducted for all cell lines (mentioned in this chapter) prior to assay. A similar method is described in Chapter 11

2.2 Bioassay of interleukin-1 (IL-1)

Bioassays for IL-1α and β commonly utilize the ability of this cytokine to induce IL-2 production by T cell-lines such as EL4.6.1 and LBRM 33, or in primary cultures of thymocytes (LAF assay). Alternatively, IL-1 can be assayed by measuring its ability to stimulate the proliferation of lines such as the subclone D10S of the murine T-helper cell line D10.64.1 (8). A subclone, NOB-1, of the murine thymoma line EL4.6.1 provides a simple and sensitive bioassay for IL-1 (see *Figure 1*). This line produces IL-2 in response to both human and murine IL-1, which is measured using the CTLL-2 cell-line based assay (see Section 2.1).

2.2.1 Maintenance of EL4/NOB-1 cell-line

Culture the cells in RPMI 1640 medium containing 5% FCS in upright 75 cm^2 flasks and feed every 2 to 3 days. Cultures are split 1:5 to 1:10 when the cell density reaches approximately 5 × 10^5 cells/ml.

Protocol 2. Bioassay of IL-1[a] using EL4/NOB-1 cell line[b]

Equipment and reagents
- EL4/NOB-1 cell culture (Section 2.2.1)
- RPMI 1640 medium containing 5% FCS
- IL-1 Standard (see Section 6 and *Table 3*)
- Test samples
- Plus equipment and reagents as *Protocol 1*

Method

1. Wash EL4/NOB-1 cells (2 to 3 days after feeding) twice in RPMI 1640 medium by centrifuging the cells at 250 *g* for 10 min and determine the viability as in step (2) of *Protocol 1*.

2. Resuspend the cells to a final concentration of 5×10^5 cells/ml in RPMI 1640 medium containing 5% FCS.

3. Distribute titrations of an IL-1 standard (see Section 6 and *Table 3*) in triplicate in 96-well microtitration plates. Start the titration of the standard at 100 pg/ml IL-1 (10 IU/ml) and make serial two-fold dilutions down to 0.09 pg/ml IL-1 (0.009 IU/ml). Make appropriate dilutions of the samples to be measured for IL-1 activity (either two-fold or ten-fold serial dilutions) in triplicate. The negative control is culture medium. Each well should contain a volume of 100 μl at this stage.

4. Add 100 μl of the washed cell suspension to each well and incubate the plates for approximately 24 h at 37°C in a humidified CO_2 incubator.

5. Remove 50 μl of the supernatant from each well and determine the IL-2 present using the CTLL-2 bioassay[c] (see *Protocol 1*). The amount of IL-2 in the supernatants will be proportional to the amount of IL-1 in the original samples. Supernatants from the EL4/NOB-1 cells can be removed and stored frozen until the IL-2 can be conveniently assayed.

[a] IL-1α and β have equal sensitivity in this assay.
[b] The EL4/NOB-1 cell line also responds to murine TNFα.
[c] Since the NOB-1 bioassay uses the CTLL-2 bioassay as a second stage, IL-2 and mIL-4 could interfere if present in the samples being tested for IL-1 activity. This can be overcome by pre-incubating the samples with the EL4/NOB-1 cell line for 4–5 h followed by thorough washing of the cells prior to steps 4 and 5.

2.3 Bioassay of interleukin-4 (IL-4)

The classical IL-4 assay involves the promotion of cell division of purified B lymphocytes co-stimulated with either anti-immunoglobulin, *Staphylococcus aureus* Cowan strain 1, or phorbol ester. Alternative assays utilize the ability of IL-4 to induce proliferation of phytohaemagglutinin (PHA) activated T lymphocytes (as in *Protocol 3*), B cell-lines such as BALM-4, HFB-1 and L4 (as described in Chapter 10), or certain haemopoietic progenitor cell-lines such as TF-1 or MO-7E (see Section 3.1). Recently, a human IL-4 responsive

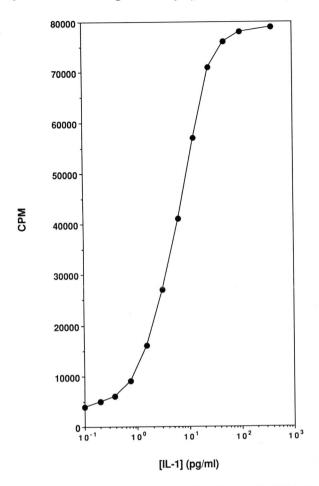

Figure 1. A standard curve for bioassay of human IL-1 using the EL4/NOB-1 and the CTLL-2 cell-lines (as described in *Protocol 2*).

cell-line has been developed by transfecting a murine CTL line with the human IL-4 receptor gene (9). This cell line proliferates in response to human IL-4 (see *Figure 2*); the assay is described in Section 2.3.1.

Protocol 3. Bioassay of IL-4 using human peripheral blood lymphocytes[a,b]

Equipment and reagents

- Fresh human blood
- Centrifuge (MSE – benchtop)
- Histopaque (Sigma) or Lymphoprep (Nyegaard)
- RPMI 1640 medium

- RPMI 1640 medium containing 5% FCS
- RPMI 1640 medium containing 5% FCS and 1% (v/v) phytohaemagglutinin (PHA; Wellcome reagent grade)
- Upright 25 cm² flasks

- 37°C, 5% CO_2, humidified incubator
- IL-2
- IL-4 standard (Section 6 and *Table 3*)
- Test samples
- [³H]thymidine (see *Protocol 1*)
- Filter mats
- Liquid scintillation counting system

Method

1. Isolate peripheral blood lymphocytes (PBL) from freshly drawn human blood by centrifugation over a cushion of suitable separation medium, e.g. histopaque (Sigma) or lymphoprep (Nyegaard Ltd).

2. Wash the isolated lymphocytes twice with warm RPMI 1640 medium by centrifuging at 400 *g* for 10 min.

3. Resuspend the cells at a concentration of 5×10^6 cells/ml in RPMI 1640 medium containing 5% FCS and 1% (v/v) PHA (Wellcome reagent grade). Incubate for 48 h in upright 25 cm² flasks at 37°C in a humidified CO_2 incubator.

4. Feed the cell suspension with an equal volume of fresh RPMI 1640 medium containing 5% FCS. Add 20 IU/ml of IL-2 to the cell suspension. Incubate for a further 48 h at 37°C in a humidified CO_2 incubator.

5. Wash cells twice with RPMI 1640 medium by centrifuging at 250 *g* for 10 min. Resuspend the cells to a concentration of 2×10^5 cells/ml in RPMI 1640 medium containing 5% FCS and 1% v/v PHA.

6. Distribute titrations of an IL-4 standard (see Section 6 and *Table 3*) in triplicate in 96-well microtitration plates. Start the titration of the IL-4 standard at 100 ng/ml (approximately 1000 U/ml) and make serial two-fold dilutions down to 0.1 ng/ml (1 U/ml) in 100 µl volumes per well. Make appropriate dilutions of the samples to be measured for IL-4 (usually 2-fold serial dilutions) in triplicate in 100 µl volumes. For a negative control, include wells containing culture medium supplemented with 1% v/v PHA but without IL-4.

7. Add 100 µl of the washed cell suspension to each well and incubate the plates for approximately 44 h at 37°C in a humidified CO_2 incubator.

8. Follow steps 5 and 6 of *Protocol 1*.

9. Plot a standard curve of c.p.m. versus concentration of IL-4. A comparison of test results with the standard curve provides an estimate of the activity in unknown samples (see Section 6).

[a] PHA stimulated PBLs will respond to IL-2, IL-4 and IL-12 and possibly other cytokines. When using peripheral blood lymphocytes for assaying IL-4 some donor to donor variation should be expected in the response of the lymphocytes to IL-4.

[b] As IL-4 is species restricted in its activity, the source of samples being measured will determine which assay is used. Assays using stimulated human B or T lymphocytes will respond to human sequence IL-4 whereas murine T cell-lines such as CTLL-2, HT-2 and AC-2 will proliferate in response to murine sequence IL-4. A blocking antibody to the murine IL-2 receptor (e.g. AMT-13) can be included to make the CTLL-2 line specific for mIL-4.

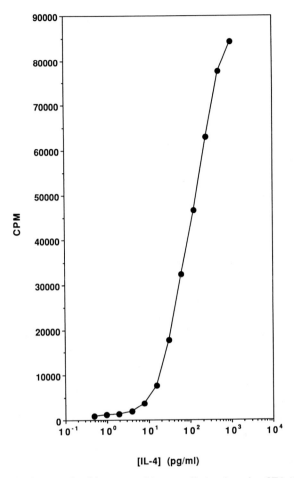

Figure 2. A standard curve for bioassay of human IL-4 using the CT.h4S cell-line (as described in *Protocol 4*).

2.3.1 Maintenance of CT.h4S cell-line

Culture the cells in RPMI 1640 medium containing 10% FCS, 2 mM L-glutamine, 1 mM sodium pyruvate, 0.05 mM 2-mercaptoethanol and 10 ng/ml (approximately 100 U/ml) human IL-4 in 25 cm^2 flasks. Cultures are split 1:4 or 1:5 every 2 to 3 days and refed with fresh medium.* Cultures are maintained at 37°C in a humidified CO$_2$ incubator.

* These cells are very sensitive to high pH and diminished 2ME concentrations so using fresh medium is important. For splitting cells, use a sterile cell scraper. Do not use trypsin or EDTA.

Protocol 4. Bioassay of IL-4 using the CT.h4S cell-line[a]

Equipment and reagents

- CT.h4S cell culture (Section 2.3.1)
- IL-4 standard (see *Table 3*)
- 96-well microtitre plates
- Sterile cell scraper (do not use trypsin or EDTA)
- RPMI 1640 medium
- Test samples
- Centrifuge (see *Protocol 1*)
- Trypan blue
- 37°C, 5% CO_2, humidified incubator
- [^{3}H]thymidine (see *Protocol 1*)
- Filter mats
- Liquid scintillation counter system

Method

1. Distribute titrations of IL-4 in 100 μl volumes in triplicate in 96-well microtitre plates. Start the titration at 10 ng/ml (approximately 100 U/ml) and serially dilute two-fold down to 1 pg/ml (0.01 U/ml). Make appropriate dilutions of the samples in 100 μl volumes in triplicate. As a negative control include medium alone.

2. Scrape cells 2 to 3 days after feeding and count. Wash cells in RPMI 1640 by centrifuging at 200 *g* for 10 min.

3. Resuspend cells to a final concentration of 1×10^5 cells/ml.

4. Add 100 μl of the cell suspension to each well and incubate plates for approximately 44 h at 37°C in a humidified CO_2 incubator.

5. Follow steps 5 and 6 of *Protocol 1*.

6. Plot a standard curve of c.p.m. versus concentration of IL-4. Compare test results with the standard curve for estimation of the activity in unknown samples (see Section 6).

[a] The assay can be adapted to 100 μl total volume of the 200 μl volume used in the Protocol above. This cell-line can respond to both murine and human sequence IL-2 and also to murine IL-4.

2.4 Bioassay for interleukin-5 (IL-5)

Bioassays for IL-5 utilize the ability of this cytokine to stimulate the formation of eosinophil colonies in an agar colony assay or by estimation of eosinophil peroxidase (see Chapter 13) (11). Alternatively, IL-5 can be assayed by its proliferative effect on the murine B-cell line, BCL-1 which is described in *Protocol 5*. Since BCL-1 cells are approximately 1000-fold less sensitive to human IL-5 than murine IL-5, a more practical means of measuring human IL-5 is by using the TF-1 cell-line (see Section 3.1).

2.4.1 Maintenance of BCL-1 line

This line is maintained in Balb/c mice by intraperitoneal injection of 10^7 cells per mouse. After approximately 4 weeks, splenic tumours appear and are removed at any time between 1 week after appearance of tumours and the death of the animals (6 to 8 weeks after inoculation). Spleen cells from tumour bearing mice (between 8×10^8–10^9 per mouse) are depleted of T cells by treatment with Thy-1.2 and anti-L3T4 monoclonal antibodies plus complement. The cells are washed thoroughly and cryopreserved in liquid nitrogen (5×10^7 cells per ampoule), until needed for assay.

Protocol 5. Bioassay for IL-5 using BCL-1 line[a]

Equipment and reagents

- Fresh or frozen BCL-1 cells (Section 2.4.1)
- IL-5 standard (Section 6 and *Table 3*)
- Test samples
- Plus equipment and reagents as *Protocol 1*

Method

1. Isolate fresh cells or thaw out an ampoule of protein cells.

2. Wash cells as in step 1 of *Protocol 1* and resuspend to a density of 2 $\times 10^5$ cells/ml in RPMI 1640 supplemented with 5% FCS.

3. Distribute a standard preparation of IL-5 (see Section 6 and *Table 3*) as a doubling dilution series in a volume of 50 μl per well in triplicate in a 96-well microtitre plate. Make appropriate dilutions of the samples in triplicate in 50 μl volumes. For a negative control, add only culture medium to the appropriate wells.

4. Add 50 μl of cell suspension to each well and incubate the plates for approximately 44 h at 37°C in a humidified CO_2 incubator.

5. Follow steps 5 and 6 of *Protocol 1*.

6. Plot a standard curve of c.p.m. versus concentration of IL-5. Estimate activity in unknown samples by comparing test results with the standard curve (see Section 6).

[a] The BCL-1 line also responds to murine IL-4 and GM-CSF.

2.5 Bioassay for interleukin-6 (IL-6)

Most bioassays for IL-6 depend upon the proliferative effect of this cytokine on IL-6 dependent murine hybridoma cell-lines such as MH60, B9 and 7TD1. (See also Chapter 10.) Alternatively, IL-6 can be assayed by its ability to enhance differentiation and IgG secretion in Epstein-Barr virus transformed human lymphoblastoid cell-lines such as CESS (as described in Chapter 10). The IL-6 dependent murine hybridoma cell-line B9 provides a reliable and sensitive assay for measuring mammalian IL-6.

2.5.1 Maintenance of B9 cell-line

Culture B9 cells in RPMI 1640 medium supplemented with 5% FCS and approximately 100 pg/ml (10 IU/ml) of IL-6 (recombinant IL-6 or crude supernatant from either IL-1 stimulated fibroblasts or LPS stimulated monocytes) in 75 cm^2 flasks. Cultures are split 1:5 to 1:10 every 2 to 3 days and refed with IL-6 when the cell density reaches approximately 5×10^5 cells/ml. Cultures are maintained at 37°C in a humidified CO_2 incubator.

Protocol 6. Bioassay for IL-6 using B9 cell-line[a]

Equipment and reagents

- B9 cells (Section 2.5.1)
- Centrifuge (see *Protocol 1*)
- IL-6 standard (Section 5 and *Table 3*)
- RPMI 1640 medium with and without 5% FCS
- Trypan blue
- 96-well microtitre plates
- 37°C, 5% CO_2, humidified incubator

- MTT tetrazolium salt (3-(4, 5-dimethyl-thiazol-2-ys)-2, 5-diphenyl tetrazolium bromide) (5 mg/ml in PBS, filter-sterilized, and stored in the dark)
- Acid SDS: 10% (w/v) SDS dissolved in 0.02 M HCl
- ELISA reader

Method

1. Wash B9 cells (2 days after feeding) as in step 1 of Protocol 1 and resuspend to a density of 5×10^4 cells/ml in RPMI 1640 medium supplemented with 5% FCS.

2. Distribute an IL-6 standard (see Section 6 and *Table 3*) as a serial two-fold dilution series in triplicate in 100 µl volumes in 96-well microtitration plates. Start the titration of the standard at 100 pg/ml (10 IU/ml) and dilute down to 0.1 pg/ml (0.01 IU/ml). Make appropriate dilutions of the samples to be measured for IL-6 activity in triplicate in 100 µl volumes. As a negative control include culture medium alone.

3. Add 100 µl of cell suspension to each well and incubate the plates for approximately 72 h[b] at 37°C in a humidified CO_2 incubator.

4. To each well add 10 µl of the tetrazolium salt MTT and return the plates to the incubator for a further 4½ h.

5. Add 25 µl of acid SDS per well. Place the plates at 37°C in a humidified CO_2 incubator overnight and determine the absorbance at a wavelength of 620 nm using an Elisa reader.

6. Plot a standard curve of absorbance versus concentration of IL-6. For determination of activity in unknown samples, compare test results with the standard curve (see Section 6).

[a] This cell line also responds to murine IL-4 and human IL-11, IL-13, and oncostatin M.
[b] Prolonging the incubation to 96 h is sometimes necessary to enhance the signal to noise ratio for this assay.

2.6 Bioassay for interleukin-7 (IL-7)

The IL-7 bioassay employs a murine pre-B cell line (clone 2b cells) which is dependent upon exogenous IL-7 for continuous growth and viability (10). The proliferation induced by human or murine IL-7 is measured by tritiated thymidine incorporation.

2.6.1 Maintenance of clone 2b cell-line

Culture clone 2b cells routinely in Iscove's modified Dulbecco's medium containing 5% FCS, 0.05 mM 2-mercaptoethanol, 2 mM L-glutamine, and 20 ng/ml (approximately 2000 U/ml) of recombinant human IL-7 in upright 75 cm^2 flasks. Cultures are split to 2×10^4 cell/ml every 3 days when the cell concentration reaches approximately $1–2 \times 10^5$ cells/ml.

Protocol 7. Bioassay for IL-7 using 2b cell-line

Equipment and reagents

- 2b cell culture (Section 2.6.1)
- Trypan blue
- Iscove's modified Dulbecco's medium (IMDM)
- IMDM containing 5% FCS
- Centrifuge (see *Protocol 1*)
- IL-7 standard (*Table 3*)
- Test samples

- 96-well microtitre plates
- 37°C, 7.5% O_2 and 5% CO_2, humidified incubator
- [^{3}H]thymidine (see *Protocol 1*)
- Filter mats
- Liquid scintillation counter system

Method

1. Wash the cells as in step 1 of *Protocol 1* and resuspend to a density of 1.25×10^3 cells/ml in medium containing 5% FCS.

2. Distribute the IL-7 standard in a two-fold dilution series in triplicate in 50 μl volumes in 96-well microtitration plates. Start the titration of the standard at 200 pg/ml (approximately 20 U/ml) and dilute down to 1 pg/ml (0.01 U/ml). Make appropriate dilutions of the samples in duplicate in 50 μl volumes. As a negative control include culture medium alone.

3. Add 50 μl of the cell suspension to each well and incubate the plates for approximately 44 h at 37°C in a humidified incubator containing 7.5% O_2 and 5% CO_2.

4. Add 2 μCi of tritiated thymidine to each well and return the plates to the incubator for approximately 4 h.

5. Follow step 6 of *Protocol 1*.

6. Plot a standard curve of c.p.m. versus concentration of IL-7. Estimate activity in unknown samples by comparison with the curve (see Section 6).

2.7 Bioassay for interleukin-9 (IL-9)

This cytokine can be assayed using long term cultured CD4[+] and CD8[+] T cell-lines and clones, or by its ability to enhance erythroid burst forming activity in the presence of erythropoietin. Murine IL-9 can stimulate human cells but not vice versa. Alternatively, the MO-7e cell-line proliferates in response to IL-9. This assay is described in *Protocol 12*.

2.8 Bioassay for interleukin-10 (IL-10)

IL-10 is commonly assayed by its ability to inhibit IFN-γ production by activated peripheral blood mononuclear cells, murine Th1 clones (11), or by its proliferative effect on the murine mast cell-line MC/9 (12). Human IL-10 is active on murine cells but the murine counterpart is inactive on human cells.

2.8.1 Maintenance of MC/9 cell-line

Culture cells in RPMI 1640 containing 10% FCS, 2 mM L-glutamine, 0.05 mM 2-mercaptoethanol, 1% spleen conditioned medium and 1 ng/ml (approximately 100 U/ml) murine GM-CSF in upright 75 cm^2 flasks. Cultures are split 1:5 to 1:7 every two to three days.

Protocol 8. Bioassay for IL-10 using the MC/9 cell-line[a]

Equipment and reagents

- MC/9 culture (Section 2.8.1)
- Centrifuge (see *Protocol 1*)
- Trypan blue
- RPMI 1640 medium
- Supplemented RPMI 1640 medium containing 10% FCS, 2 mM L-glutamine, 0.05 mM 2-mercaptoethanol
- IL-10 standard (*Table 3*)
- Test samples
- 96-well microtitre plates
- 37°C, 5% CO$_2$, humidified incubator
- [^{3}H]thymidine (see *Protocol 1*)
- Filter mats
- Liquid scintillation counter system

Method

1. Wash MC/9 cells (two to three days after feeding) as in step 1 of *Protocol 1* and resuspend to a density of 2 × 10^5 cells/ml in supplemented RPMI 1640

2. Titrate the IL-10 standard in triplicate in 96-well microtitre plates. Start the titration at 100 ng/ml (approximately 500 U/ml) and dilute down in a two-fold dilution series to 10 pg/ml (0.05 U/ml) in 10 μl volumes. Make appropriate dilutions of the samples in triplicate. For a negative control, include culture medium alone.

3. Add 10 μl of cell suspension to each well and incubate for approximately 44 h at 37°C in a humidified CO$_2$ incubator.

4. Follow steps 5 and 6 of *Protocol 1*.

---

5. Plot a standard curve of c.p.m. versus concentration of IL-10. Estimate activity in unknown samples by comparing test results with standard curve (see Section 6).

a MC/9 line also responds to human and murine IL-5 and stem cell factor (SCF).

2.9 Bioassay for interleukin-11 (IL-11)

IL-11 can be measured by its stimulatory effect on the proliferation of IL-6-dependent murine plasmacytoma lines such as T1165 (13). A subclone of this cell line, T10, has been developed; the use of this cell line is described in *Protocol 9*. Recently a subclone of the B9 line (see Section 2.5), B9–11 has been derived which shows enhanced sensitivity for IL-11 (14).

2.9.1 Maintenance of T10 cell-line

Culture the cells in RPMI 1640 medium supplemented with 10% FCS, 2 mM L-glutamine, 0.05 mM 2-mercaptoethanol and 5 ng/ml (approximately 50 U/ml) IL-11 in 75 cm^2 flasks. Cultures are split 1:4 to 1:5 every two or three days and re-fed with IL-11. T10 cells are non-adherent but do tend to cling to the bottom of the tissue culture flasks. To loosen cells, tap the flask gently. Cultures are maintained at 37°C in a humidified CO$_2$ incubator.

Protocol 9. Bioassay for IL-11 using the T10 cell-line^a

Equipment and reagents

- T10 cell culture (Section 2.9.1)
- centrifuge (see *Protocol 1*)
- Trypan blue
- RPMI 1640 medium
- Supplemented RPMI 1640 medium containing 10% FCS, 2 mM L-glutamine, 0.05 mM 2-mercaptoethanol

- IL-11 standard (*Table 3*)
- Test samples
- 37°C, 5% CO$_2$, humidified incubator
- [^{3}H]thymidine (see *Protocol 1*)
- Filter mats
- Liquid scintillation counter system

Method

1. Wash T10 cells (two to three days after feeding) as in step 1 of *Protocol 1* and resuspend the cells to a final concentration of 1 × 10^5 cells/ml in supplemented RPMI 1640.

2. Distribute titrations of IL-11 in 100 μl volume in triplicate in 96-well microtitre plates. Start the titration of the standard at 10 ng/ml (approximately 100 U/ml) and make serial dilutions down to 10 pg/ml (0.1 U/ml). Make appropriate dilutions of the unknown samples in triplicate in 100 μl volumes. As a negative control include medium alone.

3. Follow step 4 and 5 of *Protocol 4*.

Protocol 9. *Continued*

4. Plot a standard curve of c.p.m. versus concentration of IL-11. Estimate activity in unknown samples by comparing test results with standard curve (see Section 6).

[a] T10 cells respond to human and murine IL-11 and IL-6.

2.10 Bioassay for interleukin-12 (IL-12)

IL-12 can be measured by its effect on proliferation of PHA activated human lymphoblasts (15). Alternatively, it can be assayed using the IL-2-dependent human T cell leukaemic line, KIT225/K6-C (16). The murine protein is active on human cells but not vice versa.

2.10.1 Maintenance of KIT225/K6-C cell-line

Culture the cells in RPMI 1640 medium supplemented with 10% FCS, 2 mM L-glutamine, 0.05 mM 2-mercaptoethanol and 50 IU/ml IL-2 in 75 cm^2 upright flasks. Cultures are split 1:6 to 1:10 every two to three days and refed with IL-2. Cultures are maintained at 37°C in a humidified CO_2 incubator.

Protocol 10. Bioassay for IL-12 using the KIT225/K6-C cell-line

Equipment and reagents

- KIT225/K6-C cell culture (Section 2.10.1)
- Centrifuge (see *Protocol 1*)
- Trypan blue
- RPMI 1640 medium
- Supplemented RPMI 1640 containing 10% FCS, 2 mM L-glutamine, 0.05 mM 2-mercaptoethanol, 10 ng/ml (approx. 100 U/ml) rhIL-4

- IL-12 standard
- 96-well microtitre plates
- Test samples
- 37°C, 5% CO_2, humidified incubator
- [³H]thymidine (*Protocol 1*)
- Filter mats
- Liquid scintillation counter system

Method

1. Wash KIT225/K6-C cells (two to three days after feeding) as in step 1 of Protocol 1 and resuspend the cells to a final concentration of 2 × 10^5 cells/ml in supplemented RPMI 1640 medium.

2. Distribute titrations of IL-12 in 100 μl volume in triplicate in 96-well microtitration plates. Start the titration of the standard at 50 ng/ml and make serial dilutions down to 0.5 pg/ml. Make appropriate dilutions of the unknown samples in 100 μl volumes. As a negative control include medium alone.

3. Add 100 μl of the washed cell suspension to each well and incubate the plates for approximately 72 h at 37°C in a humidified CO_2 incubator.

4. Follow steps 5 and 6 of *Protocol 1*.

5. Plot a standard curve of c.p.m. versus concentration of IL-12. Estimate activity in unknown samples by comparing test results with standard curve (see Section 6).

2.11 Bioassay for interleukin-13 (IL-13)

This cytokine can be assayed either by its ability to induce expression of CD23 on B cells or IgE synthesis by purified B cells, or by proliferation of activated human B lymphocytes, as for IL-4 (Chapter 9) and IL-14 (*Protocol 11*). Alternatively, it can be measured by its proliferative effect on the human erythroleukaemic cell-line, TF-1 (described in *Protocol 13*).

2.12 Bioassay for interleukin-14 (IL-14)

IL-14 can be assayed by its ability to induce proliferation of activated B cells (17). For this, B lymphocytes are purified from peripheral blood or tonsils, using Ficoll-Hypaque separation of mononuclear cells followed by monocyte depletion by adherence and T cell depletion by E-rosetting as described in Chapter 10. The resulting B cells are then used in the co-stimulatory assay as indicated in *Protocol 11*.

Protocol 11. B cell co-stimulatory assay[a] for IL-14

Equipment and reagents

- B cells (see Chapter 10)
- RPMI 1640 supplemented with 10% FCS
- *Staphylococcus aureus* Cowan 1 strain (SAC; Immunoprecipitin, Bethesda Research Labs)
- 37°C incubator
- Ficoll-Hypaque
- Centrifuge (see *Protocol 1*)
- PBS
- IL-14 standard
- Test samples
- [³H]thymidine (*Protocol 1*)
- Filter mats
- Liquid scintillation counter system

Method

1. Prepare B cells as described in Chapter 10, and resuspend in RPMI 1640 supplemented with 10% FCS.

2. Incubate B cells with 1:25 000 dilution of *Staphylococcus aureus* Cowan 1 strain, SAC (Immunoprecipitin, Bethesda Research Laboratories, Gaithesburg, MD) in RPMI 1640 and 10% FCS for 72 h in a 37°C incubator.

3. Remove dead cells by Ficoll-Hypaque separation and centrifugation at 250 *g* for 10 min in RPMI 1640 and 10% FCS.

Protocol 11. *Continued*

4. Wash cells twice with PBS and resuspend at a concentration of 1×10^6 cells/ml.

5. Distribute titrations of an IL-14 preparation in 96-well microtitre plates in 100 μl volumes. For unknown samples, make appropriate dilutions. For a negative control, include culture medium alone.

6. Follow step 3 of *Protocol 10*.

7. Add 1 μCi of tritiated thymidine to each well and return culture plates to the incubator for a further 18 h.

8. Follow step 6 of *Protocol 1*.

9. Plot a standard curve of c.p.m. versus concentration of IL-14. For estimation of activity of unknown samples, compare test results with standard curve (see Section 6).

---

[a] This assay can be used to measure human IL-4 and human IL-13 by inclusion of an appropriate cytokine standard. For measuring activity in samples, neutralizing antibodies should be used to confirm the presence of a particular cytokine.

3. Bioassays for colony stimulating factors — interleukin-3 (IL-3), granulocyte colony stimulating factor (G-CSF), macrophage colony stimulating factor (M-CSF), granulocyte macrophage colony stimulating factor (GM-CSF), stem cell factor (SCF), and leukaemia inhibitory factor (LIF)

Classically, these factors are assayed by their ability to stimulate the formation of colonies of differentiated cells from progenitor cells of the bone-marrow in soft agar (see Chapter 15) (18). The type of colony produced depends upon the factor. IL-3 and GM-CSF stimulate the production of mixed colonies of different cell types, whereas G-CSF, M-CSF and erythropoietin (Epo) are lineage-restricted and produce predominantly granulocyte, monocyte and erythroid colonies respectively. In such assays, colonies of more than 50 cells are counted and analysed after 7–14 days by eye using a binocular microscope. The number of colonies is usually related to the specific activity or concentration of the colony stimulating factor (CSF) in the agar culture. Morphological analysis by staining of dried and fixed gels allows proper identification of the colony type (18). This bioassay is not specific and is subject to interference by contaminating factors which may enhance or inhibit colony formation. In addition, standard protocols are slow and require a long incubation period. Alternative assays for CSFs measure the prolifera-

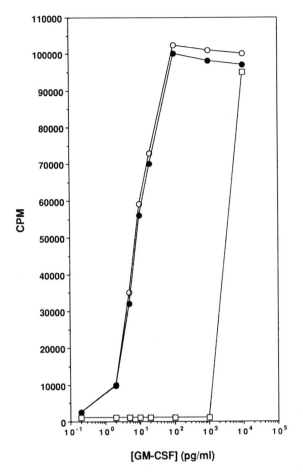

Figure 3. Dose response curve for human GM-CSF using the TF-1 cell-line bioassay. Closed circles show a standard curve for human GM-CSF (as described in *Protocol 13*). This assay responds to several other factors such as IL-3, IL-4, IL-5, and Epo. Specificity can be achieved by including a neutralizing antiserum (as shown). Open squares— titration of GM-CSF in the presence of neutralizing antiserum. Open circles—titration of GM-CSF in the presence of pre-immune serum. These sera were used at a dilution of 1:2500.

tive effect of these factors on cell lines derived from human leukaemias, e.g. AML-193, TALL-101, MO-7e, TF-1, and murine lines such as NFS-60, WEHI 3BD⁺, and 32DC1. These lines provide rapid, sensitive and reliable assays but are not specific. Specificity can be achieved by using specific neutralizing antibodies (so called 'blocking antibodies') to inhibit the appropriate activity (*Figure 3*). *Table 1* indicates the cell-lines routinely used for measuring human

Table 1. Cell-lines used in proliferation assays for measurement of human colony stimulating factors and other cytokines

Cell-line	Origin	Cytokines routinely assayed	Response to other cytokines	Reference
GNFS-60	Murine retrovirus-induced leukaemia	G-CSF	IL-6 M-CSF	19
MNFS-60	Murine retrovirus-induced leukaemia	M-CSF	IL-6 9-CSF	20
TF-1	Human erythroleukaemia	GM-CSF IL-3 IL-4 IL-5 Epo IL-13 Onco M LIF CNTF NGF	SCF	21
AML-193	Human acute myeloid leukaemia	GM-CSF G-CSF IL-3	—	22
MO-7e	Human megakaryoblastic leukaemia	GM-CSF IL-3 IL-4 IL-9 SCF	TNF-α IL-6	23
DA1.a	Murine haemopoietic cell line	LIF	G-CSF	26
BCL-1	Murine B-lymphoma	IL-5	—	9

CSFs and their cross-reactivity with other cytokines. Detailed protocols for some CSFs are described here; these protocols can be applied to measure other cross-reactive cytokines (*Table 1*) by inclusion of the appropriate cytokine standard.

For assaying CSFs of murine origin, murine cell-lines are generally employed (25). As for human lines, none of the murine IL-3 dependent lines are specific, DA1 cells also respond to IL-3, IL-4 and GM-CSF, NFS-60 cells to IL-3, IL-4, IL-6, GM-CSF, and G-CSF, FDCP cells to IL-2 and IL-4, 32 DC1 cells to IL-2, IL-4, and G-CSF and BCL-1 cells to IL-4, IL-5, and GM-CSF. Blocking antibodies can be used to determine specificity (see *Figure 3*).

3.1 Maintenance of MO-7e/TF-1 cell-lines

Culture the cells in RPMI 1640 medium supplemented with 5% FCS and GM-CSF (approximately 4 ng/ml (40 IU/ml) for MO-7e, 2 ng/ml (20 IU/ml) for TF-1) in upright 75 cm^2 flasks. Cultures are split 1:5 to 1:7 every 2 to 3 days and refed with GM-CSF when the cell density reaches approximately 5×10^5 cells/ml. Cultures are maintained at 37°C in a humidified CO_2 incubator.

Protocol 12. Bioassay using MO-7e cell-line[a]

Equipment and reagents

- MO-7e cell culture (Section 2.4.1)
- Centrifuge (see *Protocol 1*)
- Trypan blue
- RPMI 1640 medium
- RPMI 1640 supplemented with 5% FCS
- Appropriate standard (*Table 3*)

- 96-well microtitre plates
- Test samples
- 37°C, 5% CO_2, humidified incubator
- [^{3}H]thymidine (*Protocol 1*)
- Filter mats
- Liquid scintillation counter system

Methods

1. Wash MO-7e cells (2 to 3 days after feeding) as in step 1 of *Protocol 1* and resuspend cells in RPMI 1640 containing 5% FCS to a final concentration of 5×10^5 cells/ml.

2. Distribute titrations of an appropriate standard (GM-CSF, IL-3, SCF, IL-9 etc., see *Table 3*) in 100 μl volumes in triplicate in 96-well microtitration plates. Start the titration of the standard, e.g. GM-CSF, at 10 ng/ml (100 IU/ml) and make serial dilutions (two-fold or ten-fold) down to 1 pg/ml (0.1 IU/ml). Make appropriate dilutions of the samples in 100 μl volumes in triplicate. As a negative control include culture medium alone.

3. Follow steps 4 and 5 of *Protocol 4*.

4. Plot a standard curve of c.p.m. versus concentration of the particular cytokine standard. Estimate activity in unknown samples by comparison with standard curve (see Section 6).

[a] This cell line proliferates with different sensitivities, to different cytokines, the order being GM-CSF or IL-3 > IL-9 > SCF. The range of the dose response curve for IL-9 is between 10 ng/ml (approximately 100 U/ml) to 0.01 ng/ml (0.1 U/ml) and for SCF is 100 ng/ml (approximately 100 U/ml) to 0.1 ng/ml (0.1 U/ml).

Protocol 13. Bioassay using TF-1 cell-line[a]

Equipment and reagents

- TF-1 cell culture (Section 3.1)
- RPMI 1640 medium with and without 5% FCS
- Centrifuge (see *Protocol 1*)
- Trypan blue
- Appropriate standard (*Table 3*)

- Test samples
- 37°C, 5% CO_2, humidified incubator
- [^{3}H]thymidine (*Protocol 1*)
- Filter mats
- Liquid scintillation counter system

Methods

1. Wash TF-1 cells (2 to 3 days after feeding) as in step 1 of *Protocol 1* and resuspend the cells to a final concentration of 1×10^5 cells/ml in RPMI 1640 containing 5% FCS.

Protocol 13. *Continued*

2. Distribute titrations of an appropriate standard (GM-CSF, IL-3, IL-4, etc., see *Table 3*) in 100 μl volumes in triplicate. Start the titration of the standard, e.g. GM-CSF at 1 ng/ml (10 IU/ml) and make serial two-fold dilutions down to 0.1 pg/ml (0.001 IU/ml). Make appropriate dilutions of the samples in 100 μl volumes in triplicate. As a negative control include medium alone.

3. Follow steps 4 and 5 of *Protocol 4.*

4. Plot a standard curve of c.p.m. versus concentration of the particular cytokine standard. Estimate activity in unknown samples by comparison with standard curve (see Section 6).

---

[a] This cell line exhibits differential sensitivity to different cytokines, the order being GM-CSF or IL-3 > IL-5 > IL-4. For IL-4, therefore, the range of the dose response curve is between 50 ng/ml to 1 pg/ml.

3.2 Maintenance of GNFS-60 and MNFS-60 cell-lines

Culture the cells in RPMI 1640 medium supplemented with 5% FCS and 2 ng/ml (200 IU/ml) G-CSF for GNFS-60 and 17 ng/ml (1000 IU/ml) M-CSF for MNFS-60 in upright 75 cm^2 flasks. Cultures are split 1:5 to 1:10 every 2 to 3 days and re-fed with G-CSF or M-CSF when the cell density reaches approximately 5×10^5 cells/ml. Cultures should be maintained at 37°C in a humidified CO$_2$ incubator.

Protocol 14. Bioassay of G-CSF using GNFS-60 cell-line

Equipment and reagents

- GNFS-60 cell culture (Section 3.2)
- Centrifuge (see *Protocol 1*)
- Trypan blue
- RPMI 1640 medium with and without 5% FCS
- G-CSF standard (*Table 3*)

- 96-well microtitre plates
- Test samples
- 37°C, 5% CO$_2$, humidified incubator
- [^{3}H]thymidine (*Protocol 1*)
- Filter mats
- Liquid scintillation counter system

Methods

1. Wash GNFS-60 cells (2 to 3 days after feeding) as in step 1 of *Protocol 1* and resuspend cells to a final concentration of 1×10^5 cells/ml in RPMI 1640 containing 5% FCS.

2. Distribute titrations of the G-CSF standard in triplicate in 96-well microtitration plates. Start the titration of the standard at 1 ng/ml (100 IU/ml) and make serial two-fold dilutions down to 1 pg/ml (0.1 IU/ml). Make appropriate dilutions of the samples to be measured for G-CSF

activity in triplicate in 100 μl volumes. As a negative control include culture medium alone.

3. Follow steps 4 and 5 of *Protocol 4*.

4. Plot a standard curve of c.p.m. versus concentration of standard. Estimate activity in unknown samples by comparison with standard curve (see Section 6).

Protocol 15. Bioassay of M-CSF using MNFS-60 cell-line

Equipment and reagents

- MNFS-60 cell culture (Section 3.2)
- Centrifuge (see *Protocol 1*)
- Trypan blue
- RPMI 1640 medium with and without 5% FCS
- M-CSF standard (*Table 3*)

- Test samples
- 96-well microtitre plates
- 37°C, 5% CO_2, humidified incubator
- [^{3}H]thymidine (*Protocol 1*)
- Filter mats
- Liquid scintillation counter system

Method

1. Wash MNFS-60 cells (2 to 3 days after feeding) as in step 1 of *Protocol 1* and resuspend cells to a final concentration of 5×10^5 cells/ml in RPMI 1640-containing 5% FCS.

2. Distribute titrations of the M-CSF standard in 100 μl volumes in triplicate in 96-well microtitration plates. Start the titration of the standard at 100 ng/ml (6000 IU/ml) and make serial two-fold dilutions down to 10 pg/ml (0.6 IU/ml). Make appropriate dilutions of the samples to be measured for MCSF activity in triplicate in 100 μl volumes. As a negative control include culture medium alone.

3. Follow steps 4 and 5 of *Protocol 4*.

4. Plot a standard curve of absorbance versus concentration of standard. Estimate activity in unknown samples by comparison with standard curve (see Section 6).

3.3 Bioassay for leukaemia inhibitory factor (LIF)

This cytokine can be assayed by its ability to induce differentiation of the monocytic leukaemia cell-line, M1. However, a number of other assays have also been developed, e.g. bone-marrow colony assay (in which LIF synergizes with IL-3 to produce multilineage colonies) or proliferation assays employing either the murine cell-line, DA-1a or the human cell-line, TF-1.

3.3.1 Maintenance of DA-1a cell-line

Grow cells in RPMI 1640 containing 10% FCS and 1 ng/ml (approximately 10 U/ml) LIF. Cultures are split 1:5 to 1:10 every two to three days.

Protocol 16. Bioassay of LIF using DA-1a cell line[a]

Equipment and reagents

- DA-1a cell culture (Section 3.3.1)
- Centrifuge (see *Protocol 1*)
- Trypan blue
- RPMI 1640 medium with and without 10% FCS
- LIF standard (*Table 3*)

- Test samples
- 96-well microtitre plates
- 37°C, 5% CO_2, humidified incubator
- [^{3}H]thymidine (*Protocol 1*)
- Filter mats
- Liquid scintillation counter system

Methods

1. Wash DA-1a cells (2 to 3 days after feeding) as in step 1 of *Protocol 1* and resuspend to a concentration of 1×10^5 cells/ml in RPMI 1640 containing 10% FCS.

2. Titrate the LIF standard in triplicate in 96-well microtitre plates. Start the titration at 1 ng/ml (approximately 10 U/ml) and serially dilute two-fold down to 0.5 pg/ml (0.005 U/ml) in 100 μl volumes. Make appropriate dilutions of the samples. As a negative control, include culture medium alone.

3. Follow steps 4 and 5 of *Protocol 4*.

4. Plot a standard curve of c.p.m. versus concentration of LIF. Estimate activity in unknown samples by comparing test results with standard curve (see Section 6).

[a] This line also responds to human G-CSF and murine G-CSF, IL-3, IL-4, and GM-CSF. DA-1a is a clone of the DA1 cell-line.

3.4 Bioassay for stem cell factor (SCF)

The most commonly used assay for SCF is the bone-marrow colony assay; SCF synergizes with other CSFs and cytokines to form multilineage colonies. SCF can also induce proliferation in the murine mast cell-line, MC/9.[a] An easier and straightforward assay for this cytokine is based on its ability to stimulate proliferation of the megakaryocytic leukaemic cell-line, MO-7e (described in *Protocol 12*).

4. Bioassay for interleukin-8 (IL-8) and other chemokines

Most bioassays for IL-8 rely upon the stimulatory effect this cytokine has on the migration of polymorphonuclear leukocytes. The migration is assessed

---

[a] This assay is not very sensitive for human SCF. For an increased sensitivity, the use of the MO-7e cell-line is recommended.

Table 2. Bioassays for measurement of chemokines

Chemokine	Target cell types in chemotaxis assay	Other assays
IL-8	Neutrophils, basophils, T cells	Degranulation of neutrophils, respiratory burst activity and lysosomal enzyme release from neutrophils.
PF-4	Monocytes, neutrophils and fibroblasts	Inhibition of angiogenesis.
β-TG	Neutrophils, fibroblasts	
RANTES	Monocytes, eosinophils, basophils, memory T cells.	Induces histamine release from basophils.
MIP1α	Monocytes, B cells, cytotoxic T cells, CD4$^+$ T cells, basophils, eosinophils	Induces histamine release from basophils and mast cells, stem cell inhibition.
MIP1β	Monocytes, CD4$^+$ T cells (preferentially the 'naive' T cells)	Lysosomal enzyme release and degranulation of neutrophils.
MCP-1, 2, and 3	Monocytes, basophils	Induces histamine release from basophils, augments superoxide production and lysosomal enzyme release from monocytes.
GROα, β, and γ	Neutrophils	Proliferation of Hs294T melanoma cell line, induces respiratory burst activity and lysosomal enzyme release from neutrophils.
ENA-78	Neutrophils	Induces Ca^{2+} fluxes in neutrophils.
I-309	Monocytes	Induces Ca^{2+} fluxes in monocytes.
IP-10	Monocytes, T cells	
NAP-2	Neutrophils	Induces Ca^{2+} fluxes in neutrophils, elastase release from neutrophils.

either by measuring chemokinesis (an enhancement of random movement) or by chemotaxis (a directed locomotion along a concentration gradient). Chemotactic activity can be measured using a Boyden chamber or a Neuroprobe microchamber or the agarose method as described in Chapters 13 and 14. Alternative assays are based on respiratory burst activity, the elevation of intracellular Ca^{2+} levels and elastase secretion from activated neutrophils. *Table 2* lists available assays and target cell types employed in chemotaxis assays for different chemokines. The procedure used to measure release of elastase from activated neutrophils is described here.

Protocol 17. Isolation of neutrophils and measurement of elastase release by IL-8-stimulated neutrophils

Equipment and reagents

- Venous blood
- 100 mM EDTA in PBS
- PBS with and without 0.9 mM CaCl$_2$ and 0.5 mM MgCl$_2$
- BSA
- Cytochalasin B
- Histopaque-1119
- Histopaque-1077
- 50 ml centrifuge tubes
- Centrifuge

- Plate rotor
- IL-8 standard (*Table 3*)
- Test samples
- Hexadecyltrimethylammonium bromide
- Substrate solution: 0.44 mM N-t-BOC-Ala-Pro-NorVal-p-chlorothiobenzylester in 0.2 M Hepes buffer, pH 7.5, containing 0.7 M NaCl, 4% dimethyl sulphoxide, 0.75 mM 5–5'-dithiobis (2-nitrobenzoic acid (Ellman's reagent))

Method

1. Draw 50 ml of venous blood into 1 ml of 100 mM EDTA in PBS. Dilute with an equal amount of PBS (without CaCl$_2$ or MgCl$_2$).

2. Dispense 10 ml of Histopaque-1119 into a 50 ml centrifuge tube. Overlay with 10 ml of Histopaque-1077 (outlined in the Histopaque booklet).

3. Carefully layer on 20 ml aliquot of the diluted blood on to the upper gradient (i.e. Histopaque-1077).

4. Centrifuge at 700 *g* for 30 min at room temp.

5. Remove the centrifuge tube carefully. Two distinct opaque layers should be observed: the upper layer containing plasma, mononuclear cells and platelets, the middle layer containing granulocytes. The bottom of the tube contains sedimented erythrocytes.

6. Transfer granulocyte layer to another 50 ml centrifuge tube and add PBS (without CaCl$_2$ or MgCl$_2$) to the top of the tube.

7. Centrifuge at 200 *g* for 10 min.

8. Wash cells twice in PBS (without CaCl$_2$ or MgCl$_2$).

9. Resuspend cells to a concentration of 1×10^7 cells/ml in PBS with CaCl$_2$ and MgCl$_2$ containing 0.1% BSA and 5 µg/ml cytochalasin B. Incubate for 10 min at 37°C and use immediately in assay.

10. Distribute an IL-8 standard in 96-well microtitre plates. Start IL-8 titration at 500 ng/ml (500 U/ml) and serially dilute down to 5 ng/ml (5U/ml) in PBS with CaCl$_2$ and MgCl$_2$ containing 0.1% BSA, in triplicate, in 100 µl volumes. For blanks, use PBS/BSA alone. For positive control (i.e. maximum elastase release) use 0.1% hexadecyltrimethylammonium bromide. Make appropriate dilutions of the unknown samples in 100 µl volumes.

Protocol 17. *Continued*

11. Add 100 μl of pretreated cells (from step 9) and incubate for 30 min.

12. Pellet cells in plate and transfer 50 μl of the supernatant to a fresh plate.

13. To the supernatants add 50 μl substrate solution (Ellman's Reagent).

14. Allow colour to develop and measure the absorbance at 405 nm.

15. Plot a standard curve of absorbance versus IL-8. Estimate activity in unknown samples by comparison with standard curve (see Section 6).

Step 10 of assay should be prepared while the cells are being isolated so they can be used immediately.

5. Bioassays for other cytokines

5.1 Bioassay for tumour necrosis factors

The most commonly used bioassay for TNF-α or TNF-β (lymphotoxin) utilizes the cytotoxic action of these cytokines on murine fibroblasts such as L929 or L-M cells or on a human rhabdomyosarcoma cell line, KYM-1 (26) (see also biological assay on WEHI 164 cells described in Chapter 17). TNF-α and β can be distinguished from each other by inclusion of neutralizing antibodies in the assay.

5.1.1 Maintenance of L929 cells

Grow cells in RPMI 1640 containing 10% newborn calf serum in 75 cm^2 flasks. Subculture cells at weekly intervals using 0.01% trypsin and dilute 1:7 in growth medium.

Protocol 18. Bioassay for TNF-α or -β

Equipment and reagents

- Log-phase L929 cell culture (Section 5.1.1)
- Trypan blue
- RPMI 1640 medium containing 10% newborn calf serum
- Trypsin
- 96-well microtitre plates
- 37°C, 5% CO$_2$, humidified incubator
- TNF standard (Section 5 and *Table 3*)
- Actinomycin D
- Test samples
- Multichannel micropipettor

- Absorbent paper
- PBS, pH 7.2
- Naphthol blue/black stain (NBB): 0.05% NBB, 9% acetic acid (v/v), 0.1 M sodium acetate; prepared and filtered through Whatman No. 1 filter paper and stored at room temp.
- Formalin fixative: 10% formalin (v/v), 9% acetic acid (v/v), 0.1 M sodium acetate
- 50 mM NaOH
- ELISA plate reader

Methods

1. Trypsinize cells in log-phase growth and dilute to 2×10^5 cells/ml in growth medium.

2. Aliquot 100 μl of cell suspension into each well of a microtitre plate and incubate for 24 h at 37°C, 5% CO_2 humidified atmosphere.

3. Distribute titrations of TNF standard (see Section 6 and *Table 3*) in triplicate in 200 μl volumes in medium containing 2 μg/ml actinomycin D in a different microtitre plate. Start the titration of the standard at 100 IU/ml and then make serial two-fold dilutions down to 0.78 IU/ml. Make appropriate dilutions of the samples in triplicate. For a positive control, add high TNF concentration (4000 IU/ml). For the negative control, add culture medium alone.

4. Using a multichannel pipette, transfer 100 μl of standards, samples and controls (medium alone and undiluted TNF standard) to the microtitre plate containing cells.

5. Incubate plates for 24 h at 37°C, 5% CO_2 humidified atmosphere.

6. Pour out medium and blot plate by inverting on to absorbent paper.

7. Wash cells with 100 μl phosphate buffered saline pH 7.2 and invert plate on to absorbent paper.

8. Add 100 μl naphthol blue black (NBB) stain.

9. Leave plates to stain for 30 min.

10. Pour off stain and invert plates on to absorbent paper.

11. Fix cells for 15 min with 100 μl formalin fixative.

12. Wash plates with tap-water and invert plates on to absorbent paper.

13. Add 150 μl of 50 mM sodium hydroxide to each well of the plate.

14. Carefully agitate the plate until the dye is evenly dispersed throughout the wells and determine the absorbance at 620 nm using an ELISA reader.

15. Plot a standard curve of absorbance versus concentration of TNF. Estimate activity in unknown samples by comparison with standard curve (see Section 6).

5.2 Bioassay for interferons (IFNs)

Interferons are usually assayed by their ability to reduce the viral killing of target cell types by inhibiting the replication of an infecting virus. Generally, cell-lines of human origin—Hep2/C, WISH, A549 are best for assay of human interferon and cells of mouse origin, e.g. L929 for assay of mouse interferon. However, certain interferons, e.g. IFN-α may be measured using heterologous cell-lines. The challenge viruses commonly used in these assays are encephalomycarditis virus (EMCV), vesicular stomatitis virus (VSV) and semliki forest virus (SFV). These cell–virus combinations have been used in several types of assays for measuring the potency of interferon preparations. (For details of some more of these techniques see Chapter 9.) One commonly used method, the cytopathic effect reduction (CPER) assay is rapid, economic and fairly reliable. The use of the Hep2/C—EMCV combination which is sensitive to α, β, and γ forms of interferon is described. IFN-γ can also be assayed by its ability to upregulate MHC class II expression on suitable cells, e.g. Colo 205.

5.2.1 Maintenance of Hep2/C cell-line

Culture Hep2/C cells in RPMI 1640 medium supplemented with 5% FCS in 75 cm^2 flasks. Cultures are split 1:5 to 1:10 every 2 to 3 days when the flasks are confluent. Cultures are maintained at 37°C in a humidified CO_2 incubator.

Protocol 19. Bioassay for interferons using Hep2/C cell-line

Equipment and reagents

- Hep-2/C cell culture (Section 5.2.1)
- Trypan blue
- RPMI 1640 medium containing 5% FCS
- 96-well microtitre plates
- EMCV virus (Section 5.2)
- IFN sample standard (*Table 3*)
- Test samples
- Trypsin 0.01%
- 37°C, 5% CO_2, humidified incubator

- Paper towels
- Microscope
- Amido blue/black stain (ABB): 0.05% ABB in 9% acetic acid (v/v), 0.1 M sodium acetate; filtered
- Formalin fixative (*Protocol 18*)
- 0.38% NaOH
- ELISA plate reader

Method

1. Add 100 μl of RPMI 1640 containing 5% FCS to all wells in the microtitration plate.

2. Assign wells for cell and virus controls, usually lines 1 and 12 are cell controls. In line 2, rows B, C, and D are cell controls and rows E, F, G are virus controls.

3. Add IFN sample standard at approximately 500 IU/ml to the wells in duplicate in 100 μl volumes and make serial dilutions down to approximately 0.1 IU/ml. Make appropriate dilutions of the unknown samples.

4. Trypsinize Hep2/C cells and check the viability. Resuspend cells to a concentration of 5 × 10^5 cells/ml.

5. Add 100 μl of cell suspension to each well including cell and virus controls so that the final volume in each well is 200 μl.

6. Incubate the plates for 16–24 h at 37°C in a humidified CO_2 incubator.

Protocol 19. *Continued*

7. Check plates to ensure that confluent monolayers of cells are present. Remove the growth medium in the wells by flicking and blotting on a paper towel.

8. Dilute the EMCV to about 10–30 plaque forming units/cell in growth medium, or the dilution required to effect 100% cytopathic effect in unprotected virus controls. Add 200 μl of viral suspension to all wells except the cell controls. To the latter, add 200 μl of growth medium and return the plates to the CO_2 incubator.

9. The time required for maximum CPE to develop depends on the specific cell–virus pairings. In the Hep2/C-EMCV combination, 20–24 h at 37°C are adequate, the A549-EMCV combination requires 30–48 h. Examine monolayers of the virus control wells microscopically to ascertain when maximum CPE is reached. Ideally, the cells should be completely lysed.

10. Remove medium from all wells by flicking out. Add 100 μl of Amido Blue Black

11. Remove the stain solution by flicking out. Fix the cell monolayers with 100 μl of formalin fixative and leave the plates at room temp. for at least 15 min.

12. Flick off the fixative and wash plates under running tap water for 5 min. Dry plates at room temp. or at 37°C.

13. Add 100 μl of 0.38% NaOH to each well. Make sure that the contents of each well are uniformly distributed by tapping the sides of the plates.

14. Measure the absorbance at 620 nm using an ELISA reader.

15. Plot a standard curve of absorbance versus concentration of interferon. Estimate activity in unknown samples by comparison with standard curve (see Section 6).

5.3 Bioassay for TGF-β

The classical assay for TGF-β is based on its ability to induce colony growth of fibroblasts in soft agar (27). For a routine assay, the mink lung-fibroblast line MV-3D9 can be used in a proliferation assay (28); TGF-β inhibits the proliferation of this cell-line. To achieve a greater sensitivity, however, it is better to use the recently described assay based on the ability of this cytokine to inhibit IL-5 induced proliferation of the TF-1 cell-line (29). Both assays are described here.

5.3.1 Maintenance of MV-3D9 mink lung-fibroblasts

Maintain cells in RPMI 1640 containing 5% FCS in 75 cm^2 flasks. When confluent (3–4 days), cultures are split 1:7 using 0.01% trypsin.

Protocol 20. Bioassay for TGF-β using MV-3D9 cell-line

Equipment and reagents

- MV-3D9 cell culture (Section 5.3.1)
- Trypan blue
- RPMI 1640 medium containing 5% FCS
- Trypsin 0.01%
- TGF-β standard (*Table 3*)

- Test samples
- 37°C, 5% CO_2, humidified incubator
- MTT (*Protocol 6*)
- Acid SDS (*Protocol 6*)
- ELISA plate reader

Methods

1. Trypsinize MV-3D9 cells and resuspend to a density of 5×10^3 cells/ml in RPMI 1640 containing 5% FCS.

2. Distribute titrations of TGF-β standard in triplicate in a separate microtitre plate. Start this at 10 ng/ml and serially dilute two-fold down to 10 pg/ml in 100 μl volumes. Make appropriate dilutions of the samples.[a] For a negative control add culture medium alone.

3. Add 100 μl of the cell suspension to each well and incubate plates for approximately 120 h at 37°C in a humidified CO_2 incubator.

4. Follow steps 4 and 5 of *Protocol 6*.

5. Plot a standard curve of absorbance versus concentration of TGF-β (see *Figure 4*). Estimate activity in unknown samples by comparison with the standard curve (see Section 6).

Protocol 21. Bioassay for TGF-β[a] using TF-1 cell-line[b]

Equipment and reagents

- TF-1 cell culture (Section 3.1)
- Centrifuge (MSE–benchtop)
- Trypan blue
- RPMI 1640 medium with and without 5% FCS
- TGF-β standard (*Table 3*)

- IL-5
- Test samples
- 37°C, 5% CO_2, humidified incubator
- [³H]thymidine (*Protocol 1*)
- Filter mats
- Liquid scintillation counter system

Method

1. Prepare cells as described in step 1 of *Protocol 13*.

2. Distribute titrations of TGF-β standard in 100 μl volume in triplicate in 96-well microtitre plate. Start the titration at 1 ng/ml and serially dilute two-fold down to 0.1 pg/ml. Make appropriate dilutions of the test samples. As a negative control include medium alone.

3. Add 2 ng/ml (approximately 20 U/ml) of IL-5 to the prepared cell suspension at a density of 1×10^5 cells/ml.

Protocol 21. *Continued*

4. Follow steps 4 and 5 of *Protocol 4*.

5. Plot a standard curve of c.p.m. versus concentration of TGF-β (see *Figure 4*). Estimate activity in unknown samples by comparison with the standard curve (see Section 6).

> [a] Since TGF-β is initially produced in a latent form, biological samples may require acid activation by addition of 10 μl of 1.2 M HCl per 100 μl of sample for 15 min and neutralization with 25 μl of 0.5 M Hepes/0.72 M NaOH prior to inclusion in the assay.
> [b] Maintenance of the TF-1 cell line is described in Section 3.1.

5.4 Bioassay for oncostatin-M

This cytokine can be measured by its ability to inhibit the growth *in vitro* of several tumour cell-lines, e.g. the A375 melanoma line, or by stimulation of proliferation of the TF-1 cell-line (*Protocol 13*).

5.4.1 Maintenance of A375 cell-line

Culture A375 cells in Dulbecco's minimal essential medium containing 5% FCS, 2 mM L-glutamine, 0.01% non-essential amino-acids, and 1 mM sodium pyruvate. Cells are subcultured twice a week using 0.01% trypsin and seeded at 1×10^4 cells/ml.

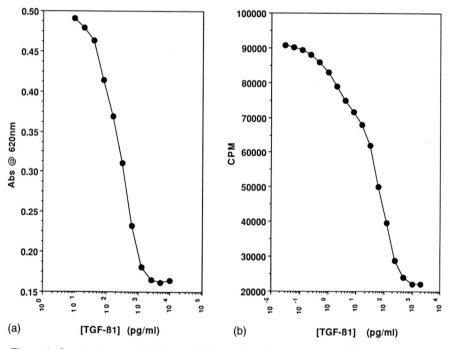

(a) [TGF-ß1] (pg/ml)   (b) [TGF-ß1] (pg/ml)

Figure 4. Standard curves for human TGF-β using (a) the MV-3D9 cell-line (see *Protocol 20*) and (b) the TF-1 cell-line (see *Protocol 21*).

Protocol 22. Bioassay for oncostatin-M using A375 cells

Equipment and reagents

- Log-phase A375 cells (Section 5.4.1)
- Dulbecco's minimal essential medium containing 5% FCS, 2 mM L-glutamine, 0.01% non-essential amino acids, 1 mM sodium pyruvate
- Trypsin 0.01%
- Trypan blue
- Oncostatin-M standard (*Table 3*)
- Test samples

- 96-well microtitre plates
- 37°C, 5% CO_2, humidified incubator
- PBS
- Methanol
- Absorbent paper
- 0.1% crystal violet
- 2% sodium deoxycholate
- ELISA plate reader

Method

1. Trypsinize cells in log-phase growth and dilute to 2×10^5 cells/ml in growth media.

2. Distribute titrations of the oncostatin-M standard in 50 μl volumes in triplicate in 96-well microtitration plates. Start the titration at 5 ng/ml (approximately 250 U/ml) and make serial two-fold dilutions down to approximately 5 pg/m (0.25 U/ml). Make appropriate dilutions of the samples to be tested for oncostatin-M activity in triplicate, in 50 μl volumes. As a negative control use culture medium alone.

3. Add 50 μl of cell suspension to each well and incubate the plates for about 72 h at 37°C in a humidified CO_2 incubator.

4. Invert and flick plates over a sink to discard supernatant.

5. Carefully wash plates three times with PBS at room temperature.

6. Add 100 μl of methanol to each well and allow to stand at room temp. for 15–20 min. Discard the methanol and blot plates on to absorbent paper.

7. Add 100 μl of 0.1% crystal violet and leave for 5 min at room temp. Wash plates gently with tap water until wash is clear, then blot on to absorbent paper.

8. Add 100 μl of 2% sodium deoxycholate and agitate plates to disperse stain.

9. Measure the absorbance at 620 nm using an ELISA reader.

10. Plot a standard curve of absorbance versus concentration of standard. Estimate activity in unknown samples by comparison with standard curve (see Section 6).

Table 3. Availability of human cytokine standards and reference reagents

Cytokine	Cytokine standards	
	Status	Source
Interleukin-1 alpha, rDNA	IS	NIBSC*
Interleukin-1 beta, rDNA	IS	NIBSC*
Interleukin-2, cell line derived	IS	NIBSC*
Interleukin-2, rDNA	RR	NIBSC
Interleukin-3, rDNA	IS	NIBSC*
Interleukin-4, rDNA	IS	NIBSC*
Interleukin-5, rDNA	RR	NIBSC*
Interleukin-6, rDNA	IS	NIBSC*
Interleukin-7, rDNA	RR	NIBSC*
Interleukin-8, rDNA	RR	NIBSC*
Rantes, rDNA	RR	NIBSC
Macrophage inflammatory protein-1 alpha, rDNA	RR	NIBSC
Interleukin-9, rDNA	RR	NIBSC
Interleukin-10, rDNA	RR	NIBSC
Interleukin-11, rDNA	RR	NIBSC
Macrophage colony stimulating factor, rDNA	IS	NIBSC*
Granulocyte colony stimulating factor, rDNA	IS	NIBSC*
Granulocyte-macrophage colony stimulating factor (GM-CSF), rDNA	IS	NIBSC*
Leukaemia inhibitory factor, rDNA	RR	NIBSC*
Stem cell factor (SCF), rDNA	RR	NIBSC
Oncostatin M, rDNA	RR	NIBSC
Transforming growth factor $\beta1$, rDNA	RR	NIBSC*
Transforming growth factor $\beta2$, rDNA	RR	NIBSC
Monocyte chemoattractant protein-1, rDNA	RR	NIBSC
Gro alpha, rDNA	RR	NIBSC
Tumour necrosis factor alpha, rDNA	IS	NIBSC*
Tumour necrosis factor beta, rDNA	RR	NIBSC
Interferon alpha, leukocyte derived	IS	NIBSC
Interferon alpha-1, rDNA	IS	NIBSC
Interferon alpha-2a, rDNA	IS	NIAID
Interferon alpha-2b, rDNA	IS	NIBSC
Interferon beta, fibroblast derived	2nd IS	NIAID
Interferon beta [Ser[17]], rDNA	IS	NIAID
Interferon gamma, leukocyte derived	BR STD	NIBSC
Interferon gamma, leukocyte derived	IS	NIAID

rDNA, recombinant DNA; IS, International Standard; RR, Reference Reagent; NIBSC, National Institute for Biological Standards and Control, South Mimms, Herts. EN6 3QG, UK; NIAID, National Institute of Allergy and Infectious Diseases, National Institutes of Health, Bethesda, MD20205, USA. * Indicates availability from BRMP, Biological Response Modifiers Program, National Cancer Institute, Frederick, Maryland 21701.

6. Analysis of results

It is essential that bioassay data is analysed correctly and statistically evaluated. This is particularly important if a valid biological potency is to be assigned to a preparation and if samples are considered to have significantly

Table 4. Availability of murine cytokine standards and reference reagents

Cytokine	Cytokine standards	
	Status	Source
Interleukin-1 alpha	RR	NIBSC
Interleukin-1 beta	RR	NIBSC
Interleukin-2	RR	NIBSC
Interleukin-3	RR	NIBSC
Interleukin-4	RR	NIBSC
Interleukin-6	RR	NIBSC
Interleukin-7	RR	NIBSC
Interleukin-9	RR	NIBSC
Tumour necrosis factor alpha	RR	NIBSC
Granulocyte macrophage colony stimulating factor	RR	NIBSC
Interferon alpha	IS	NIAID
Interferon beta	IS	NIAID
Interferon alpha/beta	2nd IS	NIAID
Interferon gamma	IS	NIAID

RR, Reference Reagent; IS, International Standard; NIBSC, National Institute for Biological Standards and Control, South Mimms, Herts. EN6 3QG, UK; NIAID, National Institute of Allergy and Infectious Diseases, National Institutes of Health, Bethesda, MD20205, USA.

different cytokine activities. A useful approach for comparison of unknown samples with standard preparations is parallel line analysis. For this, the unknown samples are titrated and then compared to the standard curve of known unitage. The parallel portions of these curves are then used to measure the displacement from the standard which is proportional to the biologically active cytokine content of the samples. These lines should be parallel if the molecule responsible for the activity in samples/standards is the same. For a detailed account of bioassay analysis see (30). Alternatively an approximate estimate can be made by taking two or three points from the titration curve and reading the values from the standard curve. 'Single point' determinations are often misleading. Inclusion of reference standards is essential; these can be obtained from National Institute of Biological Standards and Controls (NIBSC, UK) or National Cancer Institute (NCI, USA) —see *Tables 3* and *4*. An 'in-house' laboratory standard should be produced for routine use; this should be calibrated directly against the international standard or reference preparation.

Acknowledgements

We would like to thank Immunex Corporation, Schering Plough Corporation, Cetus Corporation, Bristol Myer Squibb, Collagen Corporation, R & D Systems, Glaxo Institute of Molecular Biology, Genentech Inc, Sandoz

Meenu Wadhwa et al.

Limited, Genetics Institute, Roche Products Limited, Chugai Pharmaceutical Company Limited, Dainippon Pharmaceutical Company Limited, and Amgen for their generous gift of rDNA cytokines which have been used to develop and characterize some of the assays described. We are grateful to Drs G. Rovera, L. Pegoraro, T. Kitamura, D. Urdal, L. Aarden, C. Sanderson, P. Lindquist, W. Paul, F. Moreau, and Mr J. Weaver for providing some of the cell lines described and Deborah Kirk for preparing the manuscript.

References

1. Thorpe, R., Wadhwa, M., Bird, C. R., and Mire-Sluis, A. R. (1992). *Blood Reviews*, **6**, 133.
2. Hannum, C. H., Wilcox, C. J., Arend, W. P., Joslin, F. G, Dripps, D. J., Heimdal, P. L., Armes, L. G., Sommer, A., Eisenberg, S. P., and Thompson, R. C. (1990). *Nature*, **343**, 336.
3. Seckinger, P., Zhang, J. H., Hauptmann, B., and Dayer, J. M. (1990). *Proc. Natl. Acad. Sci., USA*, **87**, 5188.
4. Ruscetti, F. W. and Palladino, M. A. (1991). *Prog. Growth Factor Res.*, **3**, 159.
5. Thorpe, R., Wadhwa, M., Gearing, A., Mahon, B., and Poole, S. (1988). *Lymphokine Res.*, **7**, 119.
6. Wadhwa, M., Thorpe, R., Bird, C. R., and Gearing, A. J. H. (1990). *J. Imm. Meths.*, **128**, 211.
7. Schrier, M. H. and Tees, R. (1981). *Immunology Methods*, **2**, 263. Acad. Press, NY, USA.
8. Orrencole, S. F. and Dinarello, C. A. (1989). *Cytokine*, **1**, 14.
9. Hu-Li, J., Ohara, J., Watson, C., Tsang, W., and Paul, W. E. (1989). *J. Immunol.*, **142**, 800.
10. Park, L. S., Friend, D. J., Schmierer, A. E., Dower, S. K., and Namen, A. E. (1990). *J. Exp. Med.*, **171**, 1073.
11. Vieira, P., de Waal-Malefyt, R., Dang, M.-N., Johnson, K. E., Kastelein, R., Fiorentino, D. F., deVries, J. E., Roncarolo, M.-G., Mosmann, T. R., and Moore, K. W. (1991). *Proc. Natl. Acad. Sci. USA*, **88**, 1172.
12. Thompson-Snipes, L., Dhar, V., Bond, M. W., Mosmann, T. R., Moore, K. W., and Rennick, D. M. (1991). *J. Exp. Med.*, **173**, 507.
13. Nordan, R. P. and Potter, M. (1986). *Science*, **223**, 566.
14. Lu, Z.-Y., Zhang, X-G., Gu, Z-J., Yasukawa, K., Etrillard, M., and Klein, B. (1994). *J. Imm. Meths.*, **173**, 19.
15. Stern, A. S., Podlaski, F. J., Hulmes, J. D., Pan, Y.-C. E., Quinn, P. M., Wolitzky, A. G., Familletti, P. C., Stremlo, D. L., Truitt, T., Chizzonite, R., and Gately, M. K. (1990). *Proc. Natl. Acad. Sci. USA*, **187**, 6808.
16. Hori, T., Uchiyama, T., Tsudo, M., Umadome, H., Ohno, H., Fukuhara, S., Kita, K., and Uchino, H. (1987). *Blood*, **70**, 1069.
17. Ambrus, J. L. Jr., Pippin, J., Joseph, A., Xu, C., Blumenthal, D., Tamayo, A., Claypool, K., McCourt, D., Srikiatchatochorn, A., and Ford, R. J. (1993). *Proc. Natl. Acad. Sci. USA*, **90**, 6330.
18. Metcalf, D. (1984). *The hemopoietic colony stimulating factors*. Elsevier Press, Amsterdam.

19. Weinstein, Y., Ihle, J.N., Lavu, S., and Reddy, E.P. (1986). *Proc. Natl. Acad. Sci.*, **83**, 5010.
20. Nakoinz, I., Lee, M., Weaver, J. F., and Ralph, P. (1989). In *7th International Congress of Immunology*, Abs. 40–23 Gustav Fischer, Stuttgart, New York.
21. Kitamura, T., Tange, T., Terasawa, T., Chiba, S., Kuwaki, T., Miyagawa, K., Piao, Y., Miyazono, K., Orabe, A., and Takaku, F. (1989). *J. Cell. Physiol.*, **140**, 323.
22. Lange, B., Valtieri, M., Santoli, D., Caracciolo, D., Mavilio, F., Gemperlein, I., Griffin, C., Emanuel, B., Finan, J., Nowell, P., and Rovera, G. (1987). *Blood*, **70**, 192.
23. Avanzi, G. C., Lista, P., Giovinazzo, B., Miniero, R., Saglio, G., Benetton, G., Coda, R., Cattonetti, G., and Pegoraro, L. (1988). *Br. J. Haematol.*, **69**, 359.
24. Moreau, J. F., Bonneville, M., Peyrat, M. A., Jacques, Y., and Soulillou, J.P. (1986). *Ann. Inst. Pasteur/Immunol.*, **137**, 25.
25. Gascan, H., Moreau, J. F., Jacques, Y., and Soulilou, J. P. (1989). *Lymphokine Res.*, **8**, 79.
26. Meager, A. (1991). *J. Immunol. Meth.*, **144**, 141.
27. Assoian, R. K., Komoriya, A., Meyers, C. A., Miller, D. M., and Sporn, M. B. (1983). *J. Biol. Chem.*, **258**, 7155.
28. Like, B. and Massague, J. (1986). *J. Biol. Chem.*, **261**, 13426.
29. Randall, L. A., Wadhwa, M., Thorpe, R., and Mire-Sluis, A. R. (1993). *J. Immunol. Meth.*, **164**, 61.
30. *European Pharmacopoeia.* (1971). **II**, 441.

RIA, IRMA, and ELISA assays for cytokines and their soluble receptors

A. MEAGER

1. Introduction

The lack of absolute specificity of cultured mammalian cells for the activities of individual cytokines and the often relatively poor reproducibility of cytokine, bioassays has created the need for more specific and reproducible, alternative assays. This need has largely been filled by the development of immunoassays for cytokines. Such assays are based on antibodies to each different cytokine, are reasonably sensitive and reliable, and are quick and easy to perform. The amount of cytokine measured in immunoassays is related to the extent of antibody binding and this in turn may be correlated to biological activity (e.g. units/ml) when suitable cytokine standards of known biological potency are available. Alternatively, homogeneous cytokines of known protein content can be used to calibrate immunoassays in terms of cytokine concentration (e.g. pg or ng/ml).

Cell surface receptors for individual cytokines have been characterized as transmembrane glycoproteins. In addition, it is now known that for the majority of cytokine receptors their glycosylated extracellular domains may be cleaved enzymatically from the cell surface to form 'soluble' receptors. These soluble receptors can enter the circulation and may act as natural physiological 'buffers' of cytokine action since they are still able to bind their respective, cognate cytokine. Alternatively, they may act to transport cytokines to certain cells, tissues or organs (e.g. to the kidneys for elimination). Levels of particular soluble cytokine receptors, e.g. soluble interleukin-2 receptor (sIL-2R), soluble tumour necrosis factor 75 kda receptor (sTNF75R), have been found to be raised in various acute and chronic pathological conditions, and can be used as disease markers. Their measurement, like cytokines themselves, can be accomplished by the use of appropriate immunoassays.

Broadly speaking, three types of immunoassay—radioimmunoassay (RIA), immunoradiometric assay (IRMA), and enzyme-linked immuno-absorbent assay (ELISA), are in regular use for cytokine quantification. All three immunoassays require the availability of monospecific antibodies, i.e.

antibodies recognizing only that cytokine to be quantified. In addition, RIA requires radiolabelled, e.g. ^{125}I, pure cytokine, and IRMA and ELISA require one radiolabelled antibody and one enzyme-linked (or biotinylated) antibody, respectively. Soluble cytokine receptors have mainly been quantified using the ELISA approach, but IRMA may also be used.

It is beyond the scope of the present chapter to describe the production and purification of anti-cytokine- and soluble receptor-immunoglobulins (Ig); these are adequately dealt with elsewhere (ref. 1 and Chapter 19 in this volume). However, methods for radiolabelling cytokines and anti-cytokine Ig, and for linking anti-cytokine Ig to suitable enzymes or biotin are detailed below.

2. Radioimmunoassays (RIA)

These require that the cytokine to be quantified is available as a pure, homogeneous protein (or glycoprotein) and that the latter can be radiolabelled to a high specific activity without untoward structural alterations being induced. There are several ways in which cytokines may be radiolabelled with ^{125}I, and the choice of method largely depends on the robustness of the cytokine to the iodination conditions. For example, chloramine T (2) will undoubtedly radiolabel cytokines to very high specific activities, but in general it is too denaturing and leads to loss of biological activity. Other methods therefore are to be preferred and these include the iodogen (Pierce/Sigma Chemical Co.), Enzymobead (Bio-rad) and Bolton-Hunter (Amersham International) methods. The suitability of these for individual cytokines should, if possible, be determined empirically. As an example, *Protocol 1* details the iodogen method as applied to radiolabelling of tumour necrosis factor alpha (TNF-α).

Protocol 1. The iodogen method for radiolabelling TNF-α

Equipment and reagents

- Eppendorf (or small glass) tube
- 1 mg/ml iodogen (1,3,4,6-tetrachloro-3α, 6α-diphenyl-glycouril) in trichloromethane
- TNF-α
- 0.25 M sodium phosphate buffer, pH 6.9
- Carrier-free [^{125}I]Na (100 mCi/ml)
- Disposable 2 ml Sephadex G-25 column
- 2 mg/ml BSA in PBS (BSA-PBS)
- Gamma counter system

Methods

1. Coat an Eppendorf (or small glass) tube with 40 μl of iodogen at 1 mg/ml in trichloromethane, by solvent evaporation.

2. Add TNF-α, 5 μg in 30 μl of 0.25 M sodium phosphate buffer, pH 6.9, together with 10 μl (1 mCi) carrier-free [^{125}I] Na, to the iodogen-coated tube. Keep on ice for 10 min.

---

3. Transfer the contents of the tube to a disposable 2 ml Sephadex G-25 column, previously equilibrated with bovine serum albumin (2 mg/ml) in phosphate-buffered saline (PBS). Wash the tube once with 40–50 µl phosphate buffer and add this to the Sephadex column.

4. Elute the column with BSA-PBS and collect twelve 200 µl fractions. Count these in a gamma counter and determine the peak of radioactivity (usually in fractions 6–8). Store radiolabelled [^{125}I] TNF-α at 4°C. It will be stable for up to 30 days.

Instructions for ^{125}I-labelling with Enzymobeads (Bio-rad) and the Bolton-Hunter reagent (Amersham) are supplied by the manufacturers. A further alternative is the *N*-bromo-succinimide (NBS) method of Reay (3) and this has been successfully applied to the radioiodination of interleukin-1β (4).

Generally speaking, most current radioimmunoassays for cytokines employ a competitive inhibition assay method. In brief, this means that variable amounts of cytokine, as serial dilutions of a standard or in samples, are incubated with a fixed amount of diluted polyclonal anti-cytokine antiserum, followed by a further incubation period with a fixed quantity of ^{125}I-labelled cytokine. Finally, antibody–cytokine complexes are removed from solution by addition of a second antibody (to the first species immunoglobulin) or other antibody-binding reagent, e.g. protein A. The amount of ^{125}I-cytokine bound therefore decreases as the concentration of unlabelled cytokine increases.

An example of this type of radioimmunoassay is given in *Protocol 2*.

Protocol 2. A cytokine radioimmunoassay

Equipment and reagents

- Cytokine standard and samples
- Assay diluent (BSA-PBS)
- Polystyrene tubes
- Rabbit polyclonal anti-cytokine
- ^{125}I-labelled cytokine tracer (e.g. 10 000 c.p.m.; 100 µCi/mg)
- 1.5% sheep anti-rabbit IgG in 4% polyethylene glycol
- Centrifuge (bench top)
- Gamma counter

Method

1. Add serial dilutions of cytokine standard or samples in 100 µl of assay diluent to polystyrene tubes containing diluted rabbit polyclonal anti-cytokine (in 300 µl assay diluent/tube) and incubate for 24–28 h at 4°C. The dilution of the rabbit polyclonal giving maximum counts per minute (c.p.m.) of bound ^{125}I-cytokine (B_o) in the absence of unlabelled cytokine should be predetermined.

2. Add ^{125}I-cytokine tracer (e.g. 10 000 c.p.m.; 100 µCi/mg) in 100 µl assay diluent and incubate for a further 20 h at 4°C.

Protocol 2. *Continued*

3. Separate free and bound cytokine by adding 1.5% sheep anti-rabbit IgG (0.5–1.0 ml) in 4% polyethylene glycol[a] (16–20 kDa) Mix and incubate for 1 h at room temp. then centrifuge at 1000 *g* for 30 min.

4. Count radioactivity (*B*) pelleted in the assay tubes and express results as a percentage of B_o ($B/B_o \times 100$; *Figure 1a*).

[a] The polyethylene glycol separation step can be replaced if the second antibody is immobilized on beads or particles. For example, the Amerlex M system (Amersham International) uses magnetic separation of second antibody-coated particles.

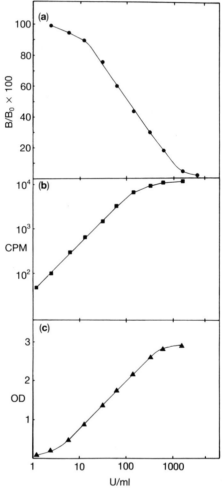

Figure 1. Typical calibration curves from (a) RIA, (b) IRMA, and (c) ELISA, based on in-house data for a human TNF-α standard (1 IU/ml = 25 pg TNF-α/ml).

3. Immunoradiometric assays (IRMA)

3.1. IRMA assays for cytokines

In these assays the concentration of cytokine in a sample is determined by the amount of ^{125}I-anti-cytokine IgG bound to cytokine captured by a first, immobilized, anti-cytokine antibody (1). Purified antibodies, but not pure cytokines, are therefore required for IRMA. For optimum performance, it is generally necessary to use two cytokine-specific antibodies, each of which recognizes a different epitope or antigenic determinant on the cytokine molecule, particularly when the cytokine is monomeric. Steric separation of the epitopes recognized is essential for the development of highly sensitive assays. Ideally, the use of two complementary anti-cytokine monoclonal antibodies (MAb) gives the most effective combination for IRMA, although polyclonal anti-cytokine IgG can also be used in many cases.

For the capture antibody, anti-cytokine IgG is purified from (a) ascitic fluid containing anti-cytokine MAb or (b) polyclonal antiserum.

Protocol 3. Purification of anti-cytokine IgG

Equipment and reagents

- Saturated ammonium sulphate solution pH 7.4
- Centrifuge (bench top)
- PBS
- Ultragel ACA44 (or Sephacryl 200)
- Elution buffer: 0.1 M sodium phosphate buffer, pH 7.2, containing 0.4 M NaCl

Methods

1. Add solid ammonium sulphate, pH 7.4, to 35% saturation to the serum or ascitic fluid and centrifuge the precipitate formed at 4°C at 5000 *g* for 10 min.
2. Dissolve the protein precipitate in a small volume of PBS and dialyse against PBS overnight at 4°C.
3. Separate anti-cytokine IgG by fractionation on Ultragel ACA44 (or Sephacryl 200) using 0.1 M sodium phosphate buffer, pH 7.2, containing 0.4 M NaCl, to elute.
4. Test fractions for anti-cytokine IgG and pool the peak fractions.

This is generally sufficient purification for the capture antibody. However, the antibody to be radiolabelled should be further purified by either:

(a) Protein A-Sepharose chromatography (5)

(b) DEAE-Affigel blue chromatography (6)

(c) HPLC (7).

Protein content may be estimated by the Lowry method (8).

Purified anti-cytokine IgG may be radiolabelled with ^{125}I using the chloramine T method (2) (cf. radioiodination of cytokines in Section 2) as described below.

Protocol 4. Chloramine T method for radiolabelling IgG

Equipment and reagents

- MAb IgG
- Eppendorf tubes
- 0.1 M sodium phosphate buffer, pH 7.4
- [^{125}I]Na (10 mCi/ml)
- 5 mg/ml chloramine T (in distilled water)
- 0.4 mg/ml L-tyrosine (in sodium phosphate buffer)
- Disposable 2 ml Dowex or Sephadex chromatography column
- BSA-PBS (*Protocol 1*)
- 10% sodium azide

Method

1. Add 10 μg of MAb IgG to an Eppendorf tube containing 10 μl of 0.1 M sodium phosphate buffer, pH 7.4 and 10 μl (100 μCi) of [^{125}I]Na in the same buffer.

2. Add 10 μl of freshly prepared chloramine T (5 mg/ml in distilled water), mix, and stir or agitate the tube contents for 45 sec.

3. Terminate reaction by the addition of 50 μl of L-tyrosine (0.4 mg/ml in sodium phosphate buffer).

4. Pass the iodination mixture through a disposable 2 ml Dowex or Sephadex chromatography column, and elute [^{125}I] MAb with BSA-PBS. Pool peak fractions and store with a drop of 10% sodium azide at 4°C.

A method for carrying out IRMA, essentially that described by Secher (9), is outlined below as applied to the estimation of TNF-α.

Protocol 5. IRMA assay for TNF-α

Equipment and reagents

- Purified capture antibody (an anti-TNF MAb)
- PBS
- Etched polystyrene balls (6.5 mm; Northumbria Biologicals)
- Diluted antibody
- Glass Universal bottles
- 0.1% BSA-PBS
- 0.5% BSA-PBS
- 100, or more, Luckham LP4 tubes
- TNF-α standard
- Assay diluent: e.g. cell growth medium containing 10% calf serum (see Section 5.2)
- Paper towels
- [^{125}I]anti-TNF-α second MAb
- Gamma counter

Method

1. Prepare immobilized antibody as follows: dilute purified capture antibody to approximately 50 μg protein/ml in PBS. Add approx. 100

398

etched polystyrene balls to 14 ml of diluted antibody in a glass Universal. Submerge the beads in the antibody solution overnight at 4°C, then aspirate the antibody solution and wash 4–5 times with 0.1% BSA-PBS. The beads may be stored under 0.1% BSA-PBS at 4°C for several weeks, if necessary.

2. Simultaneously, fill one hundred or more Luckham LP4 tubes with 0.5% BSA-PBS to block any binding to plastic surfaces, and leave overnight at 4°C. Remove tube contents by aspiration just prior to setting up the assay.

3. Prepare serial dilutions of TNF-α standard covering the range 0.1 IU (2.5 pg)/ml–1000 IU (25 ng)/ml in assay diluent. The latter should be identical, if possible, to the medium of the samples to be tested, e.g. cell growth medium containing 10% calf serum (see Section 5.2 for further discussion). Add 200 µl of TNF-α standard dilutions or samples to LP4 tubes.

4. Blot the washed antibody-coated beads to dryness on paper towels and add one bead per assay tube. The bead should be completely submerged and there should be no bubbles. Incubate assay tubes overnight at 4°C.

5. The following day, remove standard dilutions and samples by aspiration, and wash beads extensively with 0.1% BSA-PBS or simply water before addition of 200 µl of [^{125}I] anti-TNF-α second MAb to all tubes. The second MAb should be diluted in 0.1% BSA-PBS so that the resulting solution contains approximately 10^6 c.p.m./ml.

6. Leave assay tubes a further 4 h at 4°C. Remove unbound [^{125}I]-anti-TNF-α by aspiration and wash extensively with 0.1% BSA-PBS or water.

7. Count the tubes containing beads in a gamma-radiation counter. The data so obtained may be plotted as c.p.m. bound (minus negative control, i.e. non-specific binding) versus TNF-α activity/concentration, in IU/ml or pg/ml, as $\log^{10}$–$\log^{10}$ or semi-$\log^{10}$ plots to generate the calibration curve. Titres (IU/ml) or concentrations (pg/ml) of TNFα in test samples may then be simply interpolated from the calibration curve (*Figure 1b*).

3.2 IRMA assays for soluble cytokine receptors

IRMA for soluble cytokine receptors can be carried out in a manner similar to *Protocol 5*, except the capture anti-cytokine MAb is replaced by anti-soluble receptor MAb, serial dilutions of cytokine standard are replaced by serial dilutions of purified soluble receptors, preferably a reference preparation of known receptor protein content, and the ^{125}I-labelled detector anti-cytokine MAb is substituted by an appropriate ^{125}I-labelled anti-soluble receptor MAb (labelling with [^{125}I] is carried out as described in *Protocol 4*). As an alternative strategy, if capture anti-soluble receptor MAbs are available that do not interfere with the binding of cognate cytokine to soluble receptor, ^{125}I-labelled cytokine (labelling with [^{125}I] is carried out according to *Protocol 1* or similar) may be used instead of detector ^{125}I-labelled anti-soluble receptor MAb, i.e. immobilized MAb–soluble cytokine receptor-[^{125}I]cytokine. The latter guarantees that only functionally-active soluble receptors are detected and quantified.

4. Enzyme-linked immunosorbent assays (ELISA)

The principles involved in ELISA are the same as those for IRMA, except that the second anti-cytokine IgG is:

(a) conjugated to an enzyme;

(b) itself subject to detection with a third antibody-enzyme complex or variable combinations of biotinylated antibodies and streptavidin-enzyme complexes (10).

Purification of anti-cytokine IgG for ELISA is the same as that described for IRMA (see Section 3). Three enzymes, horseradish peroxidase, alkaline phosphatase, and β-galactosidase, are regularly used for conjugation to second and third antibody or streptavidin ELISA reagents. Enzyme may be directly coupled to IgG or streptavidin using a variety of methods, the simplest of which involves mixing the reagents with a low percentage solution of glutaraldehyde (11). Alternatively, antibody reagents may be biotinylated using an *N*-hydroxysuccinimide-biotin ester and detected using streptavidin reagents, which bind with high affinity to biotin molecules (12). Synthetic biotin esters with a 'spacer arm' which ensures that the protein-linked biotin is freely accessible to bind streptavidin are commercially available (e.g. Amersham International) and it is recommended that the manufacturer's instructions for biotinylation reactions are followed.

Protocol 6 illustrates how various combinations of enzyme-linked and biotinylated-reagents may be effectively used in the construction of ELISA.

Protocol 6. ELISA for cytokines

Equipment and reagents

- 96-well micro-ELISA plates
- Anti-cytokine IgG
- Enzyme-linked (horseradish peroxidase or alkaline phosphatase conjugated) second anti-cytokine IgG
- 0.1 M sodium bicarbonate buffer, pH 9.6
- 1% BSA-PBS
- 0.05% Tween-20 in PBS (Tween-20-PBS) *or* 0.1% Synperonic F68 (Serva)-PBS
- Cytokine standard
- Test samples
- 0.1 M citrate-phosphate buffer, pH 5.0

- *For horseradish peroxidase (HRT) conjugates:* 1 mg/ml orthophenylenediamine (OPD) in citrate-phosphate buffer, 0.006% hydrogen peroxide
- *For alkaline phosphatase (AP) conjugates*: 1 mg/ml *p*-nitro-phenylphosphate (pNPP)

in 1 M diethanolamine buffer, 0.5 mM MgCl$_2$, pH 9.8
- 1 M H$_2$SO$_4$ (HRP)
- 3 M NaOH (AP)
- ELISA plate reader

Methods

1. (The first step is similar in all cases.) Coat 96-well micro-ELISA plates with an anti-cytokine IgG by addition of 50–100 μl (depending on well-size) of anti-cytokine IgG (2.5–10 μg/ml) in 0.1 M sodium bicarbonate buffer, pH 9.6, to plate wells and incubate for 2 h at 37°C. Block the remaining sites in wells by addition of 150–200 μl of 1% BSA-PBS per well for at least 20 min. Seal plates and store at 4°C until required.

2. Remove unbound anti-cytokine IgG and blocking buffer by flicking out and wash the wells a few times with 0.05% Tween-20-PBS or 0.1% Synperonic F68 (Serva)—PBS (150–200 μl/well). Following the last wash, add serial dilutions of cytokine standard and undiluted/diluted samples (50–100 μl/well) in duplicate/triplicate and incubate at 37°C for 1 h. (Incubation for longer periods may improve assay sensitivity and reproducibility.)

3. Wash wells four times with 0.05% Tween-20-PBS or 0.1% Synperonic-PBS. Add enzyme-linked second anti-cytokine IgG, appropriately diluted in 1% BSA-PBS, to all wells (50–100 μl/well) and continue incubation for a further 1 h at 37°C.

4. Wash wells four times with 0.05% Tween-20-PBS or 0.1% Synperonic-PBS. Additional washes with the buffer solution used for making up the enzyme substrate are recommended before adding the substrate solution. For example, wash wells twice with 0.1 M citrate-phosphate buffer, pH 5.0, prior to the addition of horseradish peroxidase substrate, orthophenylenediamine (OPD), at 1 mg/ml in the citrate-phosphate buffer containing 0.006% hydrogen peroxide. For alkaline phosphatase conjugates, wash wells twice with water prior to the addition of *p*-nitro-phenylphosphate (pNPP) at 1 mg/ml in 1M diethanolamine buffer, 0.5 mM MgCl$_2$ pH 9.8.

5. Colour development is usually carried out in the dark for most substrates to avoid non-specific coloration occurring, and thus achieve low background optical densities. The enzyme reaction, and hence colour development, may be terminated by addition of 1 M H$_2$SO$_4$ in the case of horseradish peroxidase and 3 M NaOH in the case of alkaline phosphatase.

6. Read optical densities (OD) at the wavelengths appropriate to the colour in the wells, e.g. 492 nm for OPD and 405 nm for pNPP. Plot data as OD versus cytokine activity/concentration in a similar way to that described for IRMA (Section 3, *Figure 1c*).

Protocol 6. *Continued*

Variation 1. In step 3 addition of enzyme-linked second anti-cytokine IgG is replaced by addition of biotinylated second anti-cytokine IgG. This generates an extra step 3(a) in which, following removal of biotinylated IgG and washing wells with 0.05% Tween-20-PBS, an appropriately diluted streptavidin-enzyme conjugate is added to all wells. Since the binding of the streptavidin reagent is rapid, the incubation time after its addition is 20–30 min at 37°C, before proceeding to step 4 and continuing as previously outlined.

Variation 2. In step 3 addition of enzyme-linked second anti-cytokine IgG is replaced by addition of appropriately diluted, unconjugated second anti-cytokine IgG. Processing this ELISA then entails one or two extra steps; 3(b) addition of enzyme-linked anti-species IgG (to the second anti-cytokine IgG) or 3(c) addition of biotinylated anti-species IgG followed by further addition of a streptavidin–enzyme conjugate. Each addition is preceded by extensive washing of wells with 0.05% Tween-20-PBS.

The use of biotinylated reagents together with multivalent streptavidin–enzyme conjugates may enhance the sensitivity of an ELISA considerably, but may also tend to increase non-specific binding. It is clear that different combinations will have their advantages and disadvantages, and operators should, where possible, use the most suitable combination for a particular need. For example, samples containing large amounts of a cytokine would, in general, require a less sensitive and sophisticated ELISA than for samples containing low levels.

ELISA for soluble cytokine receptors may be carried out by following *Protocol 6,* but replacing the anti-cytokine Ig reagents by the appropriate anti-soluble receptor Ig reagents and by substituting the serial dilutions of cytokine standard with those of the relevant purified soluble receptor reference preparation. Alternatively, where anti-cytokine/anti-soluble receptor Ig reagents are available that do not interfere with cytokine binding to soluble receptor, other ELISA formats are possible, e.g. immobilized anti-soluble receptor Ig-(soluble receptor-cytokine complexes)-anti-cytokine Ig-enzyme. *Variations 1* and *2*, as outlined above, may also be applied to ELISA for soluble cytokine receptors.

5. Problems

5.1 Calibration

For all types of immunoassays, correct calibration is vital. The standard used for calibration should contain the cytokine or soluble cytokine receptor to be quantified in a molecular form(s) that it is known or predicted to be representative of the cytokine or receptor molecules present in samples. For example, a human recombinant IFN-γ standard which contains only non-glycosylated molecules should be used in the immunoassay of samples containing non-glycosylated IFN-γ; such a standard may be unsuitable for calibration when samples containing leukocyte-derived, glycosylated IFN-γ molecules are assayed. Where cytokines exist naturally as mixtures of a number of closely related molecular species, e.g. IFN-α, it may not be possible to exactly duplicate the proportions of these in a standard preparation (1).

5.2 Mixtures of cytokine and soluble receptors

In some cases, samples can contain a mixture of free cytokine and cytokine-soluble receptor complexes, and immunoassays may only detect free cytokine because relevant antigenic determinants are hidden by soluble receptor in the cytokine-soluble receptor complexes. Such samples can lead to difficulties in interpretation of results, even though the correct calibrants have been employed, i.e. it may be concluded that a cytokine is absent or present in much lower levels than actually exist. However, this problem may be overcome by utilizing anti-cytokine Ig reagents that: (a) recognize epitopes on the cytokine molecule that are not obscured by soluble receptors, or (b) are of high enough affinity to compete with and remove soluble receptors from cytokine molecules. It should be noted that such immunoassays give a measure of the total cytokine content of a sample, but give no indication of the amounts of free-(active-) and complexed-(inactive-) cytokine present.

5.3 Assay diluent or matrix

The diluent or matrix for immunoassays should be identical for the cytokine standard or soluble receptor reference preparation and for samples. 'Recognition' of cytokine/soluble receptor is often influenced by overall concentrations of other molecules, e.g. proteins, mucopolysaccharides, in the matrix, and appropriate precautions should be taken to ensure that these do not unduly influence the results. For example, biological fluids such as sera or synovia may be highly viscous when undiluted and, where possible, it is recommended that these be diluted with an appropriate buffer, or enzyme-treated, e.g. hyaluronidase for synovial fluids, to reduce viscosity before assay. For the estimation of cytokines/soluble receptors in biological fluids, the cytokine standard/soluble receptor reference preparation should be prepared by serial dilution in a comparable fluid, previously shown to give only background or baseline OD in the immunoassay.

5.4 Samples from different sources

Problems are most likely to arise in the assay of differently-sourced biological fluids containing very low levels of cytokine, and those where cytokine is actually absent. Here, there will be some variation in composition of individual samples and it will be virtually impossible to match the matrices of these to that used for standard dilution. This often leads to some scatter about the assay background OD reading and consequent difficulties in the interpretation of results. Additionally, biological fluids may contain substances other than cytokine which can 'bridge' the first and second anti-cytokine IgGs and generate false positives. For example, human sera or plasma occasionally, in varying amount, contain anti-murine IgG and this will bind to the first anti-cytokine antibody if it is a murine MAb. The second anti-cytokine IgG can then bind to the Fc portion of human anti-murine IgG to yield a positive result in the immunoassay. This kind of artefact may be reduced or eliminated by either including a few per cent mouse serum (13) in the assay diluent or by using the Fab fragments of anti-cytokine IgGs. On the other hand, biological fluids may contain cytokine inhibitors, other than soluble cytokine receptors (Section 5.2), which can bind to cytokines in a way such that epitopes recognized by anti-cytokine IgGs are blocked. This will lead to false negatives. Polyvalent substances such as heparin, which can bind to certain cytokines, e.g. fibroblast growth factors, should therefore not be used in the preparation of plasma, for example.

References

1. Meager, A. (1987). In *Lymphokines and interferons* (ed. M. J. Clemens, A. G. Morris, and A. J. H. Gearing), p. 105. IRL Press, Oxford.
2. Hunter, W. M. and Greenwood, F. C. (1962). *Nature*, **194**, 495.
3. Reay, P. (1982). *Annals of Clinical Biochemistry*, **19**, 129.
4. Poole, S., Bristow, A. F., Selkirk, S., and Rafferty, B. (1989). *Journal of Immunological Methods*, **116**, 259.
5. Ey, P. L., Prowse, S. J., and Jenkin, C. R. (1978). *Immunochemistry*, **15**, 429.
6. Bruck, C., Portelle, D., Glineur, C., and Bolton, A. (1982). *Journal of Immunological Methods*, **53**, 313.
7. Burchiel, S. W., Billman, J. R., and Albert, R. (1984). *Journal of Immunological Methods*, **69**, 33.
8. Lowry, O. H., Rosebrough, N. J., Farr, A. L., and Randall, R. J. (1951). *Journal of Biological Chemistry*, **193**, 265.
9. Secher, D. S. (1981). *Nature*, **290**, 501.
10. Kemeny, D. M. and Challacombe, S. J. (ed.) (1989). *ELISA and other solid phase immunoassays*, pp. 1–29. J. Wiley & Sons, Chichester, UK.
11. Avrameas, S. (1969). *Immunochemistry*, **6**, 43.
12. Guesdon, J.-L., Ternynck, T., and Avrameas, S. (1979). *Journal of Histochemistry and Cytochemistry*, **27**, 1131.
13. Boscato, L. M. and Stuart, M. C. (1986). *Clinical Chemistry*, **32**(8), 1491.

A1

Suppliers of specialist items

Affinity Research Products Ltd, GPT Business Park, Nottingham NG9 2ND, UK.

Ambion, AMS Biotechnology UK Ltd, 5 Thorney Leys Park, Witney, Oxon OX8 7GE, UK.

Ambis Inc., 3939 Ruffin Road, San Diego, CA 92123, USA.

American Radiolabelled Chemicals Inc., 11624 Bowling Green Drive, St Louis, MO 63146, USA.

Amersham Buchier GmbH & Co KG, Life Science, Gieselweg 1, D-38110 Braunschweig, Germany; Amersham International plc, Amersham Place, Little Chalfont, Bucks HP7 9NA, UK; Amersham North America, 2636 Clearbrook Drive, Arlington Heights, Illinois 60005, USA.

Amicon Inc., 72 Cherry Hill Drive, Beverly, MA 21915, USA.

Applied Biosystems—Division of Perkin-Elmer, 850 Lincoln Center Drive, Foster City, CA 94404, USA.

Bachem Feinchemikalien AG, Haupstrasse 144, CH-4416, Bubendorf, Germany.

BDH Ltd, Poole, UK.

Beckton-Dickinson (UK) Ltd, Between Towns Road, Cowley, Oxford OX4 3LY, UK; Becton-Dickinson, Microbiology Systems, PO Box 243, Cockeysville, MD 21030-0243, USA; Becton-Dickinson, Sunnyvale, California, USA.

Biochrom KG, Leonorenstr. 2-6, D-12247, Berlin, Germany.

Bio 101 Inc., American Bioanalytical, Natick, MA 01760, USA; Bio 101 Inc., Stratech Scientific Ltd., Luton, Beds LU2 0NP, UK; Bio 101 Inc., La Jolla, California, USA.

Bio Image, 777 East Eisenhower Parkway, Suite 950, Ann Arbor, MI 48108, USA; Bio Image, UK, Radway Green Venture Park, Radway Green, Crewe, Cheshire CW2 5PK, UK.

Biological Industries Ltd, Media House, Donswood Road, Ward Park South, Cumbernauld, UK.

Biomol Feinchemikalien GmbH, Waidmannstr. 35, D-200 Hamburg 50, Germany.

Biol Research Laboratories Inc., Biomol, 5166 Campus Drive, Plymouth Meeting, PA 19462, USA.

Bio-Rad Laboratories Ltd, Bio-Rad House, Maylands Avenue, Hemel

Hempstead, Herts HP2 7TD, UK; Bio-Rad Laboratories, Life Science Group, 2000 Alfred Nobel Drive, Hercules, CA 94547, USA.

BioRad, Richmond, California, USA.

Bio-Whittaker, Walkersville, MD 21793, USA.

Boehringer Mannheim Corporation, 9115 Hague Road, Indianapolis, IN 46250-0414, USA; Boehringer Mannheim GmbH, Sandhofer Strasse 116, D-6800 Mannheim 31, Germany; Boehringer Mannheim UK, Bell Lane, Lewes, E. Sussex, UK.

Brinkmann Instruments Inc., Cantiague Road, Westbury, New York 11590, USA.

Cambridge Bioscience, 25 Signet Court, Newmarket Road, Cambridge CB5 8LA, UK.

Campden Instruments, 185 Campden Hill Road, London W8 1TH, UK.

Cappelen Laboratory Technics (CAPP), Kallerupvej 26, PO Box 824, DK-5230 Odense M, Denmark.

Carl Zeiss, D-7082, Oberkochen, Germany.

CellPro, Europe, St Pietersplein 11/12 Parvis-St. Pierre, 1970 Werembeekoppem, Belgium.

CLONTECH Laboratories Inc., 4030 Fabian Way, Palo Alto, CA 94303, USA.

Collaborative Biomedical Products, Collaborative Research Inc., Two Oak Park, Bedford, MA 01730, USA.

Costar/Nucleopore, 1 Alewife Center, Cambridge, Massachusetts 02140, USA; Costar/Nucleopore, 10 Valley Center, Gordon Road, High Wycombe, Bucks HP13 GEK, UK.

CP Labs Ltd, PO Box 22, Bishops Stortford, Herts CM23 3DH, UK.

Cresent Chemical Company, Distributor for Serva, 1324 Motor Parkway, Hauppauge, NY 11788, USA.

Dako Ltd, 16 Manor Courtyard, Hughenden Avenue, High Wycombe, Bucks HP13 5RE, UK.

David Kopf, Tujunga, California, USA.

Difco, PO Box 1058, Detroit, Michigan 48232, USA; Difco Laboratories Ltd, PO Box 13B, Central Avenue, West Molesey, Surrey KT8 0SE, UK.

Drummond Scientific Co., 500 Parkway, Broomal, PA 19008, USA.

DuPont NEN Research Products, 549 Albany Street, Boston, MA 02118, USA.

Dynal (UK) Ltd, 10 Thursby Road, Croft Business Park, Bromborough, Wirral, Merseyside L62 3PW, UK.

Dynell, 24–6 Grove Street, New Ferry, Bromborough, Wirral, UK; Dynell, New York, USA.

E. Merck, Frankfurther Straße 150, D-6100 Darmstadt, Germany.

Eppendorf, 2000 Hamburg 65, Postfach 650670, Germany.

Eurogenetics (UK), 111–13 Waldegrave Road, Teddington, Middx TW11 8LL, UK.

E-Y Labs, PO Box 1787, San Mateo, CA 94401, USA.

Fisher Scientific, Fair Lawn, NJ 07410, USA.

Flow ICN USA—ICN Pharmaceuticals Inc., 3300 Hyland Avenue, Costa Mesa, CA 92626, USA; Flow-ICN UK—ICN Pharmaceuticals Ltd, Thame Park Business Centre, Wenman Road, Thame, Oxfordshire OX9 3XA, UK; Flow Laboratories Ltd, Woodcock Hill Industrial Estate, Harefield Road, Rickmansworth, Herts WD3 1PQ, UK.

FMC Bio Products, 191 Thomasten Street, Rockland, ME 04841, USA; Risingevej 1, OK-2665 Vallensbaek Strand, Denmark.

FOTODYNE Inc., 950 Walnut Ridge Drive, Hartland, WI 53029, USA.

Fresenius AG, Bad Hamburg, Germany.

Genzyme, 50 Gibson Drive, Kings Hill, West Malling, Kent ME19 6HG, UK; Genzyme Corp, Genzyme Diagnostics Division, 1 Kendall Square, Cambridge, MA 02139, USA.

Gesellschaft für Biotechnologische Forschung, Mascheroder Weg 1, 38124 Braunschweig, Germany.

GIBCO, Grand Island, New York, NY 07410, USA; GIBCO BRL UK—Life Technologies Ltd, PO Box 35, Trident House, Renfrew Road, Paisley, Renfrewshire, Scotland PA3 4EF, UK; GIBCO BRL USA—Life Technologies Inc., 8400 Helgerman Court, PO Box 6009, Gaithersburg, MD 20884-9980, USA.

Harleco, 480 Democrat Road, Gibbstown, New Jersey 08027, USA.

Heat Systems Co., 60 Broad Hollow Road, Melville, New York, USA.

Herolab, 69168 Wiesloch, Germany.

Hoeffer Pharmacia Biotech, 654 Minnesota Street, Box 777387, San Francisco, CA 94197, USA; PO Box 351, Newcastle-under-Lyme, Staffs ST5 0TT, UK.

Hyclone Lab, 1725 So. State Highway 89–91, Logan UH, USA.

IBF, Villeneuve la Garenne, France.

IBI/Eastman Kodak Co., PO Box 9558, New Haven, CT 06335, USA.

Ilford Scientific Products, Mobberly, Knutsford, Cheshire WA16 7HA, UK.

Industrial Chemicals, EM Industries Inc., 5 Skyline Drive, Hawthorne, NY 10532, USA.

Intracel Ltd, Unit 4, Station Road, Shepreth, Royston, Herts SG8 6PZ, UK.

Lab-Line Instruments Inc., 15th & Bloomingdale Avenues, Melrose Park, IL 60160, USA.

Labtech International Ltd, Woodside, Easons Green, Uckfield, E. Sussex TN22 5RE, UK; Life Sciences, 2900 72nd Street North, St Petersburg, Florida 33710, USA.

Life Sciences International Europe Ltd (Savant), 93/96 Chadwick Road, Astmoor Industrial Estate, Runcorn, Cheshire WA7 1PR, UK.

Life Technologies Ltd (GibcoBRL), Paisley PA3 4EF, Renfrewshire, Scotland, and Gaithersburg, MD 20884-9980, USA.

Luckham, Victoria Gardens, Burgess Hill, West Sussex RH15 9QN, UK.

Marathon Laboratory Supplies (Falcon), Unit 6, 55–7 Park Royal Road, London NW10 7LP.

Matheson, Coleman, and Bell, Norwood, USA.

Merck/BDH, Merck Ltd, Merck House, Poole, Dorset BH15 1TD, UK.

Micro Instruments Ltd, 18 Hanborough Park, Long Hamborough, Witney, Oxon OX8 8LH, UK.

Millipore, Intertech, PO Box 255, Bedford, MA 017130, USA; Millipore (UK) Ltd, The Boulevard, Ascot Road, Croxley Green, Herts WD1 8YW, UK.

Molecular Dynamics, 880 E. Arques Ave., Sunnyvale, CA 94066, USA.

Molecular Probes Inc., 4849 Pitchford Avenue, PO Box 22010, Eugene, Oregon 97402, USA.

Nariishige Scientific Instrument Laboratory, 9.28 Kasuya, 4 Chome Seta-gayaku, Tokyo, Japan.

Nasco Co., 101 Janesville Avenue, Ft. Atkinson, Wisconsin 53538, USA.

National Blood Transfusion Centre, Colindale Avenue, Colindale, London, UK.

Neuroprobe, Cabin John, Madison, USA; Neuroprobe, 7621 Cabin Road, Bethesda, Maryland 20034, USA; Neuroprobe, Porvair Filtronics Ltd, Shepperton, Middx, UK.

New England Biolabs, 67 Knoll Place, Wilbury Way, Hitchin, Herts SG4 0TY, UK; New England Biolabs (NBL), 32 Tozer Road, Beverley, MA 01915-5510, USA.

Nikon (UK) Ltd, Instrument Division, Haybrook, Halesfield 9, Telford, Shropshire TF7 4EW, UK.

Nordic Immunological Laboratories b.v., Langestraat 55–61, PO Box 22, NL-5000 AA Tilburg, The Netherlands.

Nuclepore, Filtration Products, 7035 Commerce Circle, Pleasanton, California, CA 94566-3294, USA.

Nunc, Naperville, Illinois, USA; Nunc A/S, Life Technologies A/S, PO Box 198, Industrivej 1, DK-4000 Roskilde, Denmark.

Nycomed (UK) Ltd, Nycomed House, 211 Coventry Road, Sheldon, Birmingham, UK; Nycomed Pharma, Oslo, Norway.

Olympus Optical Co. (Europa) GMBH, Postfach 104908, Wendenstrasse 14–16, 2 Hamburg 1, Germany.

Oxford GlycoSystems, 75-G Wiggins Avenue, Bedford, MA 01730, USA; Oxford GlycoSystems, Hitching Court, Blacklands Way, Abingdon, OX14 1RG, UK.

Packard Instrument Company, 2200 Warrenville Road, Downers Grove, Illinois 60515, USA; 800 Research Pkway, Meriden, CT 06450, USA.

Perkin Elmer Cetus, Beaconsfield, Buckinghamshire HP9 1QA, UK; Perkin Elmer Cetus, Norwalk, CT 06859, USA.

Pharmacia, 23 Grosvenor Road, St Albans, Herts AL1 3AW, UK; Pharmacia, 800 Centennial Avenue, PO Box 1327, Piscataway, NJ 08854, USA;

Pharmacia Biotech, Bjorkgatan 30, S-751 82 Uppsala, Sweden; Pharmacia Biotech Europe, Procordia Eurocentre, Rue de la Fusee 62, B1130 Brussels, Belgium; Pharmacia P-L Biochemicals, Uppsala, Sweden; Pharmacia UK-Pharmacia Biotech Ltd, Davy Avenue, PO Box 100, Knowlhill, Milton Keynes MK5 8PB, UK.

Pharmingen, 10975 Torreyana Road, San Diego, CA 92121, USA; Pharmingen, Cambridge BioScience, 25 Signet Court, Newmarket Road, Cambridge CB5 8LA, UK.

Pierce Chemical Co., PO Box 117, Rockford, Illinois 61105, USA; Pierce Chemical Co. Inc., 3747 North Meridian Road, Rockford, IL 61105, USA.

Pierce & Warriner (UK) Ltd, 44 Upper Northgate Street, Chester CH1 4EF, UK.

Pillar Engineering Supplies Ltd, 103/109 Waldegrave Road, Teddington, Middx TW11 8LL, UK.

Promega, Southampton, UK; Promega Biotec, Madison, Wisconsin, USA; Promega Corporation, 2800 Woods Hollow Road, Madison, WI 53711-5399, USA.

Dr Hugh Pross, Dept of Radiation Oncology and Microscopy and Immunology, Queen's University, Kingston, Ontario, Canada.

R&D Systems, 614 McKinley Place N.E., Minneapolis, MN 55413, USA; R&D Systems Europe, 4–10 The Quadrant, Barton Lane, Abingdon, Oxon OX14 3YS, UK.

Robbins Scientific Co., 814 San Aleo Avenue, Sunnyvale, CA 94086, USA.

Savant Instruments Inc., 110-103 Bi-County Blvd, Farmingdale, NY 11735, USA.

Schleicher & Schuell GmbH, PO Box 4, D-3354 Dassel, Germany; 10 Optical Avenue, PO Box 2012 Keene, NH 03431, USA.

Serotec, 22 Bankside, Station Approach, Kidlington, Oxford OX5 1JE, UK.

Serva Feinbiochemica GmbH & Co., KG, Carl-Benz-Str. 7, D69115 Heidelberg, Germany.

Sigma, 3050 Spruce Street, St Louis, Missouri 63103, USA; Sigma Chemical—Sigma–Aldrich Co. Ltd, Fancy Road, Poole, Dorset BH17 7NH, UK; Sigma Chemical Co., Sigma–Aldrich Techware, PO Box 14508, St Louis, MO 63178-9916, USA.

Sigma–Aldrich Chemie GmbH, Grünwalder Weg 30, D-82041 Deisenhofen, Germany.

Skatron Inc., PO Box 530, Sterling, VA 22170, USA; Skatron UK, Unit 11, Studlands Park Ave., Newmarket, Suffolk CD8 7DR, UK.

Steralin Ltd, Lampton House, Lampton Road, Hounslow, Middx, UK.

Stratagene, 11011 N. Torrey Pines Road, La Jolla, CA 92037, USA; Stratagene Limited, Cambridge Innovation Centre, 140 Cambridge Science Park, Milton Road, Cambridge CB4 4GF, UK.

Stratech Scientific Ltd, 61–3 Dudley Street, Luton LU2 0HP, UK.

Takeda Chemical Industry Ltd, 17–85 Jusohonmachi 2-Chome, Yodogawa-ku, Osaka 532, Japan.

Techne, Duxford, Cambridge CB4 4PZ, UK.

Terry Fox Laboratory, British Columbia Cancer Research Centre, 6012 West 10th Avenue, Vancouver, B.C., Canada V5Z 1L3.

TIB MolBiol, Emser Str. 103, 12051 Berlin, Germany.

Tissue Culture Services—TCS Biologicals Ltd, Botolph, Claydon, Bucks MK18 2LR, UK.

Tocris Neuramin Ltd, Churchill Building, Langford House, Landford, Bristol BS18 7DY, UK.

Wallac. EG+G, 20 Vincent Avenue, Crownhill Business Park, Crownhill, Milton Keynes, MK8 0AB, UK.

Wellcome Reagents Ltd, Beckenham, Kent BR3 3BS, UK.

Wild Leitz UK Limited, Davey Avenue, Knowlhill, Milton Keynes, MK5 8LB, UK.

Worthington Biochemical Corporation, Halls Mill Road, Freehold, NJ 07728, USA.

Index

A375 cell line 387–8
A549 cells 383
activation of B cells 159–62
activation of macrophages 269–78
activation of T cells 192–3
adhesion assay 332–3
adjuvant 332–3
AET-treated SRBC 150
African green monkey cells 10
agarose assay for chemotaxis 215–20
agarose gel electrophoresis 22–3
agarose plugs 21
aggregation of platelets 238
alkaline phosphatase 304
alkaline phosphatase assays 282–3, 400–2
Alsevers solution 150
AML-193 cells 376
antibodies to cytokines 327–38
   adjuvant 332–3
   guinea pigs 335–6
   heterobifunctional cross-linkers 330–2
   homobifunctional cross-linkers 329–30
   immunization schedules 333–7
   immunogen 327–9
   mice 333–4
   purification 337–8
   screening 336
   small peptide analogues 328–9
antigen retrieval for paraffin embedded
   tissues 346–7
antisense asymmetric RNA probes 47–8
antiviral activity-measurement 129–41,
   383–5
   microtitre plate assay 135–7
   MTT staining procedure 134, 137
   multi-well assays 132–3
   plaque assay 137–40
   virus stocks 140–1
   virus yield reduction assay 137–40
APAAP technique 341–2
apoptosis 283–5, 294–5
autoradiography of labelled nuclei 287–8

B9 cells 365–6
bacillus Calmette–Guerin (BCG) 328
BALM-4 cells 360
BCDF 148
BCGF 148
BCL-1 cells 367
beads, anti IgM 146
β-galactosidase 400

BHK cells 152
biological assays for cytokines 183–8, 357–91
   analysis 389
   colony stimulating factors 372–7
   interferons 129–41, 382–4
   interleukin-1 359–60
   interleukin-2 358–9
   interleukin-3 372–7
   interleukin-4 360–4
   interleukin-5 364–5
   interleukin-6 365–6
   interleukin-7 367
   interleukin-8 379–80
   interleukin-9 368
   interleukin-10 368–9
   interleukin-11 369–70
   interleukin-12 370–1
   interleukin-13 371
   interleukin-14 377–8
   leukaemia inhibitory factor (LIF) 377–8
   oncostatin-M 386–9
   standards 388–9
   stem cell factor (SCF) 378
   TGF-β 384–6
   tumour necrosis factors 381–2
blastocyst injection 319–22
B lymphocyte methods 143–77
   activation 159–60
   B cell lines 158–9, 166–7
   co-stimulation assays 166–7
   density gradient fractionation 153
   differentiation assays 166–9
   E-rosetting 150–1
   heavy and light B cells 153
   immunofluorescence analysis 156–8
   immunoglobulin assays 166–8, 171–6
   MACS cell sorter 154–5
   monoclonal antibody separation 153–5
   monocyte depletion 151–3
   Percoll preparation 145
   peripheral blood mononuclear cells 148
   preparation of B cells 148–55
   proliferation assays 162–6
   reagents for assay 144
   responses to cytokines 144
   RNA synthesis 160
   subpopulations 153–5
   surface antigens 160–2
   T cell replacing factor assays 169–71
   tonsillar mononuclear cells 148
Bolton–Hunter reagent 397
bone-marrow colony growth 248–50

bone-marrow preparation
  human 249–50
  murine 248–9
Boyden chamber assay 230–1
bromodeoxyuridine assay 291–2

calibration of immunoassays 402–3
cDNA libraries 2–10
cDNA synthesis 3–5, 58–9, 72
cell counting 279–82
cell cycle analysis 285–92
cell lysis assays 283–5
cell mediated cytotoxicity assays 197–213
  chromium release assay 208–11
  cryopreservation of tumours and
    lymphocytes 207
  generation of lymphokine activated killers
    204–7
  generation of T cell killers 205–6
  isolation of effector cells 199–204
  lactate dehydrogenase assay 212
  single cell CMC assays 211–12
  tumour infiltrating lymphocytes (TIL)
    203–4
cell signalling 93–110
cellulose ester filters 221
centrifugal elutriation 259–61
ceramide 94, 104–5
CFC assays 247–66
charring densitometry 104
chemokine assays 360
chemokines 1, 215–24, 360, 380–1
chemotaxis assays 215–24, 229–33
  agarose assay 215–20
  agarose plate preparation 216–17
  Boyden chamber assay 230–1
  granulocyte preparation 217–18
  in vivo assay 222–4
  microchamber assay 221–2
  micropore filters 220–2
chimaeric mice 318–22
chloramine T method 398
CHO cells 11
chromium release assay 208–11
clonal assays for bone marrow 247–53
clone 2b cells 367
cloning of cytokine/receptor genes 1–18,
  85–8
  G-CSF 8
  GM-CSF 7, 8
  IFN-α/β receptors 86–8
  IFN-γ receptor 85–8
  IL-2 13
  IL-3 7, 8
  IL-5 12–14
  IL-6 9, 12–14
  M-CSF 8

TNF 8–9
colony forming cells 247–53
colony stimulating factor assays 247–66,
  374–9
  biological assays using cell lines 372–7
  centrifugal elutriation 259–61
  clonal assays 250–3
  enrichment of progenitors 257–66
  FACs sorting 256–9, 263–6
  growth factor-dependent lines 254–6
  human bone-marrow 249–50
  immunoadsorption methods 261–2
  murine bone-marrow 248–9
  red cell separation 250
  serum-free culture 253–4
computer based analysis of 2-D gels 123
Coomassie staining 119–23
COS cells 10–11
co-stimulation assays 166–7
Coulter counter 280–1
covalent cross-linking of ligand to receptor
  82–4
cross-linking DNA 25
cross-linking of ligand to receptor 101–2
cryopreservation of tumours and
  lymphocytes 207
crystal violet 283–5
CTLL cells 359
Cunningham chamber 351
cytofluorimetry 290
cytostasis assays 279–95
cytotoxicity assays 208–12
cytotoxic lymphocytes 197–213
  cryopreservation 207
  cytotoxicity assays 208–12
  generation 204–7
  isolation 199–204

DA-1a cell line 374–8
1,2, DAG 93–5
dark field microscopy 49
DEAE–Affigel Blue chromatography 303
degranulation assay for PAF 238–9
2-D gel electrophoresis 111–28
  analysis 123
  applications 123–6
  identification of proteins 119–23
  isolation of proteins 115–18
  PDQUEST system 123
  preparation of samples 112–15
  preparative gel applications 124–5
diacylglycerol (DAG) 94–110
differential screening 9–10
discontinuous density gradient centrifugation
  201
DNA fragmentation assays 293–5

egg transfer for transgenic mice 302–3
EL4 cell line 361
elastase 381
ELISA assays 400–2
elispot assay 351–3
elutriator 259–62
embryonic stem cells 308–11
encephalomyocarditis virus (in anti-viral
    assays) 130–1, 133, 138, 140
endothelial cells 225–45
  adhesion assays 241–5
  Boyden chamber assay 230–1
  chemotaxis 229–33
  culture 226–7
  HUVEC 226, 230, 233
  microchemotaxis 231–2
  mouse PmT-transformed lines 227–9
  platelet activating factor assays 235–40
  procoagulant activity 233–5
  thromboplastin activity 233–5
  transmigration assays 242–5
enzymatic digestion of tissues 176, 183
enzyme-linked immunoabsorbant assays for
    cytokines 185, 400–2
E-rosetting 150–1, 181–2
ES cells, *see* embryonic stem cells
expression of cytokine genes 1–18
  bacterial systems 13–14
  lower eukaryotes 11
  mammalian cells 10–11
  permanent expression 10
  transient expression 10

FACS analysis and sorting 162–3, 201–2
FACS analysis of cell cycle 290
FACS sorting of bone-marrow 256–9, 263–6
Ficoll separation medium 148–9

Gey's haemolytic solution 150
gibbon T cell line 8
GNFS-60 cell line 376–7
goat antibodies to cytokines 336
granulocyte adhesion 241–3
granulocyte preparation 217–18
guinea-pig antibodies to cytokines 335–6

H₂O₂ release 271
haemocytometer 280
haemopoietic growth factor assays 247–68
  centrifugal elutriation 259–61
  clonal assays 250–3
  enrichment of progenitors 257–66
  FACS sorting 256–9, 263–6
  growth factor-dependent lines 254–6
  human bone-marrow 249–50

immunoadsorption methods 261–2
  murine bone-marrow 248–9
  red cell separation 250
  serum-free culture 253–4
haemopoietic growth factor-dependent cell
    lines 254–6
haptenization of peptides 328–9
HeLa cells 130
Hep2 cells 130, 383
heterobifunctional cross-linkers 330–2
heterologous expression of cytokine genes
    10–13
homobifunctional cross-linkers 329–30
human bone-marrow preparation 252–3
HUVEC 225–45
hydroxyethyl starch 217

Ig assays 171–6
immune assays for cytokines and soluble
    receptors 393–404
  calibration 402–3
  chloramine T method 398
  diluent 403
  ELISA 400–2
  iodogen method of labelling 394–5
  IRMA 397–400
  matrix 403
  problems 402–4
  purification of anti-cytokine IgG 397
  RIA 394–6
  soluble cytokine receptors 393, 399–400,
    402
immunization schedules for cytokine
    antibodies 333–7
immunoglobulin secretion assays 171–6
immunolocalization of cytokines in cells
    339–56
  APAAP 341–2
  ELISPOT assay 351–3
  reverse haemolytic plaque assay 347–51
  streptavidin-biotin (AB) 343–4
individual cell assays for cytokines 339–6
influenza virus 147, 169–70
inositol phosphates 94
in situ hybridization to cytokine mRNA 35,
    46–56, 184
  antisense probes 47–8
  autoradiography 48, 54
  controls 48–9
  counterstaining 54
  dark field microscopy 49–50
  DNA probes 47
  hybridization 52
  interpretation 54–5
  nick translation 47
  post-hybridization 53
  pre-hybridization 50

in situ (*cont.*)
  probe choice 47
  probe labelling 47–8, 51–2
  sample handling 46
  sample preparation 44, 49–50
  sense probes 49
  slide preparation 49
interferon-α
  assays 129–41, 324–6
  B cells 144
  2-D gel analysis of induced proteins 124
interferon-γ
  B cells 61, 144
  2-D gel electrophoresis of induced
    proteins 92
  ELISPOT assay 351
  macrophages 272
  primer for RT-PCR 63
  receptors 77–91
  RHPA 247
interferons, biological assays for 129–41,
    383–5
interleukin-1 (IL-1)
  antibodies 333–4
  assays 361–2
  B cells 144
  chemotaxis 215, 223
  2-D gel analysis of induced proteins 124
  endothelial cells 225
  LAK cells 206
  primer for RT-PCR 63
  T cells 183–6
interleukin-2 (IL-2)
  assay 358–60
  B cells 144
  cloning 13
  LAK cells 206
  primer for RT-PCR 63
  release bioassay 194
  T cells 179, 183–6
interleukin-3 (IL-3)
  assay 374–9
  B cells 144
  cloning 7–8
  primer for RT-PCR 63
interleukin-4 (IL-4)
  assay 364–6
  B cells 144
  cloning 5–6
  macrophages 272
  primer for RT-PCR 63
interleukin-5 (IL-5)
  assay 364–5
  B cells 144
interleukin-6 (IL-6)
  antibodies 334
  assays 365–6
  B cells 144

cloning 9, 13
ELISPOT assay 351
endothelial cells 215
expression 11
primer for RT-PCR 63
interleukin-7 (IL-7)
  assay 367
  B cells 144
  LAK cells 206
interleukin-8 (IL-8)
  assay 215, 378, 381
  chemotaxis 215, 223
  endothelial cells 225
  immunocytochemistry 343
interleukin-9, assay 368
interleukin 10, assay 368
interleukin-11, assay 369–70
interleukin-12, assay 370–1
interleukin-13, assay 371
interleukin-14, assay 371–2
iodination of proteins 394–5, 398
iodogen method 394–5
IRMA assays 397–400
isotype switching 173–4

keyhole limpet haemocyanin (KLH) 328
KIT225/K6-C cell line 370–1
knockout mice 308–25
  blastocyst injection 319–22
  chimaeric mice generation 318–22
  embryonic stem (ES) cells 308–11
  gene targeting ES cells 311–13
  identification of targeted cells 317–18
  picking and expansion of ES cells 314–15
  storage and recovery of ES cells 315–17
  transfection of ES cells 313–14

L4 cells 363
L-929 bioassay 134, 382–3
L-929 cells 134, 382–3
lactate dehydrogenase assay 212
LAF assay 361
LAK cells, *see* lymphokine activated killer
    cells
large granular lymphocytes 199–201
leucocyte adhesion and transmigration
    assays 241–2
leukaemia inhibitory factor (LIF) assay
    379–80
LGL, *see* large granular lymphocytes
limiting dilution cloning 190–1
lipid extraction 77–99

Index

L-M cells 381
locomotion assays 214, 224, 229, 233
lymphokine activated killer cells 197–213
  cryopreservation 207
  cytotoxicity assays 208–12
  density purification 201
  FACS sorting 201–2
  generation with other cytokines 206–7
  generation with IL-2 204–5
  isolation of PBMC, PBL, LGL 199–201
  magnetic bead sorting 202–3

macrophage activation assays 269–78
  cytokine assays 276–7
  mannosyl receptor assays 273–6
  MHC class II expression 272–3
  nitric oxide 269, 271–2
  respiratory burst activity 271–2
  single cell assays 276
  superoxide anion release assay 270–1
MACS cell sorter 154–5
magnetic cell sorting 154–5
mannosyl receptor assay 273–6
matrix effects 403
MC/9 cell line 370–1
Mengo virus 130–1, 133, 138
metabolic cell labelling 96–7, 100
MHC class II assay in macrophages 272–3
microchemotaxis assay 231–2
microexudate plates 152–3
micropore filters for chemotaxis assays
  220–2
mitotic index assay 292
MNFS-60 cells 373–7
MO-7e cells 373–7
monoclonal antibodies
  in B cell purification 140, 153–5
  to cytokines 332, 393–404
monocyte chemotactic protein 223, 225
monocyte depletion 151–3
mononuclear cell preparation 145–6, 179–81
MT-1 cells 358
MTT-type assay 283
murine antbodies to cytokines 333–4
murine bone-marrow colony assays 250–2
murine bone-marrow preparation 248
MV-3D9 cells 385–6

natural killer cells 197–213
nick-translation 27
nitric oxide assay 269, 271–2
nitrobluetetrazolium 271
NK cells, see natural killer cells
NOB-1 cells 359–60

Northern analysis of cytokine mRNA 35–42,
  184
  blotting 39–40
  electrophoresis 38–9
  handling RNA 36
  hybridization 41–2
  preparation of RNA 36–7
  probe labelling 40–1
  spun column 4

oligo(dT)-cellulose 2
oncostatin-M assay 386–7
oocyte recovery 300

PDECGF immunostaining 341
PDQUEST analysis of 2-D gels 123
Percoll preparation 145–6
phosphatidyl choline 99
phosphatidylcholine-specific phospholipase C
  93–5, 102
phospholipase D 101
phosphorylation of interferons 78–91
plaque assay 138–9
platelet activating factor 235–41
  cell cultures for assay 235–6
  degranulation assay 238–9
  extraction 236–7
  identification 237
  quantitation 239–41
platelet aggregation assay 238
platelet contamination of blood cells 200
poly(A)$^+$ fraction of RNA 3
polyclonal antibodies to cytokines 327–38
polymerase chain reaction for cytokine
  mRNA 57–68, 69–75
  cDNA synthesis 59
  comparative 69–75
  control fragment design 60–2
  DNAse treatment 71
  internal DNA control standard 57–8
  multispecific fragment 62
  primers 63–5
  quantitation 62–8, 73–5
  reaction 59–60, 72–3
  RNA extraction for 70
poly(U)-Sepharose 3
polyvinyl pyrrolidine-polycarbonate
  membranes 221–2
primers for cytokine RT-PCR 63, 65
priming for macrophages 269
procoagulant activity assay 233–5
progenitor cell enrichment 256–8
proliferation assays 279–83
pulsed field gel electrophoresis 23–4

rabbit antibodies to cytokines 334–5
radioimmune assays for cytokines 393–404
radiolabelling
  antibodies 393–404
  cytokines 394
  phosphorylation 78–9
  probes 27–9, 40
random priming 27, 40–1
receptor-binding studies 77–91
  binding of ligand 79–82
  covalent crosslinking 82–4
  radiolabelling interferon 78–9
  Scatchard analysis 80–1
receptor signalling 93–110
red cell removal from bone-marrow 250
respiratory burst activity 271–2
restriction enzyme digest of DNA 21–2
restriction fragment length polymorphisms
  19
reverse haemolytic plaque assay 347–51
reverse transcription 58–9, 72
RIA, *see* radioimmune assay
ribonuclease protection 35, 42–6
  analysis on sequencing gel 45–6
  cleaning protected fragment 45
  controls 42–3
  hybridization 44
  probe preparation 43–4
  template DNA 43
RNA
  conversion to cDNA 3–4, 58–9, 72
  Northern blotting 35–42
  preparation 2–3, 36–8, 58, 70
RT-PCR, *see* polymerase chain reaction

Saccharomyces cerevisiae 12
Scatchard plot 79–81
screening of cDNA libraries 4–9
  direct expression in bacterial cells 6
  direct expression in mammalian cells 7
  direct expression by *in vitro* transcription 5
  hybrid selection 4
  oligonucleotide screening 8
  screening on basis of protein sequence
    data 9
  screening of biological activity 4
second messenger sytems 61
Semliki Forest virus 130–1, 133
sense RNA probes 47–9
serum-free culture of bone-marrow 253–4
sheep antibodies to cytokines 336
Sindbis virus 130, 133
single cell assays for cytokines 339–56
single cell CMC assays 211–12
slot blots of tail DNA 306–7
small peptide analogues for immunization
  328–9

Southern blotting of cytokine genes 19–34
  agarose plugs 21
  artefacts 30–1
  background sources 31–2
  contamination with plasmid DNA 30
  controls 24
  cross-linking of DNA 24
  degraded DNA 28
  gel electrophoresis 22–4
  interpretation 30–2
  membrane hybridization 29–30
  nick-translation 27–29
  oligolabelling 25
  partial digests 28
  preparation of DNA 19–21
  probes 27–9
  pulsed field gel electrophoresis 23
  quantitation 33
  random priming 27
  restriction enzyme digest 21–2
  size markers 24
  transfer of DNA 24–7
SP6 polymerase 5
sphingomyelin 94
sphingomyelinases 94, 102–8
spun column 41
standards for cytokines 363
Staphylococcus aureus Cowan 1 strain 147
stem cell factor (SCF) assay 378–9
streptavidin-biotin (AB) method 343–5
superoxide anion release 270–1

tail DNA preparation 304–5
T cell depletion 150–1
T cell replacing factor 169–71
tetanus antigen-specific T cell lines 189–90
TF-1 cell line 377, 386
TGF-β, *see* transforming growth factor β
Th₁/Th₂ cells 186–7
thin-layer chromatography 99, 101, 103
thromboplastin activity assay 233–5
thymidine uptake assay 285–8
T lymphocyte methods 179–96
  autoreactive T cells 192–3
  cloning 192
  cytokine effects on T lymphocytes 188–95
  cytokine production 183–8
  IL-2 release bioassay 194–5
  isolation for cytokine production 179–83
  isolation from peripheral blood 180–1
  isolation from solid tissue 183
  isolation from synovial fluid 182–3
  limiting dilution 190–1
  one step ficoll 180–1
  SRBC rosette formation 181–2
  T cell lines and clones 189–92
  tetanus antigen specific T cell lines 189–90

tonsillar mononuclear cells 148–9
transformation of *E. Coli*
transforming growth factor-β
  assay 384–6
  B cells 144
  T cells 183
transgenic mice 297–308, 322–4
  animals 298–9
  egg transfer 302–3
  identification of transgene progeny 304–8
  integration/expression transgenes 303–4
  microinjection 301–2
  PCR screening 307–8
  pronuclear injection of DNA 297–301
  purification of DNA 299
  recovery of oocytes 300
  slot blots of tail DNA 306–7
  tail DNA preparation 304–5
transmigration assay 242–5
trypsin 262
tumour infiltrating lymphocytes 203–4
tumour necrosis factors (TNFs)
  antiviral activity 130
  APAAP technique 341
  assay 283–4, 382–3
  B cells 144
  cloning 9

  cytostasis/cytotoxicity 279–95
  2-D gel analysis of induced proteins 125
  expression 14
  IRMA assays 394–6
  LAK cells 206
  macrophages 276–7
  primer for RT-PCR 63–5
  radiolabelling 394–5
  T cells 184, 193
two dimensional gel electrophoresis, *see* 2-D
  gel

versene 281
virus stock preparation 140
virus yield reduction assay 138–40
VSV 130–1, 133, 138, 140

WEHI cells 283–4, 382
WISH cells 383

Xenopus laevis oocytes 5–6, 76, 81–2

zoo blots 19